Bayesian Tensor Decomposition for Signal Processing and Machine Learning

Lei Cheng · Zhongtao Chen · Yik-Chung Wu

Bayesian Tensor Decomposition for Signal Processing and Machine Learning

Modeling, Tuning-Free Algorithms, and Applications

Springer

Lei Cheng
College of Information Science
and Electronic Engineering
Zhejiang University
Hangzhou, China

Zhongtao Chen
Department of Electrical and Electronic
Engineering
The University of Hong Kong
Hong Kong, China

Yik-Chung Wu
Department of Electrical and Electronic
Engineering
The University of Hong Kong
Hong Kong, China

ISBN 978-3-031-22437-9 ISBN 978-3-031-22438-6 (eBook)
https://doi.org/10.1007/978-3-031-22438-6

This Springer imprint is published by the registered company Springer Nature Switzerland AG
The registered company address is: Gewerbestrasse 11, 6330 Cham, Switzerland

Preface

Our world is full of data, and these data often appear in high-dimensional structures, with each dimension describing a unique attribute. Examples include data in social sciences, medicines, pharmacology, and environmental monitoring, just to name a few. To make sense of the multi-dimensional data, advanced computational tools, which directly work with tensor rather than first converting a tensor to a matrix, are needed to unveil the hidden patterns of the data. This is where tensor decomposition models come into play. Due to the remarkable representation capability, tensor decomposition models have led to state-of-the-art performances in many domains, including social network mining, image processing, array signal processing, and wireless communications.

Previous research on tensor decompositions mainly approached from an optimization perspective, which unfortunately does not come with the capability of tensor rank learning and requires heavy hyper-parameter tuning. While these two tasks are important in complexity control and avoiding overfitting, they are often overlooked or downplayed in current research, and assumed can be achieved by trivial operations, or somehow can be obtained from other methods. In reality, estimating the tensor rank and a proper set of hyper-parameters usually involve exhaustive search. This requires running the same algorithm many times, effectively increasing the computational complexity in actual model deployment.

Another path for model learning is Bayesian methods. They provide a natural recipe for the integration of tensor rank learning, automatic hyper-parameter determination, and tensor decomposition. Due to this unique capability, Bayesian models and inference trigger a recent interest in tensor decompositions for signal processing and machine learning. From these recent works, Bayesian models show comparable or even better performance than optimization-based counterparts.

However, Bayesian methods are very different from optimization methods, with the former learning distributions of the unknown parameters, and the latter learning a point estimate. The process of building the models and inference algorithm derivations are fundamentally different as well. This leads to a barrier between the two groups of researchers working on similar problems but starting from different perspectives. This book aims to distill the essentials of Bayesian modeling and inference in tensor research, and present a unified view of various models. The book addresses the needs of postgraduate students, researchers, and practicing engineers whose interests lie in tensor signal processing and machine learning. It can be used as a textbook for short courses on specific topics, e.g., tensor learning methods, Bayesian learning, and multi-dimensional data analytics. Demo codes can be downloaded from https://github.com/leicheng-tensor/Reproducible-Bayesian-Tensor-Matrix-Machine-Learning-SOTA. It is our hope that by lowering the barrier to understanding and entering the Bayesian landscape, more ideas and novel algorithms can be stimulated and facilitated in the research community.

This book starts by reviewing the basics and classical algorithms for tensor decompositions, and then introduces their common challenge on rank determination (Chap. 1). To overcome this challenge, this book develops models and algorithms under the Bayesian sparsity-aware learning framework, with the philosophy and key results elaborated in Chap. 2. In Chaps. 3 and 4, we use the most basic tensor decomposition format, Canonical Polyadic Decomposition (CPD), as an example to elucidate the fundamental Bayesian modeling and inference that can achieve automatic rank determination and hyper-parameter learning. Both parametric and non-parametric modeling and inference are introduced and analyzed. In Chap. 5, we demonstrate how Bayesian CPD is connected with stochastic optimization in order to fit large-scale data. In Chap. 6, we show how the basic model can incorporate additional nonnegative structures to achieve enhanced performances in various signal processing and machine learning tasks. Chapter 7 discusses the extension of Bayesian methods to complex-valued data, handling orthogonal constraints and outliers. Chapter 8 uses the direction-of-arrival estimation, which has been one of the focuses of array signal processing for decades, as a case study to introduce the Bayesian tensor decomposition under missing data. Finally, Chap. 9 extends the modeling idea presented in previous chapters to other tensor decomposition formats, including tensor Tucker decomposition, tensor-train decomposition, PARAFAC2 decomposition, and tensor SVD.

The authors sincerely thank the group members, Le Xu, Xueke Tong, and Yangge Chen, at The University of Hong Kong for working on this topic together over the years. This project is supported in part by the NSFC under Grant 62001309, and in part by the General Research Fund from the Hong Kong Research Grant Council under Grant 17207018.

Hangzhou, China — Lei Cheng
Hong Kong, China — Zhongtao Chen
Hong Kong, China — Yik-Chung Wu
August 2022

Contents

Chapter 1
Tensor Decomposition: Basics, Algorithms, and Recent Advances

Abstract In this chapter, we will first introduce the preliminaries on tensors, including terminologies and the associated notations, related multi-linear algebra, and more importantly, widely used tensor decomposition formats. Then, we link the tensor decompositions to the recent representation learning for multi-dimensional data, showing the paramount role of tensors in modern signal processing and machine learning. Finally, we review the recent algorithms for tensor decompositions, and further analyze their common challenge in rank determination.

1.1 Terminologies and Notations

1.1.1 Scalar, Vector, Matrix, and Tensor

Plain letters (e.g., x) are used to denote *scalars*. The boldface lowercase (e.g., $\mathbf{x}$) and uppercase letters (e.g., $\mathbf{X}$) are used for *vectors* and *matrices*, respectively. For *tensors*, they are denoted by boldface calligraphic letters $\boldsymbol{\mathcal{X}}$.

In multi-linear algebra, the term *order* measures the number of indices used to assess each data element (in scalar form). Specifically, vector $\mathbf{x} \in \mathbb{R}^I$ is of order 1 since its element $\mathbf{x}_i$ can be assessed via only one index. Matrix $\mathbf{X} \in \mathbb{R}^{I \times J}$ is of order 2, because two indices are enough to traverse all of its elements $\mathbf{X}_{i,j}$. As a generalization, tensors are of order three or higher. An Nth order tensor $\boldsymbol{\mathcal{X}} \in \mathbb{R}^{I_1 \times \cdots \times I_N}$ utilizes N indices to address its elements $\boldsymbol{\mathcal{X}}_{i_1,\ldots,i_N}$. For illustration, we depict scalar, vector, matrix, and tensor in Fig. 1.1.

For an Nth order tensor $\boldsymbol{\mathcal{X}}$, addressing each element requires N indices, and each index corresponds to a *mode*, which is used to generalize the concepts of rows and columns in matrices. For example, for a third-order tensor $\boldsymbol{\mathcal{X}} \in \mathbb{R}^{I_1 \times I_2 \times I_3}$, given indices i_2 and i_3, the vectors $\boldsymbol{\mathcal{X}}_{:,i_2,i_3}$ are termed as *mode-1 fibers*.

L. Cheng et al., *Bayesian Tensor Decomposition for Signal Processing and Machine Learning*, https://doi.org/10.1007/978-3-031-22438-6_1

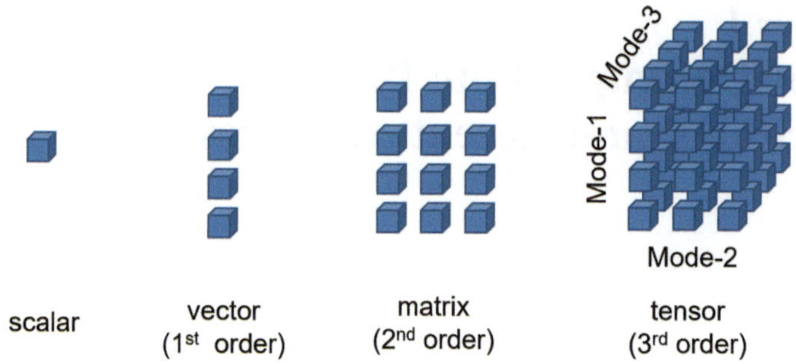

Fig. 1.1 Illustration of scalar, matrix, and tensor

1.1.2 Tensor Unfolding/Matricization

Tensor unfolding/matricization aims to re-organize the fibers in one mode into a matrix. For an Nth order tensor $\mathcal{X} \in \mathbb{R}^{I_1 \times \cdots \times I_N}$, since it has N modes, there are N types of unfolding, each termed as *mode-n unfolding*. We formally define it as follows and illustrate it in Fig. 1.2.

Definition 1.1 (*Mode-n Unfolding*) Given a tensor $\mathcal{X} \in \mathbb{R}^{I_1 \times \cdots \times I_N}$, its mode-$n$ unfolding gives a matrix $\mathbf{X}^{(n)} \in \mathbb{R}^{I_n \times \prod_{k=1,k\neq n}^{N} I_k}$. Each tensor element $\mathcal{X}_{i_1,\ldots,i_N}$ is mapped to the matrix element $\mathbf{X}^{(n)}_{i_n,j}$ with $j = i_1 + (i_2 - 1)I_1 + \cdots + (i_{n-1} - 1)I_1 \cdots I_{n-2} + (i_{n+1} - 1)I_1 \cdots I_n + \cdots + (i_N - 1)\, I_1 \cdots I_{N-1}$.

Tensor unfolding/matricization is one of the most important operators in tensor-based data analytics, since it gives a "matrix" view to describe an Nth order tensor data, such that fruitful results in linear algebra can be utilized. As will be seen in the later sections, most tensor algorithms involve basic operations on the matrices provided by "unfolding/matricization", and the special tensor/matrix products introduced in the following subsection.

1.1.3 Tensor Products and Norms

Tensor decompositions and products are essentially built on matrix products. We introduce the most widely used ones in this subsection. For a full list of matrix/tensor products, readers can refer to [1]. All the tensor operations in this subsection have been implemented in Matlab tensor toolbox [2].

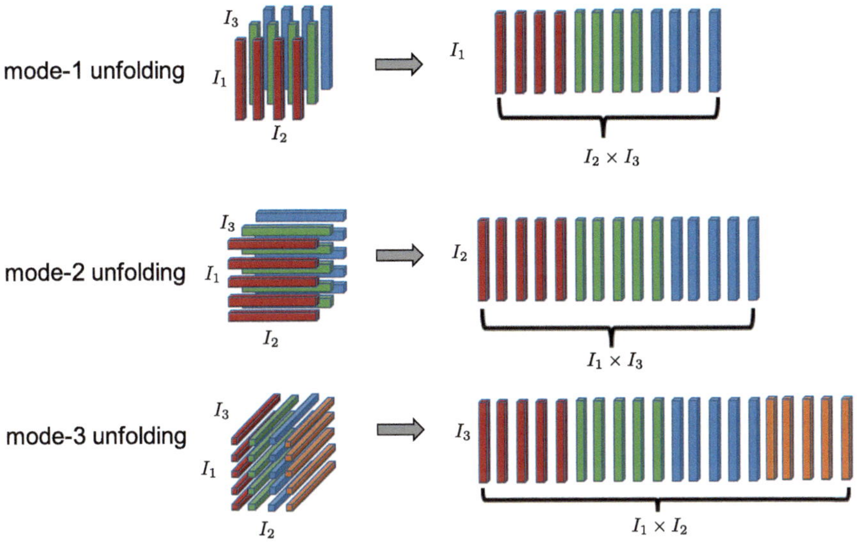

Fig. 1.2 Illustration of tensor unfolding/matricization for different modes

Definition 1.2 (*Kronecker Product*) Given two matrices $\mathbf{A} \in \mathbb{R}^{I_1 \times I_2}$ and $\mathbf{B} \in \mathbb{R}^{J_1 \times J_2}$, their Kronecker product is defined as

$$\mathbf{A} \odot \mathbf{B} = \underbrace{\begin{bmatrix} a_{11}\mathbf{B} & \cdots & a_{1I_2}\mathbf{B} \\ \vdots & \ddots & \vdots \\ a_{I_11}\mathbf{B} & \cdots & a_{I_1I_2}\mathbf{B} \end{bmatrix}}_{\triangleq \mathbf{C}} \in \mathbb{R}^{I_1J_1 \times I_2J_2}. \tag{1.1}$$

As seen in (1.1), the Kronecker product between $\mathbf{A}$ and $\mathbf{B}$ results in a matrix $\mathbf{C}$ with enlarged dimensions. From another angle, the Kronecker product provides an effective way to represent a large matrix $\mathbf{C}$ (if it satisfies (1.1)) by two smaller matrices $\{\mathbf{A}, \mathbf{B}\}$. This product will be useful for tensor Tucker decompositions, as will be elaborated in later sections.

Next, using the Kronecker product, we define another important matrix product.

Definition 1.3 (*Khatri–Rao Product*) Given two matrices $\mathbf{A} \in \mathbb{R}^{I \times R}$ and $\mathbf{B} \in \mathbb{R}^{J \times R}$, their Khatri–Rao product is defined as:

$$\mathbf{A} \diamond \mathbf{B} = \left[\mathbf{A}_{:,1} \odot \mathbf{B}_{:,1}, \mathbf{A}_{:,2} \odot \mathbf{B}_{:,2}, \ldots, \mathbf{A}_{:,R} \odot \mathbf{B}_{:,R}\right] \in \mathbb{R}^{IJ \times R}. \tag{1.2}$$

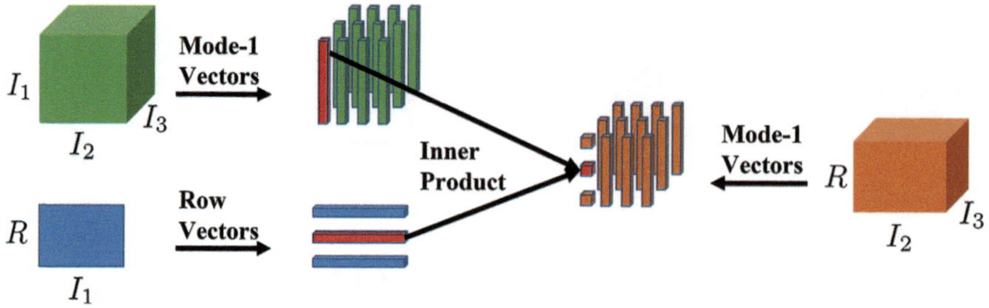

Fig. 1.3 Illustration of 1-mode product

From (1.2), it is easy to see that the Khatri–Rao product performs the column-wise Kronecker product between two matrices $\{\mathbf{A}, \mathbf{B}\}$. The Khatri–Rao product is one of the most critical operators in tensor canonical polyadic decomposition, which will be elucidated in later sections.

The Hadamard product, which performs element-wise product between two matrices $\{\mathbf{A}, \mathbf{B}\}$, is defined as follows.

Definition 1.4 (*Hadamard Product*) Given two matrices $\mathbf{A} \in \mathbb{R}^{I_1 \times I_2}$ and $\mathbf{B} \in \mathbb{R}^{I_1 \times I_2}$, their Hadamard product is:

$$\mathbf{A} \circledast \mathbf{B} = \begin{bmatrix} a_{11}b_{11} & \cdots & a_{1I_2}b_{1I_2} \\ \vdots & \ddots & \vdots \\ a_{I_1 1}b_{I_1 1} & \cdots & a_{I_1 I_2}b_{I_1 I_2} \end{bmatrix} \in \mathbb{R}^{I_1 \times I_2}. \tag{1.3}$$

Then we define several tensor products.

Definition 1.5 (*n-mode Product*) The n-mode product between a tensor $\boldsymbol{\mathcal{X}} \in \mathbb{R}^{I_1 \times \cdots \times I_N}$ and a matrix $\mathbf{M} \in \mathbb{R}^{R \times I_n}$ results in a tensor $(\boldsymbol{\mathcal{X}} \times_n \mathbf{M}) \in \mathbb{R}^{I_1 \times \cdots \times I_{n-1} \times R \times I_{n+1} \times \cdots \times I_N}$, with each element being:

$$(\boldsymbol{\mathcal{X}} \times_n \mathbf{M})_{i_1,\ldots,i_{n-1},r,i_{n+1},\ldots,i_N} = \sum_{i_n=1}^{I_n} \mathbf{M}_{r,i_n} \boldsymbol{\mathcal{X}}_{i_1,\ldots,i_N}. \tag{1.4}$$

An illustration of 1-mode product between a 3D tensor $\boldsymbol{\mathcal{X}}$ and a matrix $\mathbf{M}$ is given in Fig. 1.3. Furthermore, the n-mode product can also be expressed as a matrix product in terms of the mode-n unfolding,

$$(\boldsymbol{\mathcal{X}} \times_n \mathbf{M})^{(n)} = \mathbf{M} \times \mathbf{X}^{(n)}. \tag{1.5}$$

If the tensor $\mathcal{X} \in \mathbb{R}^{I_1 \times I_2}$ is a matrix (or a 2D tensor), its 1-mode product with $\mathbf{M}$ reduces to a matrix product, i.e., $\mathcal{X} \times_1 \mathbf{M} = \mathbf{M} \times \mathcal{X}$, where $\mathbf{M} \in \mathbb{R}^{R \times I_1}$. Similarly, $\mathcal{X} \times_2 \mathbf{M} = \mathcal{X} \times \mathbf{M}^T$, where $\mathcal{X} \in \mathbb{R}^{I_1 \times I_2}, \mathbf{M} \in \mathbb{R}^{R \times I_2}$. Another generalization from vector/matrix algebra is the generalized inner product.

Definition 1.6 (*Generalized Inner Product*) For a tensor $\mathcal{X} \in \mathbb{R}^{I_1 \times \cdots \times I_N}$ and a tensor $\mathcal{Y} \in \mathbb{R}^{I_1 \times \cdots \times I_N}$, their generalized inner product is defined as:

$$< \mathcal{X}, \mathcal{Y} >= \sum_{i_1=1}^{I_1} \sum_{i_2=1}^{I_2} \cdots \sum_{i_n=1}^{I_N} \mathcal{X}_{i_1,\ldots,i_n} \mathcal{Y}_{i_1,\ldots,i_n}. \tag{1.6}$$

In data analytic tasks, the l_p norm, which was defined for vectors and matrices, frequently appears in the designs of cost functions and regularizations. For tensors, we can generalize its definition as follows.

Definition 1.7 (l_p *tensor norm*) For a tensor $\mathcal{X} \in \mathbb{R}^{I_1 \times \cdots \times I_N}$, its l_p norm is:

$$||\mathcal{X}||_p = \left(\sum_{i_1,\ldots,i_N} |\mathcal{X}_{i_1,\ldots,i_N}|^p \right)^{1/p}. \tag{1.7}$$

For $p = 0$, the l_0 norm $||\mathcal{X}||_0$ gives the number of non-zero elements (strictly speaking l_0 does not satisfy the usual norm properties), and thus acts as a measure of sparsity. As its tightest convex surrogate, the l_1 norm $||\mathcal{X}||_1$ computes the sum of absolute values of tensor $\mathcal{X}$, and also can be treated as a convenient measure of sparsity. The most widely used one is the l_2 norm $||\mathcal{X}||_2$, which is also called the *Frobenius norm* and denoted by $||\mathcal{X}||_F$.

1.2 Representation Learning via Tensors

Multi-dimensional data from various applications can be naturally represented as tensors. To understand these data, representation learning aims at extracting low-dimensional yet informative parameters (in terms of smaller tensors, matrices, and vectors) from the tensor data. It is hoped that the extracted parameters can preserve the structures endowed by the physical phenomenon and reveal hidden interpretations. To achieve this goal, tensor decompositions with various structural constraints are developed, as illustrated in Fig. 1.4. In the following subsections, we introduce three widely used tensor decomposition formats with increasing complexity in modeling, namely canonical polyadic decomposition (CPD), Tucker decomposition (TuckerD), and tensor train decomposition (TTD).

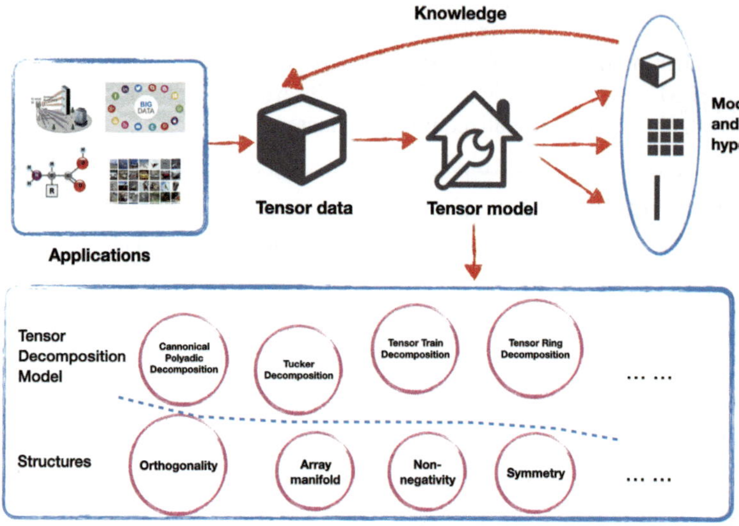

Fig. 1.4 Illustration of representation learning via tensors

1.2.1 Canonical Polyadic Decomposition (CPD)

As illustrated in Fig. 1.5, the CPD, also known as PARAFAC [3], decomposes tensor data $\mathcal{X} \in \mathbb{R}^{I_1 \times \cdots \times I_N}$ into a summation of R rank-1 tensors [3]:

$$\mathcal{X} = \sum_{r=1}^{R} \underbrace{\mathbf{u}_r^{(1)} \circ \cdots \circ \mathbf{u}_r^{(N)}}_{\text{rank-1 tensor}}, \tag{1.8}$$

where $\circ$ denotes the vector outer product. Equation (1.8) states that the tensor $\mathcal{X}$ consists of R rank-1 component tensors. If we put the vectors $\mathbf{u}_1^{(n)}, \ldots, \mathbf{u}_R^{(n)}$ into a factor matrices $\mathbf{U}^{(n)} \in \mathbb{R}^{I_n \times R}$ defined as:

$$\mathbf{U}^{(n)} = \left[\mathbf{u}_1^{(n)}, \ldots, \mathbf{u}_R^{(n)}\right], \tag{1.9}$$

Equation (1.8) can be expressed as another equivalent form $\mathcal{X} = \sum_{r=1}^{R} \mathbf{U}_{:,r}^{(1)} \circ \cdots \circ \mathbf{U}_{:,r}^{(N)} := [\![\mathbf{U}^{(1)}, \ldots, \mathbf{U}^{(N)}]\!]$, where $[\![\cdots]\!]$ is known as the Kruskal operator. Notice that the minimum number R that makes (1.8) hold is termed as *tensor rank*, which generalizes the notion of matrix rank to high-order tensors.

Tensor CPD has been found in various data analytic tasks due to its appealing uniqueness property. Here, we present one of the most widely used sufficient conditions for CPD uniqueness. For other conditions that take additional structures (e.g., nonnegativity, orthogonality) into account, interested readers can refer to [1, 3].

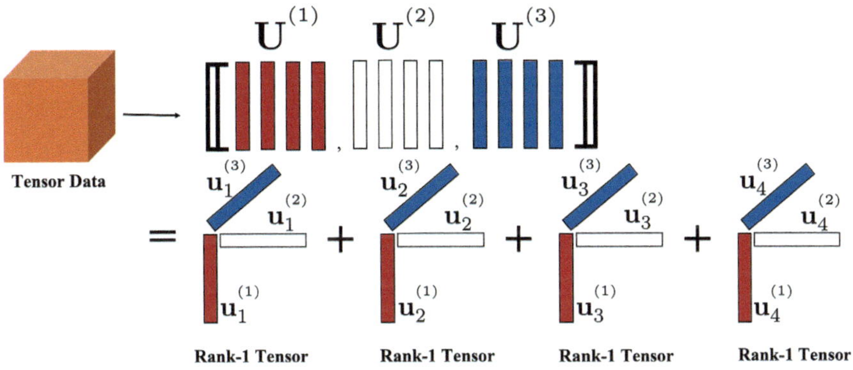

Fig. 1.5 Illustration of a CPD for a third-order tensor

Property 1.1 *(Uniqueness condition for CPD* [1]) Suppose $[\![\mathbf{U}^{(1)}, \mathbf{U}^{(2)}, \ldots, \mathbf{U}^{(N)}]\!] = [\![\Xi^{(1)}, \Xi^{(2)}, \ldots, \Xi^{(N)}]\!]$, and $\sum_{n=1}^{N} k_n \geq 2R+(N-1)$, where k_i denotes the k-rank of matrix $\mathbf{U}^{(i)}$ and R is the tensor rank. Then the following equations hold: $\Xi^{(1)} = \mathbf{U}^{(1)}\mathbf{\Delta}\mathbf{\Lambda}^{(1)}$, $\Xi^{(2)} = \mathbf{U}^{(2)}\mathbf{\Delta}\mathbf{\Lambda}^{(2)}$, ..., $\Xi^{(N)} = \mathbf{U}^{(N)}\mathbf{\Delta}\mathbf{\Lambda}^{(N)}$ where $\mathbf{\Delta}$ is a permutation matrix and the diagonal matrix $\mathbf{\Lambda}^{(n)}$ satisfies $\prod_{n=1}^{N} \mathbf{\Lambda}^{(n)} = \mathbf{I}_R$.

In **Property 1.1**, the k-rank of matrix $\mathbf{A}$ is defined as the maximum value k such that any k columns are linearly independent [1]. **Property 1.1** states that under mild conditions, tensor CPD is unique up to trivial scaling and permutation ambiguities. This is one of the major differences between tensor CPD and low-rank matrix decomposition, which is, in general, not unique unless some constraints are imposed. This nice property has made CPD an important tool in the *blind source separation* and *data clustering*-related tasks, as will be demonstrated in the following chapters.

1.2.2 Tucker Decomposition (TuckerD)

The CPD disregards interactions among the columns of factor matrices and requires the factor matrices to have the same number of columns. To achieve a more flexible tensor representation, tensor TuckerD was introduced to generalize CPD by allowing different column numbers of factor matrices and introducing a core tensor $\mathcal{G} \in \mathbb{R}^{R_1 \times \cdots \times R_N}$. Particularly, tensor TuckerD is defined as [1, 4]:

$$\mathcal{X} = \mathcal{G} \times_1 \mathbf{U}^{(1)} \times_2 \mathbf{U}^{(2)} \times_3 \cdots \times_N \mathbf{U}^{(N)}, \tag{1.10}$$

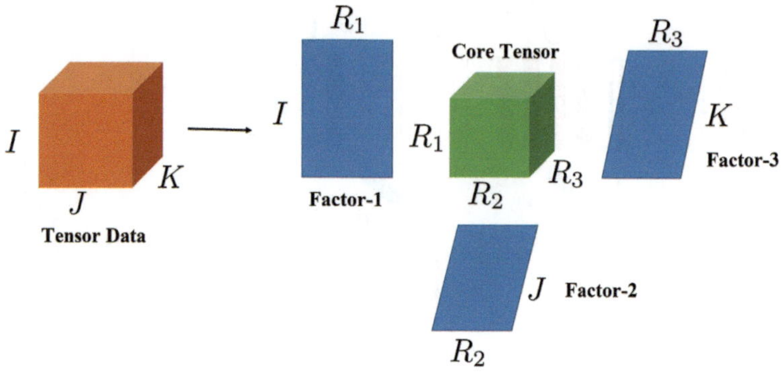

Fig. 1.6 Illustration of a TuckerD for a third-order tensor

where each factor matrix $\mathbf{U}^{(n)} \in \mathbb{R}^{I_n \times R_n}, \forall n$. The tuple $(R_1, \ldots, R_N)$ is known as *multi-linear rank*. An illustration of TuckerD is provided in Fig. 1.6. Note that when the core tensor $\mathcal{G}$ is super-diagonal and $R_1 = \cdots = R_N$, TuckerD reduces to CPD. Using the Kruskal operator, Tucker D can be compactly denoted by:

$$\mathcal{X} = [\![\mathbf{G}; \mathbf{U}^{(1)}, \ldots, \mathbf{U}^{(N)}]\!]. \tag{1.11}$$

Although TuckerD provides flexibilities for data representation, it is not unique in general [1, 4]. Therefore, it is frequently used in the *data compression*, *basis function learning*, and *feature extraction*-related tasks, where uniqueness is not the most important consideration.

1.2.3 Tensor Train Decomposition (TTD)

The TTD decomposes tensor data $\mathcal{X} \in \mathbb{R}^{I_1 \times \cdots \times I_N}$ into a set of core tensors $\{\mathcal{G}^{(n)} \in \mathbb{R}^{R_n \times I_n \times R_{n+1}}\}$ such that [5]

$$\mathcal{X}_{i_1,\ldots,i_N} = \mathcal{G}^{(1)}_{:,i_1,:} \times \cdots \times \mathcal{G}^{(N)}_{:,i_N,:}. \tag{1.12}$$

In (1.12), each core tensor slice $\mathcal{G}^{(n)}_{:,i_n,:} \in \mathbb{R}^{R_n \times R_{n+1}}$. Since each $\mathcal{X}_{i_1,\ldots,i_N}$ is a scalar, R_1 and R_{N+1} are both required to be 1. The tuple $(R_1, \ldots, R_{N+1})$ is termed as *TT-rank*. In quantum physics, TTD is known as a matrix-product state [5]. The TTD for a third-order tensor is illustrated in Fig. 1.7.

Due to its flexibility, TTD with appropriately chosen TT-rank has shown superior performance in a variety of data analytic tasks, including *image completion*, *classification*, and *neural network compression*.

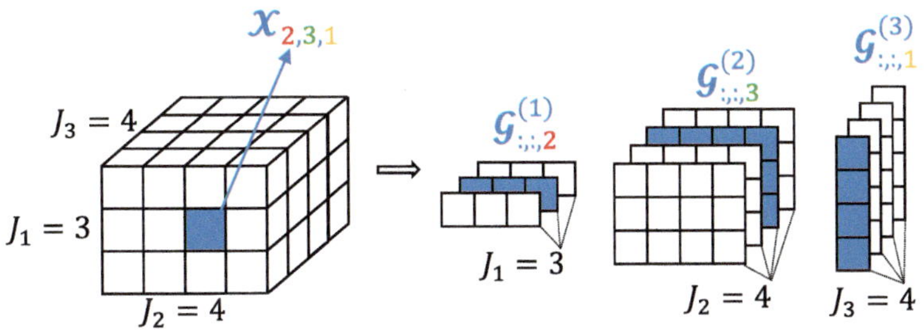

Fig. 1.7 Illustration of a TTD for a third-order tensor

1.3 Model Fitting and Challenges Ahead

Given the tensor decomposition models introduced in the last section, the next task is to estimate the model parameters and hyper-parameters from the observed tensor data. One straightforward approach is to formulate the learning problem as an optimization problem (see Fig. 1.8). Specifically, in different application contexts, cost functions can be designed to encode our knowledge of the tensor data and the tensor model. Constraints on model parameters can be further added to embed the side information. The problem formulation generally appears in the form:

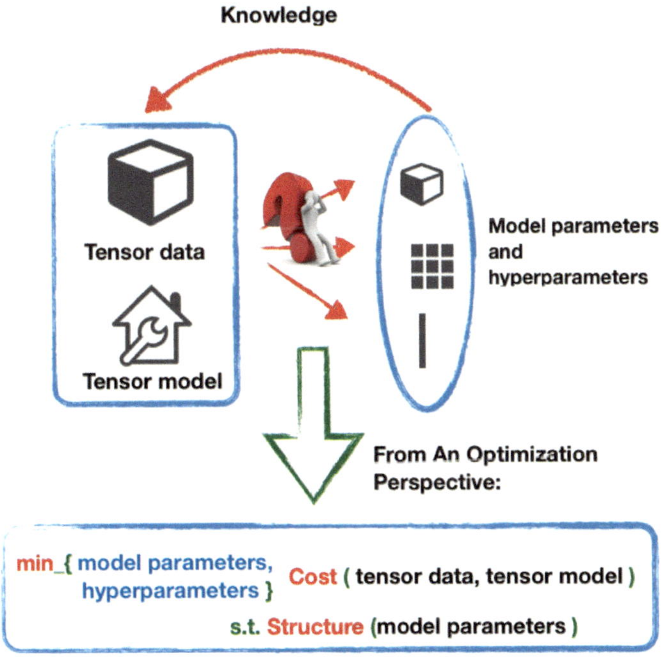

Fig. 1.8 Tensor-based representation learning from an optimization perspective

$$\begin{aligned}
&\min_{\boldsymbol{\Theta},\eta} c(\boldsymbol{\mathcal{Y}}; \boldsymbol{\Theta}, \eta) \\
&\text{s.t. } g_i(\boldsymbol{\Theta}) = \mathbf{0}, i = 1, \ldots, I, \\
&\qquad f_j(\boldsymbol{\Theta}) \geq \mathbf{0}, j = 1, \ldots, J,
\end{aligned} \tag{1.13}$$

where $c(\cdot)$ is the cost function, e.g., least-squares function; $g(\cdot)$ denotes the function for equality constraints; and $f(\cdot)$ is the function for inequality constraints. $\boldsymbol{\mathcal{Y}} \in \mathbb{R}^{I_1 \times \cdots \times I_N}$ is the observed tensor data. $\boldsymbol{\Theta}$ includes all the model parameters, e.g., factor matrices of CPD, and η denotes the hyper-parameters, e.g., tensor rank or regularization parameters.

In the following, we first provide a concrete example for the tensor CPD format, and then discuss the optimizations for tensor TuckerD and TTD.

1.3.1 Example: Tensor CPD

For the tensor CPD, as defined in Sect. 1.2.1, the model parameters $\boldsymbol{\Theta} = \{\mathbf{U}^{(n)} \in \mathbb{R}^{I_n \times R}\}_{n=1}^N$ and the hyper-parameter η is the tensor rank R. Adopting the least-squares cost function, the problem can be formulated as

$$\min_{\{\mathbf{U}^{(n)} \in \mathbb{R}^{I_n \times R}\}_{n=1}^N, R} ||\boldsymbol{\mathcal{Y}} - [\![\mathbf{U}^{(1)}, \ldots, \mathbf{U}^{(N)}]\!]||_F^2. \tag{1.14}$$

In problem (1.14), optimizing R, which is an integer, is known to be non-deterministic polynomial-time hard (NP-hard) [6]. Trial-and-errors or cross-validations are widely used in practice to tune this hyper-parameter for the best data interpretation. Even given a tensor rank R, the optimization of $\{\mathbf{U}^{(n)} \in \mathbb{R}^{I_n \times R}\}$ is still a non-convex problem. To tackle the non-convexity, alternating optimization is commonly used. Particularly, after fixing factor matrices other than $\mathbf{U}^{(k)}$, the problem can be equivalently formulated as

$$\min_{\mathbf{U}^{(k)}} ||\mathbf{Y}^{(k)} - \mathbf{U}^{(k)} \left(\mathop{\diamond}_{n=1,n\neq k}^{N} \mathbf{U}^{(n)} \right)^T ||_F^2, \tag{1.15}$$

where $\mathop{\diamond}_{n=1,n\neq k}^{N} \mathbf{U}^{(n)} = \mathbf{U}^{(N)} \diamond \cdots \diamond \mathbf{U}^{(k+1)} \diamond \mathbf{U}^{(k-1)} \diamond \cdots \diamond \mathbf{U}^{(1)}$, and $\diamond$ stands for the Khatri–Rao product introduced in **Definition 1.3**. Problem (1.15) is a convex problem with respect to variable $\mathbf{U}^{(k)}$. After taking the gradient and then letting it to be zero, it is easy to obtain:

$$\hat{\mathbf{U}}^{(k)} = \mathbf{Y}^{(k)} \left[\left(\mathop{\diamond}_{n=1,n\neq k}^{N} \mathbf{U}^{(n)} \right)^T \right]^{\dagger}, \tag{1.16}$$

Algorithm 1 CPD-ALS($\mathcal{Y}$, R)

Initializations: initialize $\mathbf{U}^{(n,0)} \in \mathbb{R}^{I_n \times R}$, for $n = 1, \ldots, N$.
Iterations: for the tth iteration ($t \geq 0$),
for the kth factor matrix $\mathbf{U}^{(k,0)}$, $k = 1, \ldots, N$,

$$\mathbf{U}^{(k,t+1)} = \mathbf{Y}^{(k)} \left(\mathop{\diamond}_{n=1,n\neq k}^{N} \mathbf{U}^{(n,s)} \right)^T \left(\mathop{\circledast}_{n=1,n\neq k}^{N} \mathbf{U}^{(n,s)T} \mathbf{U}^{(n,s)} \right)^{\dagger},$$

where $s = t + 1$ when $s \leq k$, and otherwise $s = t$.
Until Convergence

where $\dagger$ denotes the Moore–Penrose pseudo inverse. Due to the property of the Khatri–Rao product, Eq. (1.16) can be simplified as:

$$\hat{\mathbf{U}}^{(k)} = \mathbf{Y}^{(k)} \left(\mathop{\diamond}_{n=1,n\neq k}^{N} \mathbf{U}^{(n)} \right)^T \left(\mathop{\circledast}_{n=1,n\neq k}^{N} \mathbf{U}^{(n)T} \mathbf{U}^{(n)} \right)^{\dagger}. \tag{1.17}$$

Using (1.17), we only need to compute the pseudo inverse of a small matrix with size $R \times R$, rather than a large matrix with size $\prod_{n\neq k} I_n \times R$ in (1.16).

By alternatively updating factor matrix $\mathbf{U}^{(k)}$, $\forall k$, using (1.17) until convergence, we arrive at the workhorse algorithm for a given R: alternating least-squares (ALS) method [7], which is summarized in **Algorithm 1**.

The alternating optimization framework still applies even if additional structural constraints are imposed on the factor matrices. For example, if nonnegativeness or orthogonality constraints are added, the subproblem (1.15) becomes:

$$\begin{aligned} &\min_{\mathbf{U}^{(k)}} \left\| \mathbf{Y}^{(k)} - \mathbf{U}^{(k)} \left(\mathop{\diamond}_{n=1,n\neq k}^{N} \mathbf{U}^{(n)} \right)^T \right\|_F^2, \\ &\text{s.t. } \mathbf{U}^{(k)} \geq 0 \ \text{(non-negativeness)}; \\ &\text{or s.t. } \mathbf{U}^{(k)T}\mathbf{U}^{(k)} = \mathbf{I}_R \ \text{(orthogonality)}. \end{aligned} \tag{1.18}$$

As long as each subproblem (1.18) can be solved, CPD with additional structures can be learned. Recent advances in large-scale non-convex optimizations have tackled problems in the form of (1.18) [8, 9].

1.3.1.1 Optimizations for Tensor TuckerD and TTD

Following the general problem formulation (1.13), the learning problem of tensor TuckerD is usually given as

Algorithm 2 TuckerD-HOOI($\mathcal{Y}$, $\{R_n\}_{n=1}^N$)

Initializations: initialize $\mathbf{U}^{(n,0)} \in \mathbb{R}^{I_n \times R_n}$, for $n = 1, \ldots, N$.
Iterations: for the tth iteration ($t \geq 0$),
for the kth factor matrix $\mathbf{U}^{(k,0)}$, $k = 1, \ldots, N$,

$$\mathbf{C}_{(k)}^t = \mathbf{Y}^{(k)} \left(\mathbf{U}^{(N,t-1)} \odot \cdots \odot \mathbf{U}^{(k+1,t-1)} \odot \mathbf{U}^{(k-1,t)} \odot \cdots \odot \mathbf{U}^{(1,t)} \right). \tag{1.20}$$

$\mathbf{U}^{(k,t)} \leftarrow R_k$ leading left singular vectors of $\mathbf{C}_{(k)}^t$.
Until Convergence

Algorithm 3 TT-SVD($\mathcal{Y}$, ϵ)

Initializations: compute truncation parameter $\delta = \frac{\epsilon}{\sqrt{N-1}} \|\mathcal{Y}\|_F$. Set $C = \mathcal{Y}$, $R_0 = 1$.
For $n = 1$ **to** $N-1$

$\mathbf{C} = reshape\left(C, \left[R_{n-1} I_n, \frac{\prod_{n=1}^N I_n}{R_{n-1} I_n}\right]\right)$.

Compute δ-truncated SVD: $\mathbf{C} = \mathbf{USV} + \mathbf{E}$, $\|\mathbf{E}\| \leq \epsilon$, $R_n = \text{rank}_\delta(\mathbf{C})$.

$\mathcal{G}^{(n)} = reshape(\mathbf{U}, [R_{n-1}, I_k, R_n])$.

$\mathbf{C} = \mathbf{SV}^T$.

End
$\mathcal{G}^{(N)} = \mathbf{C}$, $R_N = 1$.
Return $\{\mathcal{G}^{(n)}\}_{n=1}^N$, $\{R_n\}_{n=1}^N$.

$$\min_{\mathbf{G}, \{\mathbf{U}^{(n)}\}_{n=1}^N, \{R_n\}_{n=1}^N} \| \mathcal{Y} - [\![\mathbf{G}; \mathbf{U}^{(1)}, \ldots, \mathbf{U}^{(N)}]\!] \|_F$$
$$\text{s.t. } \mathbf{U}^{(n)T} \mathbf{U}^{(n)} = \mathbf{I}_{R_n}, \tag{1.19}$$

where the orthogonal constraint is imposed on each factor matrix. Once again, we can use the ALS approach to solve problem (1.19). That is, in each iteration, we optimize one factor matrix (or core tensor) at a time while fixing other unknown parameters to the most recent updated values. The resulting algorithm, also known as higher order orthogonal iteration (HOOI), is summarized in **Algorithm 2**. Interested readers can find the detailed derivation and convergence analysis in [10].

Similarly, for the TTD format, the learning problem becomes:

$$\min_{\{\mathcal{G}^{(n)}\}_{n=1}^N, \{R_n\}_{n=1}^N} \| \mathcal{Y} - \mathcal{X}(\{\mathcal{G}^{(n)}\}_{n=1}^N) \|_F, \tag{1.21}$$

where $\mathcal{X}(\{\mathcal{G}^{(n)}\}_{n=1}^N)$ is a TT-format tensor following (1.12). TT-SVD, which iteratively optimizes the TT cores by truncated SVDs, can yield TT cores $\{\mathcal{G}^{(n)}\}_{n=1}^N$ such that $\left\| \mathcal{Y} - \mathcal{X}(\{\mathcal{G}^{(n)}\}_{n=1}^N) \right\|_F \leq \epsilon \|\mathcal{Y}\|_F$, where ϵ is the prescribed accuracy. The TT-SVD algorithm is summarized as **Algorithm 3** [5].

1.3.2 Challenges in Rank Determination

From **Algorithms 1, 2 and 3**, it can be seen that in addition to the tensor data $\mathcal{Y}$, a number of hyper-parameters are required to be set, including tensor rank R in **Algorithm 1**, multi-linear rank $\{R_n\}_{n=1}^N$ in **Algorithm 2**, and the prescribed accuracy ϵ in **Algorithm 3**. Notice that in **Algorithm 3**, the prescribed accuracy ϵ determines TT-rank $\{R_n\}_{n=1}^N$. Setting these ranks to large values would lead to overfitting, while setting them too small would lead to inflexibility of the modeling abilities. Manually tuning these hyper-parameters is computationally costly. In many previous works (e.g., [8]), research effort has been put into designing fast algorithms for each tensor decomposition format. Although the reduction of running time for various tensor decomposition formats has been witnessed, executing the algorithm numerous times (for different combinations of hyper-parameters) is still inevitable in order to select the best hyper-parameters. Compared to CPD, which has only a single number as the rank, the rank determination issue for TuckerD and TTD is much more challenging, since they have more than one hyper-parameter. Exhaustively testing different combinations of hyper-parameter values would result in a prohibitive computation burden. This challenge raises an immediate question: Could the hyper-parameters automatically be learned from the training data along with the factor matrices/core tensors by running an algorithm only one time?

This thought seems arduous since it is known that the optimization of tensor rank is NP-hard. However, recent advances in Bayesian modeling and inference [11] have provided viable solutions, see, e.g., [12–18], each with a comparable computational complexity to the optimization-based methods [8] with fixed ranks. Their common idea is to first assume an over-parameterized model by letting the associated hyper-parameters take large values. It then leverages the Bayesian sparsity-aware modeling and Occam's razor principle to prune out redundant parameters [11]. Consequently, in addition to the posterior distributions of unknown factor matrices/core tensors, which provide additional uncertainty information compared to the optimization-based counterparts, the Bayesian learning process also unveils the most suitable values of hyper-parameters that can avoid overfitting and further improve the model interpretability. This book aims to give a unified view and insight into recent Bayesian tensor decompositions.

References

1. T.G. Kolda, B.W. Bader, Tensor decompositions and applications. SIAM Rev. **51**(3), 455–500 (2009)
2. B.W. Bader, T.G. Kolda, et al., Matlab tensor toolbox version 3.1 (2019)
3. J.H.d.M. Goulart, M. Boizard, R. Boyer, G. Favier, P. Comon, Tensor cp decomposition with structured factor matrices: algorithms and performance. IEEE J. Selected Topics Signal Process. **10**(4), 757–769 (2015)
4. V. Bhatt, S. Kumar, S. Saini, Tucker decomposition and applications. Mater. Today: Proc. (2021)

5. I.V. Oseledets, Tensor-train decomposition. SIAM J. Sci. Comput. **33**(5), 2295–2317 (2011)
6. H. Johan, Tensor rank is np-complete. J. Algor. **4**(11), 644–654 (1990)
7. J.D. Carroll, J.-J. Chang, Analysis of individual differences in multidimensional scaling via an n-way generalization of "eckart-young" decomposition. Psychometrika **35**(3), 283–319 (1970)
8. X. Fu, N. Vervliet, L. De Lathauwer, K. Huang, N. Gillis, Computing large-scale matrix and tensor decomposition with structured factors: a unified nonconvex optimization perspective. IEEE Signal Process. Mag. **37**(5), 78–94 (2020)
9. B. Yang, A.S. Zamzam, N.D. Sidiropoulos, Large scale tensor factorization via parallel sketches. IEEE Trans. Knowl. Data Eng. (2020)
10. C.A. Andersson, R. Bro, Improving the speed of multi-way algorithms: part i. tucker3. Chemom. Intell. Lab. Syst. **42**(1–2), 93–103 (1998)
11. S. Theodoridis, *Machine Learning: a Bayesian and Optimization Perspective*, 2nd edn. (Academic, Cambridge, 2020)
12. L. Cheng, X. Tong, S. Wang, Y.-C. Wu, H.V. Poor, Learning nonnegative factors from tensor data: probabilistic modeling and inference algorithm. IEEE Trans. Signal Process. **68**, 1792–1806 (2020)
13. L. Xu, L. Cheng, N. Wong, Y.-C. Wu, Overfitting avoidance in tensor train factorization and completion: prior analysis and inference, in *International Conference on Data Mining (ICDM)* (2021)
14. L. Cheng, Y.-C. Wu, H.V. Poor, Scaling probabilistic tensor canonical polyadic decomposition to massive data. IEEE Trans. Signal Process. **66**(21), 5534–5548 (2018)
15. L. Cheng, Y.-C. Wu, H.V. Poor, Probabilistic tensor canonical polyadic decomposition with orthogonal factors. IEEE Trans. Signal Process. **65**(3), 663–676 (2016)
16. Y. Zhou, Y.-M. Cheung, Bayesian low-tubal-rank robust tensor factorization with multi-rank determination. IEEE Trans. Pattern Anal. Mach. Intell. 62–76 (2019)
17. Z. Zhang, C. Hawkins, Variational bayesian inference for robust streaming tensor factorization and completion, in *Proceeding of the IEEE International Conference on Data Mining (ICDM)* (2018), pp. 1446–1451
18. Q. Zhao, L. Zhang, A. Cichocki, Bayesian CP factorization of incomplete tensors with automatic rank determination. IEEE Trans. Pattern Anal. Mach. Intell. **37**(9), 1751–1763 (2015)

Chapter 2
Bayesian Learning for Sparsity-Aware Modeling

Abstract In this chapter, we will first introduce the Bayesian philosophy, under which the two essential merits, namely uncertainty quantification and model selection, are highlighted. These merits shed light on the design of sparsity-promoting prior for automating the model pruning in recent machine learning models, including deep neural networks, Gaussian processes, and tensor decompositions. Then, we introduce the variational inference framework for algorithm development and discuss its tractability in different Bayesian models.

2.1 Bayes' Theorem

Bayes' theorem is the first profound triumph of statistical inference [1], since it elegantly establishes a framework for combining prior experience with the data observations.

In a typical inverse problem [2], for a given dataset $\mathcal{D}$, we hope to extract the knowledge from $\mathcal{D}$ via a machine learning model parameterized by a vector $\boldsymbol{\theta}$. In the Bayesian framework, a prior distribution $p(\boldsymbol{\theta}|\boldsymbol{\xi}_p)$ is assumed for the model parameters $\boldsymbol{\theta}$, encoding our belief before observing the data, where $\boldsymbol{\xi}_p$ denotes an unknown yet deterministic hyper-parameter vector of the prior distribution. After collecting the data $\mathcal{D}$, the likelihood function $p(\mathcal{D}|\boldsymbol{\theta}, \boldsymbol{\xi}_l)$, where $\boldsymbol{\xi}_l$ is the hyper-parameters, is used to model the *forward problem*. While $p(\mathcal{D}|\boldsymbol{\theta}, \boldsymbol{\xi}_l)$ links the data and the model, our ultimate task is to learn the parameters $\boldsymbol{\theta}$ from data $\mathcal{D}$, which is encoded in the posterior distribution $p(\boldsymbol{\theta}|\mathcal{D}, \boldsymbol{\xi}_l, \boldsymbol{\xi}_p)$. Bayes' theorem rigorously formulates such a process [1]:

$$p(\boldsymbol{\theta}|\mathcal{D}, \boldsymbol{\xi}) = \frac{p(\mathcal{D}|\boldsymbol{\theta}, \boldsymbol{\xi}_l)p(\boldsymbol{\theta}|\boldsymbol{\xi}_p)}{\int p(\mathcal{D}|\boldsymbol{\theta}, \boldsymbol{\xi}_l)p(\boldsymbol{\theta}|\boldsymbol{\xi}_p)d\boldsymbol{\theta}}, \tag{2.1}$$

where $\boldsymbol{\xi}$ includes both the hyper-parameters contained in prior (i.e., $\boldsymbol{\xi}_p$), and hyper-parameters of the likelihood function $\boldsymbol{\xi}_l$ (e.g., the noise power in the dataset $\mathcal{D}$).

L. Cheng et al., *Bayesian Tensor Decomposition for Signal Processing and Machine Learning*, https://doi.org/10.1007/978-3-031-22438-6_2

Bayes' theorem stated in (2.1) reveals how to link the *inverse problem* (posterior distribution $p(\boldsymbol{\theta}|\mathcal{D}, \boldsymbol{\xi})$) to the *forward problem* (likelihood function $p(\mathcal{D}|\boldsymbol{\theta}, \boldsymbol{\xi}_l)$) via incorporating the *prior* $p(\boldsymbol{\theta}|\boldsymbol{\xi}_p)$. As the likelihood is usually easy to obtain from the task objective or the measurement process, the most critical step lies in seeking a suitable prior, which will be covered in the next section.

The denominator of (2.1), which can be expressed as $p(\mathcal{D}|\boldsymbol{\xi})$, plays an important role. This term is known as "model evidence" since it measures the "fitness" of data $\mathcal{D}$ given the model hyper-parameters $\boldsymbol{\xi}$, which is crucial for model selection. Concretely, rather than manually tuning those hyper-parameters on a separated validation dataset, we can learn their optimal values directly from the training dataset by solving the following problem:

$$\max_{\boldsymbol{\xi}} \log p(\mathcal{D}|\boldsymbol{\xi}). \tag{2.2}$$

This problem is known as the model *evidence maximization* problem [2].

Bayes' theorem returns the posterior distribution of model parameters, $p(\boldsymbol{\theta}|\mathcal{D}, \boldsymbol{\xi})$. Unlike the discriminative approach (or cost function optimization approach), a distribution (rather than a single point estimate) of the model parameter is obtained, thus conveying much richer information for the downstream process of unseen data [2]. One of the most important pieces of information possessed is the uncertainty of model parameters, which shows the extent of our belief in the learned model. This information is invaluable for many mission-critical applications like auto-driving [3] and medical diagnosis [4].

2.2 Bayesian Learning and Sparsity-Aware Learning

Bayesian learning is a class of machine learning methods that rely on Bayes' theorem [2, 5]. Following the principle of Bayes' theorem introduced in the last section, Bayesian learning involves two crucial stages: (1) prior design for the model parameters [6]; and (2) the inference algorithm development that learns both the posterior distribution and the hyper-parameter values from the training (or observed) data.

In this monograph, we focus on sparsity-aware learning (SAL), which aims at leveraging the explicit or implicit sparsity pattern of data to enhance the understanding of data [7, 8]. The history of SAL dates back to the early 1970s when estimating the reflected signals from seismic measurements [6]. At that time, the notion of "sparsity", which means that the signal to be estimated only has a few elements being non-zero, was exploited to enhance the signal estimation performance. After the 2000s, driven by the bloom of compressive sensing, SAL has become a focus of signal processing and machine learning research, giving rise to many strong theoretical results and successful algorithms in various domains [7, 8].

There is a common misunderstanding of "sparsity" that floats around newcomers of signal processing and machine learning: "sparsity" is oftentimes superficially understood as nulling the values of parameters. However, if the model parameters are uniformly shrunk to be all-zeros, the machine learning model collapses and cannot learn interpretable results on the data. Consequently, in addition to favoring small values, the "sparsity" notion also promotes *a few large* values of model parameters. These few but significant model parameters (learned via SAL) distill the essential information from data, thus effectively avoiding overfitting and delivering enhanced interpretability. More discussions can be found in [2].

With SAL being a desirable property, the next question is: how do we properly design the prior distribution to encode the sparsity structure? Answering this question is the first step toward the Bayesian SAL, which has recently enabled automatic model structure pruning in deep learning [9], Gaussian processes [10], and tensor decompositions [11]. We elaborate on its design next.

2.3 Prior Design for Sparsity-Aware Modeling

There are two main streams for modeling sparsity in recent Bayesian studies, namely the parametric and non-parametric ways. Here, we introduce the parametric prior modeling relying on the heavy-tailed distributions. Interested readers on non-parametric modeling could refer to [2].

To see why the heavy-tailed distributions successfully model the sparsity, we present the Laplacian distribution and Gaussian distribution for $\boldsymbol{\theta} \in \mathbb{R}^2$ in Fig. 2.1. It can be seen that the Laplacian distribution is heavy-tailed compared to the Gaussian

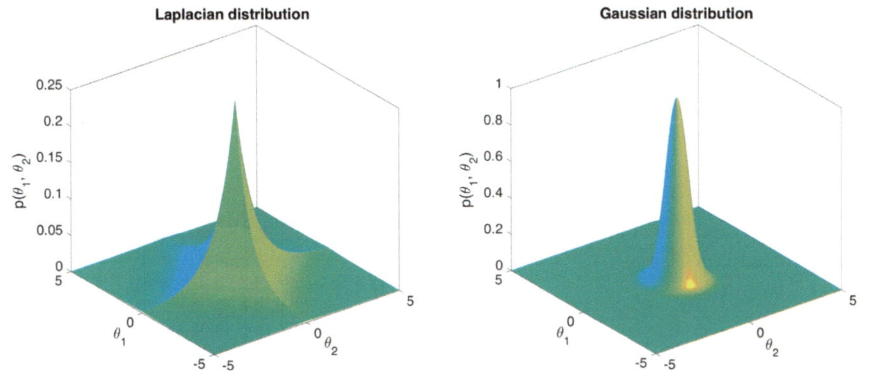

(a) (b)

Fig. 2.1 Joint probability distribution of the model parameters in 2D space. Subfigure **a** shows the Laplacian distribution and subfigure **b** shows the Gaussian distribution. The heavy-tailed Laplacian distribution peaks sharply around zero and falls slowly along the axes, thus promoting sparse solutions in a probabilistic manner. On the contrary, the Gaussian distribution decays more rapidly along both dimensions than the Laplacian distribution

distribution. In other words, for Gaussian distributed parameters, the probability that they take non-zero values goes to zero very fast, and most of the probability mass concentrates around zero. This is not ideal for sparsity modeling since we want most of the values to be (close to) zero, but still, some of the parameters have large values. In contrast, observe that for the Laplacian distributed parameters, although most of the probability mass is close to zero, there is still a high enough probability for non-zero values. More importantly, this probability mass is concentrated along the axes, where one of the parameters is zero. This is how Laplacian prior promotes sparsity.

In statistics, Laplacian is not the only distribution having heavy tails. In the sequel, we introduce an important Gaussian scale mixture (GSM) family which has heavy tails and thus can be used as sparsity-promoting priors. For a vector $\boldsymbol{\theta} = [\theta_1, \ldots, \theta_L]$, the main idea of the GSM prior is to assume that (a) the parameters, $\theta_l,\ l = 1, 2, \ldots, L$, are mutually statistically independent; (b) each one of them follows a Gaussian prior with zero mean; and (c) the respective variances, $\zeta_l,\ l = 1, 2, \ldots, L$, are also random variables, each one following a prior $p(\zeta_l|\xi_p)$, where ξ_p is a set of tuning hyper-parameters associated with the prior. Thus, the GSM prior for each θ_l is expressed as

$$p(\theta_l|\xi_p) = \int \mathcal{N}(\theta_l|0, \zeta_l)p(\zeta_l|\xi_p)d\zeta_l. \tag{2.3}$$

By varying the functional forms of $p(\zeta_l|\xi_p)$, the marginalization performed in (2.3) induces different prior distributions on θ_l. For example, if $p(\zeta_l|\xi_p)$ is an inverse Gamma distribution, (2.3) induces a Student's t distribution [12]; if $p(\zeta_l|\xi_p)$ is a Gamma distribution, (2.3) induces a Laplacian distribution [12]. Table 2.2 summarizes different heavy-tailed distributions in the GSM family, including Normal-Jefferys, generalized hyperbolic, and horseshoe distributions, among others. To illustrate the sparsity-promoting property of the GSM family, in addition to the Laplacian distribution shown in Fig. 2.1, we depict two more representative GSM prior distributions, namely the Student's t distribution and the horseshoe distribution, in Fig. 2.2 (Table 2.1).

Table 2.1 Examples of GSM prior. Abbreviations: Ga = Gamma, IG = inverse Gamma, GIG = generalized inverse Gaussian, C^+ = Half Cauchy

GSM prior $p(\theta_l\|\xi_p)$	Mixing distribution $p(\zeta_l\|\xi_p)$
Student's t	Inverse Gamma: $p(\zeta_l\|\xi_p = \{a, b\}) = \mathrm{IG}(\zeta_l\|a, b)$
Normal-Jefferys	Log-uniform: $p(\zeta_l\|\xi_p = \{\ \}) \propto \frac{1}{\|\zeta_l\|}$
Laplacian	Gamma: $p(\zeta_l\|\xi_p = \{a, b\}) = \mathrm{gamma}(\zeta_l\|a, b)$
Generalized hyperbolic	Generalized inverse Gaussian: $p(\zeta_l\|\xi_p = \{a, b, \lambda\}) = \mathrm{GIG}(\zeta_l\|a, b, \lambda)$
Horseshoe	$\zeta_l = \tau_l \upsilon_l,\ \xi_p = \{a, b\}$ Half Cauchy: $p(\tau_l) = C^+(0, a)$ $p(\upsilon_l) = C^+(0, b)$

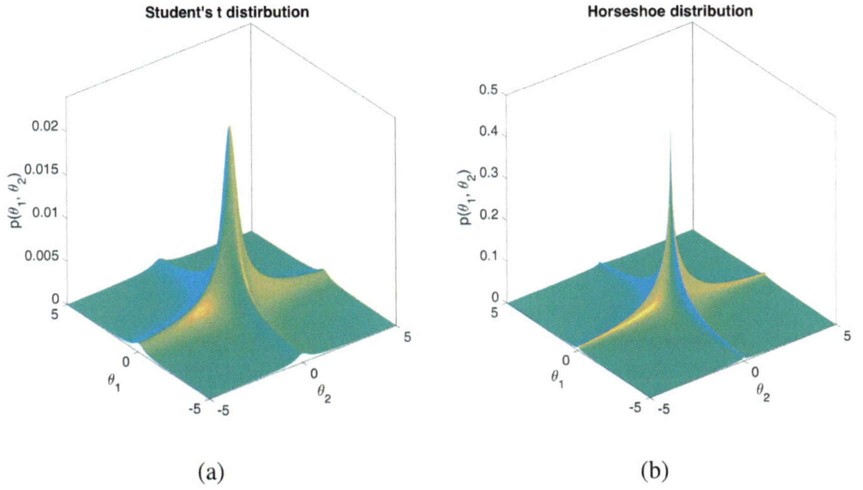

Fig. 2.2 Representative GSM prior distributions in 2D space. Subfigures **a** and **b** show the Student's t distribution and the horseshoe distribution, respectively. It can be seen that these two distributions show different heavy-tailed profiles but are both sparsity-promoting

Table 2.2 Examples of exponential family distributions

Exponential family distribution	Natural parameter	Sufficient statistic
Univariate Gaussian distribution $\mathcal{N}(x\|\mu, \sigma)$	$\left[\frac{\mu}{\sigma^2}; \frac{-1}{2\sigma^2}\right]$	$\left[x; x^2\right]$
Multivariate Gaussian distribution $\mathcal{N}(\mathbf{x}\|\boldsymbol{\mu}, \boldsymbol{\Sigma})$	$\left[\boldsymbol{\Sigma}^{-1}\boldsymbol{\mu}; -\frac{1}{2}\text{vec}(\boldsymbol{\Sigma}^{-1})\right]$	$\left[\mathbf{x}; \text{vec}(\mathbf{x}\mathbf{x}^T)\right]$
Gamma distribution gamma$(x\|a, b)$	$\left[-b; a-1\right]$	$\left[x; \log x\right]$

2.4 Inference Algorithm Development

Given a signal processing or machine learning task with the associated likelihood function $p(\mathcal{D}|\boldsymbol{\theta}, \boldsymbol{\xi}_l)$ and a sparsity-promoting prior $p(\boldsymbol{\theta}|\xi_p)$, the goal of Bayesian SAL is to infer the posterior distribution $p(\boldsymbol{\theta}|\mathcal{D}, \boldsymbol{\xi})$ and to obtain the model hyperparameters $\boldsymbol{\xi}$ by maximizing the evidence $p(\mathcal{D}|\boldsymbol{\xi})$.

In most cases, the multiple integrations required in computing the evidence are analytically intractable. Inspired by the ideas of the *Minorize-Maximization* (also called Majorization-minimization (MM)) optimization framework, we can seek for a tractable lower bound that minorizes the evidence function (see discussion around (2.2)), and maximize the lower bound iteratively until convergence. It has been shown, see, e.g., [2], that such an optimization process can obtain a stationary point of the evidence function. More concretely, the logarithm of the evidence function is lower bounded as follows:

$$\log p(\mathcal{D}|\boldsymbol{\xi}) \geq \mathcal{L}(Q(\boldsymbol{\theta}); \boldsymbol{\xi}), \tag{2.4}$$

where the lower bound

$$\mathcal{L}(Q(\boldsymbol{\theta}); \boldsymbol{\xi}) \triangleq \int Q(\boldsymbol{\theta}) \log \frac{p(\mathcal{D}, \boldsymbol{\theta}|\boldsymbol{\xi})}{Q(\boldsymbol{\theta})} d\boldsymbol{\theta}, \tag{2.5}$$

is called *evidence lower bound* (ELBO), and $Q(\boldsymbol{\theta})$ is known as the variational distribution. The tightness of the ELBO is determined by the closeness between the variational distribution $Q(\boldsymbol{\theta})$ and the posterior $p(\boldsymbol{\theta}|\mathcal{D}, \boldsymbol{\xi})$, measured by the Kullback–Leibler (KL) divergence, KL $(Q(\boldsymbol{\theta})||p(\boldsymbol{\theta}|\mathcal{D}, \boldsymbol{\xi}))$. In fact, the ELBO becomes tight (i.e., the lower bound becomes equal to the evidence) if and only if $Q(\boldsymbol{\theta}) = p(\boldsymbol{\theta}|\mathcal{D}, \boldsymbol{\xi})$ or equivalently KL $(Q(\boldsymbol{\theta})||p(\boldsymbol{\theta}|\mathcal{D}, \boldsymbol{\xi})) = 0$. This is easy to see if we expand (2.5) and reformulate it as

$$\log p(\mathcal{D}|\boldsymbol{\xi}) = \mathcal{L}(Q(\theta); \boldsymbol{\xi}) + \mathrm{KL}(Q(\theta)||p(\boldsymbol{\theta}|\mathcal{D}|\boldsymbol{\xi})). \tag{2.6}$$

As the KL divergence is nonnegative, the equality in (2.4) holds if and only if the KL divergence is equal to zero.

Since the ELBO in (2.5) involves two arguments, namely $Q(\boldsymbol{\theta})$ and $\boldsymbol{\xi}$, solving the maximization problem

$$\max_{Q(\boldsymbol{\theta}), \boldsymbol{\xi}} \mathcal{L}(Q(\boldsymbol{\theta}); \boldsymbol{\xi}) \tag{2.7}$$

can provide both an estimate of the model hyper-parameters $\boldsymbol{\xi}$ and the variational distributions $Q(\boldsymbol{\theta})$. These two blocks can be optimized in an alternating fashion. Different strategies for optimizing $Q(\boldsymbol{\theta})$ and $\boldsymbol{\xi}$ result in different inference algorithms. For example, the variational distribution $Q(\boldsymbol{\theta})$ can be optimized either via functional optimization, or via the Monte Carlo method, while the hyper-parameters $\boldsymbol{\xi}$ can be optimized via various non-convex optimization methods.

We have argued that the ELBO will be maximized when KL divergence equals zero or equivalently $Q(\boldsymbol{\theta}) = p(\boldsymbol{\theta}|\mathcal{D}, \boldsymbol{\xi})$. But this brings us back to the intractable multiple integration problem we face in the first place. Therefore, in variational inference, certain restriction will be applied to $Q(\boldsymbol{\theta})$. A widely adopted restriction is the mean-field approximation [13], which will be introduced in the following section.

2.5 Mean-Field Variational Inference

2.5.1 General Solution

Since optimizing $Q(\boldsymbol{\theta})$ in (2.7) given $\boldsymbol{\xi}$ is usually intractable, a widely adopted approximation to the variational pdf $Q(\boldsymbol{\theta})$ is the mean-field family $Q(\boldsymbol{\theta}) =$

$\prod_{k=1}^{K} Q(\boldsymbol{\theta}_k)$, where $\bigcup \boldsymbol{\theta}_k = \boldsymbol{\theta}$ and $\bigcap \boldsymbol{\theta}_k = \emptyset$. That is, the unknown set $\boldsymbol{\theta}$ is partitioned into exhaustive but non-overlapping subsets $\{\boldsymbol{\theta}_k\}_{k=1}^{K}$. With the mean-field equality constraint being incorporated into the objective function, problem (2.7) is written as ($\boldsymbol{\xi}$ is fixed);

$$\max_{\{Q(\boldsymbol{\theta}_k)\}_{k=1}^{K}} -\ln p\left(\mathcal{D}\right) + \mathbb{E}_{\prod_{k=1}^{K} Q(\boldsymbol{\theta}_k)}\left[\ln p\left(\boldsymbol{\theta}, \mathcal{D}\right)\right] - \mathbb{E}_{\prod_{k=1}^{K} Q(\boldsymbol{\theta}_k)}\left[\ln \prod_{k=1}^{K} Q\left(\boldsymbol{\theta}_k\right)\right]. \tag{2.8}$$

Although problem (2.8) is not jointly convex with respect to variational pdfs $\{Q(\boldsymbol{\theta}_k)\}_{k=1}^{K}$, it is convex with respect to a single variational pdf $Q(\boldsymbol{\theta}_k)$ when the others $\{Q(\boldsymbol{\theta}_j)\}_{j\neq k}$ are fixed. This inspires the use of the coordinate descent algorithm in seeking the optimal $\{Q^*(\boldsymbol{\theta}_k)\}_{k=1}^{K}$. That is, with $\{Q(\boldsymbol{\theta}_j)\}_{j\neq k}$ being fixed, the optimal $Q^*(\boldsymbol{\theta}_k)$ is obtained by solving;

$$\begin{aligned} &\min_{Q(\boldsymbol{\theta}_k)} \int Q\left(\boldsymbol{\theta}_k\right)\Big(-\mathbb{E}_{\prod_{j\neq k} Q(\boldsymbol{\theta}_j)}\left[\ln p\left(\boldsymbol{\theta}, \mathcal{D}\right)\right] + \ln Q\left(\boldsymbol{\theta}_k\right)\Big) d\boldsymbol{\theta}_k \\ &\text{s.t.} \int Q(\boldsymbol{\theta}_k) d\boldsymbol{\theta}_k = 1\,,\ Q(\boldsymbol{\theta}_k) \geq 0. \end{aligned} \tag{2.9}$$

For this convex problem, the Karush–Kuhn–Tucker (KKT) condition gives the optimal variational pdf $Q^*\left(\boldsymbol{\theta}_k\right)$ as

$$Q^*\left(\boldsymbol{\theta}_k\right) = \frac{\exp\Big(\mathbb{E}_{\prod_{j\neq k} Q(\boldsymbol{\theta}_j)}\left[\ln p\left(\boldsymbol{\theta}, \mathcal{D}\right)\right]\Big)}{\int \exp\Big(\mathbb{E}_{\prod_{j\neq k} Q(\boldsymbol{\theta}_j)}\left[\ln p\left(\boldsymbol{\theta}, \mathcal{D}\right)\right]\Big) d\boldsymbol{\theta}_k}. \tag{2.10}$$

In Eq. (2.10), it is seen that the computation of variational pdf $Q^*\left(\boldsymbol{\theta}_k\right)$ relies on the statistics of other variables in $\boldsymbol{\theta}$. Therefore, each $Q^*\left(\boldsymbol{\theta}_k\right)$ is updated alternatively. Since $Q^*\left(\boldsymbol{\theta}_k\right)$ in (2.10) exactly solves the problem (2.9) at each iteration, a stationary point of the KL divergence is reached after convergence.

2.5.2 *Tractability of MF-VI*

While the general rule of mean-field VI seems simple, it is seen from (2.10) that the exact functional form of $Q^*\left(\boldsymbol{\theta}_k\right)$ is not explicit unless the detailed probabilistic model $p\left(\boldsymbol{\theta}, \mathcal{D}\right)$ is specified and the expectation computations are carried out. Unfortunately, the calculation in (2.10) might not be easy, as it involves a possibly intractable integration in the denominator. This variability of mean-field VI generally poses great difficulty in probabilistic modeling and algorithm development.

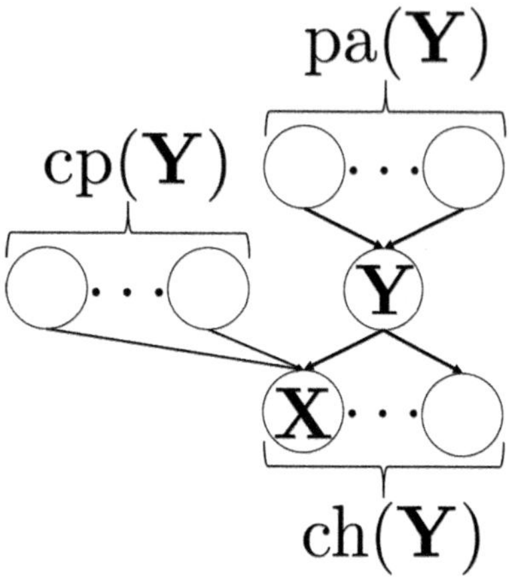

Fig. 2.3 Part of a Bayes network showing a node **Y**, the parents and children of **Y**, and the co-parents **Y** with respect to a child node **X** [14]

However, there are special cases in which the optimal variational pdfs $\{Q^*(\boldsymbol{\theta}_k)\}_{k=1}^K$ in (2.10) follow predictable patterns. To better illustrate these cases, we firstly provide readers with background knowledge of Bayes networks. A Bayes network represents a set of random variables and their conditional probabilities via a directed graph without cycles. In particular, the joint distribution of all random variables $\mathbf{X} = \{\mathbf{X}_i\}_{i=1}^N$ is

$$p(\mathbf{X}) = \prod_{i=1}^{N} p(\mathbf{X}_i|\text{pa}(\mathbf{X}_i)), \tag{2.11}$$

where $\text{pa}(\mathbf{X}_i)$ denotes the parents of node $\mathbf{X}_i$. An illustration of parents, children, and co-parents is presented in Fig. 2.3.

Next, we introduce exponential family distribution and conjugacy, which is essential for yielding closed-form optimal variational pdfs for Bayes networks. Examples of exponential family distributions are listed in Table 2.2.

Definition 2.1 (*Exponential family distribution*) A random variable $\mathbf{x}$ is in exponential family distribution if its distribution admits the following form:

$$p(\mathbf{x}|\boldsymbol{\eta}) = h(\mathbf{x}) \exp\left(\mathbf{n}(\boldsymbol{\eta})^T \mathbf{t}(\mathbf{x}) - a(\boldsymbol{\eta})\right), \tag{2.12}$$

where $\boldsymbol{\eta}$ is a vector parameterizing the distribution, $h(\mathbf{x})$ is the normalizing constant, $\mathbf{n}(\boldsymbol{\eta})$ is called the natural parameter, $\mathbf{t}(\mathbf{x})$ is the sufficient statistic, and $a(\boldsymbol{\eta})$ is the log-partition function.

Definition 2.2 (*Conjugacy*) A prior probability distribution $p(\boldsymbol{\eta})$ is conjugate to the likelihood function $p(\mathcal{D} \mid \boldsymbol{\eta})$, if the posterior distribution $p(\boldsymbol{\eta} \mid \mathcal{D})$ shares the same parametric form as that of $p(\boldsymbol{\eta})$.

In a Bayes network, if the prior distribution is an exponential family distribution, and the likelihood function is an exponential form containing a linear term with respect to the sufficient statistic in the prior distribution, the prior distribution is conjugate to the likelihood function. This is formally stated in **Proposition 2.1**.

Proposition 2.1 Given the prior distribution $p(\mathbf{Y}|pa(\mathbf{Y}))$ in exponential family,

$$p(\mathbf{Y}|\text{pa}(\mathbf{Y})) = h_{\mathbf{Y}}(\mathbf{Y}) \exp\Big(\mathbf{n}_{\mathbf{Y}}(\text{pa}(\mathbf{Y}))^T \mathbf{t}_{\mathbf{Y}}(\mathbf{Y}) - a_{\mathbf{Y}}(\text{pa}(\mathbf{Y}))\Big). \tag{2.13}$$

If the likelihood function $p(\mathbf{X}|\mathbf{Y}, \text{cp}_{\mathbf{Y}})$ can be expressed as

$$p(\mathbf{X}|\mathbf{Y}, \text{cp}(\mathbf{Y})) = \exp\Big(\mathbf{n}_{\mathbf{X},\mathbf{Y}}(\mathbf{X}, \text{cp}(\mathbf{Y}))^T \mathbf{t}_{\mathbf{Y}}(\mathbf{Y}) - \lambda(\mathbf{X}, \text{cp}(\mathbf{Y}))\Big), \tag{2.14}$$

the prior distribution $p(\mathbf{Y}|\text{pa}(\mathbf{Y}))$ is conjugate to the likelihood function $p(\mathbf{X}|\mathbf{Y}, \text{cp}(\mathbf{Y}))$.

Proof The posterior distribution $p(\mathbf{Y}|\mathbf{X}, \text{pa}(\mathbf{Y}), \text{cp}(\mathbf{Y}))$ is calculated as

$$\begin{aligned} & p(\mathbf{Y}|\mathbf{X}, \text{pa}(\mathbf{Y}), \text{cp}(\mathbf{Y})) \\ \propto\ & p(\mathbf{X}|\mathbf{Y}, \text{cp}(\mathbf{Y})) p(\mathbf{Y}|\text{pa}(\mathbf{Y})) \\ \propto\ & h_{\mathbf{Y}}(\mathbf{Y}) \exp\bigg(\Big(\mathbf{n}_{\mathbf{Y}}(\text{pa}(\mathbf{Y})) + \mathbf{n}_{\mathbf{X},\mathbf{Y}}(\mathbf{X}, \text{cp}(\mathbf{Y}))\Big)^T \mathbf{t}_{\mathbf{Y}}(\mathbf{Y}) - a_{\mathbf{Y}}(\text{pa}(\mathbf{Y})) - \lambda(\mathbf{X}, \text{cp}(\mathbf{Y}))\bigg), \end{aligned} \tag{2.15}$$

which shares the same parametric form as that of prior distribution $p(\mathbf{Y}|\text{pa}(\mathbf{Y}))$, but conditional on a new natural parameter $\mathbf{n}_{\mathbf{Y}}(\text{pa}(\mathbf{Y})) + \mathbf{n}_{\mathbf{X},\mathbf{Y}}(\mathbf{X}, \text{cp}(\mathbf{Y}))$. Therefore, the prior (2.13) is conjugate to the likelihood function (2.14). □

With the knowledge of exponential family and conjugacy, we are now ready to discuss tractable classes of mean-field VI. The earliest known case that yields closed-form optimal variational pdf is the conjugate exponential family (CEF) model in [15], which is a two-layer Bayes network as shown in Fig. 2.4a. In this model, the unknown variable is $\boldsymbol{\theta} = \{\{\mathbf{z}_n\}_{n=1}^N, \boldsymbol{\eta}\}$, consisting of the local variable $\mathbf{z}_n$ that is only associated with data $\mathcal{D}_n$, and the global variable $\boldsymbol{\eta}$ that controls all the data. Two conditions are assumed in this model. (1) The joint likelihood function $p(\{\mathcal{D}_n, \mathbf{z}_n\}_{n=1}^N|\boldsymbol{\eta})$ is a member of the exponential family distribution parameterized by $\boldsymbol{\eta}$, and (2) the prior

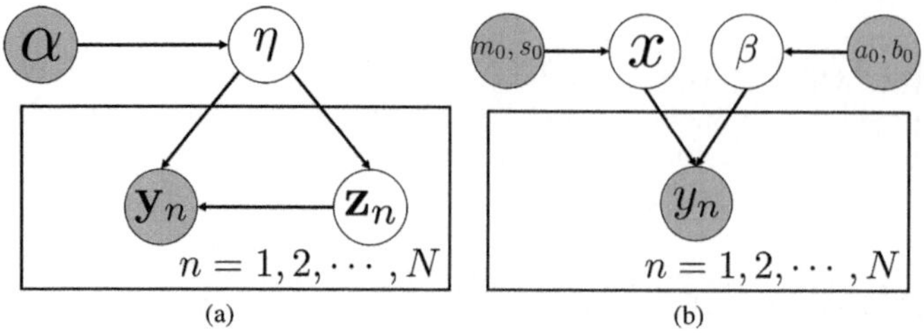

Fig. 2.4 **a** Conjugate exponential family (CEF) model. **b** Univariate Gaussian model, which belongs to CEF. In this figure, the observed variables are denoted by shaded circles while the unknown variables are denoted by unshaded circles

distribution $p(\boldsymbol{\eta}|\boldsymbol{\alpha})$ is conjugate to $p(\{\mathcal{D}_n, \mathbf{z}_n\}_{n=1}^N|\boldsymbol{\eta})$ with a fixed hyper-parameter $\boldsymbol{\alpha}$. Due to this conjugacy, it was shown in [15] that closed-form optimal variational pdfs of (2.10) exist in this model.

Example 2.1 (*Univariate Gaussian model*) To better illustrate the CEF, we use a concrete example of univariate Gaussian model, as shown in Fig. 2.4b. In particular, given data $\mathcal{D} = \{y_n \in \mathbb{R}\}_{n=1}^N$, univariate Gaussian model assumes that each observation y_n is independently drawn from a univariate Gaussian distribution $\mathcal{N}(y_n|x, \beta^{-1})$. Bayesian modeling further assigns the mean variable x with a univariate Gaussian prior $\mathcal{N}(x|m_0, s_0^{-1})$ and the precision (i.e., the inverse of variance) variable β with a gamma prior gamma$(\beta|a_0, b_0)$. Therefore, the joint probability distribution $p(\{y_n\}_{n=1}^N, x, \beta|m_0, s_0, a_0, b_0)$ can be read from Fig. 2.4b using the definition of Bayes network (2.11),

$$\begin{aligned} p(\{y_n\}_{n=1}^N, x, \beta|m_0, s_0, a_0, b_0) &= \prod_{n=1}^N p(y_n|\text{pa}(y_n))p(x|\text{pa}(x))p(\beta|\text{pa}(\beta)) \\ &= \prod_{n=1}^N \mathcal{N}(y_n|x, \beta^{-1})\mathcal{N}(x|m_0, s_0^{-1})\text{gamma}(\beta|a_0, b_0). \end{aligned} \tag{2.16}$$

To verify such a model is within CEF, we need to check the two conditions in the definition of CEF. For the first condition of the CEF model, we need to prove that $p(\{y_n\}_{n=1}^N|x, \beta)$ is in the exponential family. To show this, we note that the joint likelihood function is

$$
\begin{aligned}
p(\{y_n\}_{n=1}^N|x,\beta) &= \prod_{n=1}^{N} \mathcal{N}(y_n|x,\beta^{-1}) \\
&= \exp\left(\begin{bmatrix} \beta x \\ -\frac{\beta}{2} \end{bmatrix}^T \begin{bmatrix} \sum_{n=1}^N y_n \\ \sum_{n=1}^N y_n^2 \end{bmatrix} + \frac{1}{2}N(\ln\beta - \beta x^2 - \ln 2\pi)\right).
\end{aligned}
\tag{2.17}
$$

We can identify that $p(\{y_n\}_{n=1}^N|x,\beta)$ takes the form of (2.12), with natural parameter $\mathbf{n}(x,\beta) = \left[\beta x; -\frac{\beta}{2}\right]$, sufficient statistic $\mathbf{t}(\{y_n\}_{n=1}^N) = \left[\sum_{n=1}^N y_n; \sum_{n=1}^N y_n^2\right]$, and $a(x,\beta) = -\frac{1}{2}N(\ln\beta - \beta x^2 - \ln 2\pi)$. Therefore, $p(\{y_n\}_{n=1}^N|x,\beta)$ is in the exponential family and the first condition of CEF is satisfied.

Next, we verify the second condition of CEF, which is to show the priors of x and β are both conjugate to the likelihood function in (2.17). To prove the prior is conjugate to the likelihood function, we can utilize **Proposition 2.1**, which states that the conjugacy holds if the prior is in the exponential family and the likelihood function admits the form in (2.14). In particular, for random variable x, the univariate Gaussian prior is in the exponential family,

$$
p(x|m_0,s_0) = \exp\left(\begin{bmatrix} m_0 s_0 \\ -\frac{s_0}{2} \end{bmatrix}^T \begin{bmatrix} x \\ x^2 \end{bmatrix} + \frac{1}{2}(\ln s_0 - s_0 m_0^2 - \ln 2\pi)\right),
\tag{2.18}
$$

with sufficient statistic $\mathbf{t}_x(x) = \left[x; x^2\right]$, as summarized in Table 2.2. We rewrite the joint likelihood function (2.17) with respect to random variable x,

$$
p(\{y_n\}_{n=1}^N|x,\beta) = \exp\left(\begin{bmatrix} \beta\sum_{n=1}^N y_n \\ -\frac{N\beta}{2} \end{bmatrix}^T \begin{bmatrix} x \\ x^2 \end{bmatrix} + \frac{1}{2}\sum_{n=1}^{N}(\ln\beta - \beta y_n^2 - \ln 2\pi)\right).
\tag{2.19}
$$

It is seen that (2.19) is in the form of (2.14), where the natural parameter $\mathbf{n}_{\{y_n\}_{n=1}^N,x}(\{y_n\},\beta) = \left[\beta\sum_{n=1}^N y_n; -\frac{N\beta}{2}\right]$. Its exponent contains the linear product of sufficient statistic $\mathbf{t}_x(x)$ and natural parameter $\mathbf{n}_{\{y_n\}_{n=1}^N,x}(\{y_n\},\beta)$. Therefore, by **Proposition 2.1**, the prior $\mathcal{N}(x|m_0,s_0^{-1})$ is conjugate to the likelihood function $p(\{y_n\}_{n=1}^N|x,\beta)$.

Similarly, for random variable β, the gamma prior is an exponential family distribution with sufficient statistic $\mathbf{t}(\beta) = \left[\beta; \ln\beta\right]$,

$$
p(\beta|a_0,b_0) = \exp\left(\begin{bmatrix} -b_0 \\ a_0 - 1 \end{bmatrix}^T \begin{bmatrix} \beta \\ \ln\beta \end{bmatrix} + a_0\ln b_0 - \Gamma(a_0)\right).
\tag{2.20}
$$

The joint likelihood function (2.17) can rewritten with respect to β as

$$p(\{y_n\}_{n=1}^N|x,\beta) = \exp\left(\begin{bmatrix} -\sum_{n=1}^N \frac{1}{2}y_n^2 + x\sum_{n=1}^N y_n - \frac{N}{2}x^2 \\ \frac{N}{2} \end{bmatrix}^T \begin{bmatrix} \beta \\ \ln\beta \end{bmatrix} - \frac{1}{2}N\ln 2\pi\right), \tag{2.21}$$

which coincides with (2.14), with the natural parameter $\mathbf{n}_{\{y_n\}_{n=1}^N,\beta}(\{y_n\}_{n=1}^N, x) = \left[-\sum_{n=1}^N \frac{1}{2}y_n^2 + x\sum_{n=1}^N y_n - \frac{N}{2}x^2; \frac{N}{2}\right]$ and $\lambda(\{y_n\}_{n=1}^N, x) = \frac{1}{2}N\ln 2\pi$. Therefore, the prior gamma$(\beta|a_0, b_0)$ is conjugate to likelihood function by **Proposition 2.1**, and thus the second condition of CEF is satisfied.

To see how the MF-VI framework yields closed-form optimal variational pdfs for the univariate Gaussian model, here we demonstrate the computation of $Q^*(x)$ and $Q^*(\beta)$ using (2.10) under the mean-field $Q(x,\beta) = Q(x)Q(\beta)$. To derive the optimal variational pdf $Q^*(x)$, we extract the terms related to x in the joint probability distribution (2.16),

$$\begin{aligned} &p(\{y_n\}_{n=1}^N|x,\beta)p(x|m_0,s_0) \\ &\propto \exp\left(\begin{bmatrix} \beta\sum_{n=1}^N y_n \\ -\frac{N\beta}{2} \end{bmatrix}^T \begin{bmatrix} x \\ x^2 \end{bmatrix} + \frac{1}{2}\sum_{n=1}^N (\ln\beta - \beta y_n^2 - \ln 2\pi)\right) \\ &\times \exp\left(\begin{bmatrix} m_0 s_0 \\ -\frac{s_0}{2} \end{bmatrix}^T \begin{bmatrix} x \\ x^2 \end{bmatrix} + \frac{1}{2}(\ln s_0 - s_0 m_0^2 - \ln 2\pi)\right) \\ &\propto \exp\left(\begin{bmatrix} \beta\sum_{n=1}^N y_n + m_0 s_0 \\ -\frac{N\beta}{2} - \frac{s_0}{2} \end{bmatrix}^T \begin{bmatrix} x \\ x^2 \end{bmatrix}\right), \end{aligned} \tag{2.22}$$

where we utilize (2.19) in the second line and (2.18) in the third line. By substituting (2.22) into (2.10), it can be derived that the optimal variational pdf

$$Q^*(x) \propto \exp\left(\begin{bmatrix} \mathbb{E}[\beta]\sum_{n=1}^N y_n + m_0 s_0 \\ -\frac{N\mathbb{E}[\beta]}{2} - \frac{s_0}{2} \end{bmatrix}^T \begin{bmatrix} x \\ x^2 \end{bmatrix}\right), \tag{2.23}$$

which is a univariate Gaussian distribution, since its sufficient statistic is $[x; x^2]$. Furthermore, by matching the natural parameter in Table 2.2, the variance of $Q^*(x)$ is $(N\mathbb{E}[\beta] + s_0)^{-1}$ and the mean is $(N\mathbb{E}[\beta] + s_0)^{-1}(\mathbb{E}[\beta]\sum_{n=1}^N y_n + m_0 s_0)$.

For random variable β, by utilizing (2.20) and (2.21), the terms related to β in (2.16) are

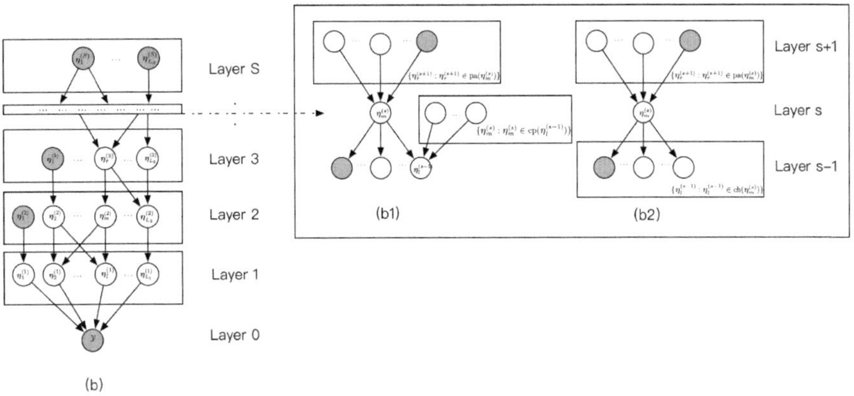

Fig. 2.5 **b** Multilayer hierarchical model. **b1** The conjugate relationship for the hierarchical model and **b2** the conjugate relationship in the MPCEF model. In this figure, the observed variables are denoted by shaded circles while the unknown variables are denoted by unshaded circles

$$
\begin{aligned}
& p(\{y_n\}_{n=1}^N|x,\beta)p(\beta|a_0,b_0) \\
& \propto \exp\left(\begin{bmatrix} -\sum_{n=1}^N \frac{1}{2}y_n^2 + x\sum_{n=1}^N y_n - \frac{N}{2}x^2 \\ \frac{N}{2} \end{bmatrix}^T \begin{bmatrix} \beta \\ \ln\beta \end{bmatrix} - \frac{1}{2}N\ln 2\pi\right) \\
& \quad \times \exp\left(\begin{bmatrix} -b_0 \\ a_0 - 1 \end{bmatrix}^T \begin{bmatrix} \beta \\ \ln\beta \end{bmatrix} + a_0\ln b_0 - \Gamma(a_0)\right) \\
& \propto \exp\left(\begin{bmatrix} -\sum_{n=1}^N \frac{1}{2}y_n^2 + x\sum_{n=1}^N y_n - \frac{N}{2}x^2 - b_0 \\ \frac{N}{2} + a_0 - 1 \end{bmatrix}^T \begin{bmatrix} \beta \\ \ln\beta \end{bmatrix}\right).
\end{aligned} \tag{2.24}
$$

Putting (2.24) into (2.10), the optimal variational pdf $Q^*(\beta) = \text{gamma}(\beta|a,b)$, where $a = a_0 + \frac{N}{2}$ and $b = b_0 + \frac{1}{2}(\sum_{n=1}^N y_n^2 - 2\mathbb{E}[x]\sum_{n=1}^N y_n + N\mathbb{E}[x^2])$. ■

CEF model is a two-layer model. More recently, a multilayer hierarchical model [14] is also found to have closed-form VI property. Figure 2.5b1 shows this hierarchical model, which considers a variable $\boldsymbol{\eta}_m^{(s)}$ in the sth layer. Its parent variables are grouped in the set $\text{pa}(\boldsymbol{\eta}_m^{(s)})$, and the variable $\boldsymbol{\eta}_l^{(s-1)}$ is one of the children variables of $\boldsymbol{\eta}_m^{(s)}$, with other co-parent variables denoted by the set $\text{cp}(\boldsymbol{\eta}_l^{(s-1)})$. In [14], it was shown that when the conditional pdfs $p(\boldsymbol{\eta}_l^{(s-1)}|\text{cp}(\boldsymbol{\eta}_l^{(s-1)}), \boldsymbol{\eta}_m^{(s)})$ and $p(\boldsymbol{\eta}_m^{(s)}|\text{pa}(\boldsymbol{\eta}_m^{(s)}))$ are conjugate pdf pairs in the exponential family, closed-form variational pdfs also exist.

Although the CEF model [15] and the hierarchical model in [14] cover many popular models, such as the factor model, hidden Markov model, and Boltzmann machine, there are many recent models that do not fall into these two classes. For example, variational relevance vector machine (RVM), low-rank matrix estimation, blind separation and decomposition, interference mitigation in wireless communications, sparse and massive channel estimation, and tensor canonical polyadic decomposition do not obey the conjugacy required by [15] or [14]. But interestingly, after tedious derivations using (2.10), the optimal variational pdfs in these works can all be derived in closed form. One may wonder if this is just a coincidence or if there is something common among these models that gives closed-form optimal variational pdfs $\{Q^*(\boldsymbol{\theta}_k)\}_{k=1}^K$. This subsection will reveal the latter is true.

In fact, the models in these works belong to another class of multilayer models. This multilayer model also employs conjugate pdf pairs in the exponential family so that closed-form VI updates can be guaranteed. However, it takes a different form of conjugacy compared to the model in [14]. In particular, this new multilayer model, shown in Fig. 2.5b2, assumes that for a variable $\boldsymbol{\eta}_m^{(s)}$ in the sth layer, the conditional pdf $p(\mathrm{ch}(\boldsymbol{\eta}_m^{(s)})|\boldsymbol{\eta}_m^{(s)})$ is conjugate to the pdf $p(\boldsymbol{\eta}_m^{(s)}|\mathrm{pa}(\boldsymbol{\eta}_m^{(s)}))$, where $\mathrm{pa}(\boldsymbol{\eta}_m^{(s)})$ and $\mathrm{ch}(\boldsymbol{\eta}_m^{(s)})$ stand for the parent set and children set of $\boldsymbol{\eta}_m^{(s)}$, respectively. Since this model only involves partial conjugacy among pdfs in adjacent layers, we term it as multilayer partial conjugate exponential family (MPCEF) model. In the next subsection, we introduce the formal definition of the MPCEF model.

2.5.3 *Definition of MPCEF Model*

The MPCEF model satisfies the following conditions:

Condition 1. For each variable $\boldsymbol{\eta}_l^{(1)}$ in Layer 1, with the remaining unknown variables $\{\boldsymbol{\eta}_j^{(1)}\}_{j\neq l}$ in Layer 1 being fixed, the likelihood function $p(\mathcal{D}|\boldsymbol{\eta}_l^{(1)}, \{\boldsymbol{\eta}_j^{(1)}\}_{j\neq l})$ with respect to $\boldsymbol{\eta}_l^{(1)}$ lies in the exponential family, and its expression can be written as[1]

$$p(\mathcal{D}|\boldsymbol{\eta}_l^{(1)}, \{\boldsymbol{\eta}_j^{(1)}\}_{j\neq l}) = \exp\left\{\mathbf{n}(\mathcal{D}, \{\boldsymbol{\eta}_j^{(1)}\}_{j\neq l})^T \mathbf{t}(\boldsymbol{\eta}_l^{(1)}) - \lambda(\mathcal{D}, \{\boldsymbol{\eta}_j^{(1)}\}_{j\neq l})\right\}. \quad (2.25)$$

Furthermore, the prior distribution $p(\boldsymbol{\eta}_l^{(1)}|\mathrm{pa}(\boldsymbol{\eta}_l^{(1)}))$ conditional on its parents $\mathrm{pa}(\boldsymbol{\eta}_l^{(1)})$ is in the exponential family and it takes the following form:

$$p(\boldsymbol{\eta}_l^{(1)}|\mathrm{pa}(\boldsymbol{\eta}_l^{(1)})) = \exp\left\{\mathbf{n}(\mathrm{pa}(\boldsymbol{\eta}_l^{(1)})^T \mathbf{t}(\boldsymbol{\eta}_l^{(1)}) - \lambda(\mathrm{pa}(\boldsymbol{\eta}_l^{(1)})\right\}. \quad (2.26)$$

That is, the prior distribution $p(\boldsymbol{\eta}_l^{(1)}|\mathrm{pa}(\boldsymbol{\eta}_l^{(1)}))$ is conjugate to the likelihood function $p(\mathcal{D}|\boldsymbol{\eta}_l^{(1)})$ (see **Proposition 2.1**).

Condition 2. For each variable $\boldsymbol{\eta}_m^{(s)}$ in Layer $S > s > 1$ with at least one parent, the distribution of its children variables $\mathrm{ch}(\boldsymbol{\eta}_m^{(s)})$ conditioned on itself is an exponential family distribution, and can be expressed as

[1] For simplicity, we use the overloaded notation for the function $\mathbf{n}(\cdot)$, $\mathbf{t}(\cdot)$, and $\lambda(\cdot)$. Their detailed functional forms depend on their arguments.

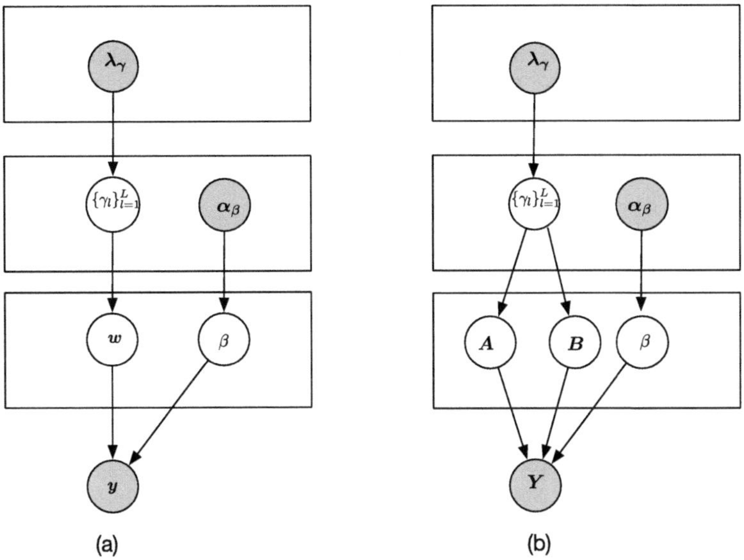

Fig. 2.6 Probabilistic model of **a** relevance vector machine and **b** matrix decomposition

$$p(\mathrm{ch}(\boldsymbol{\eta}_m^{(s)})|\boldsymbol{\eta}_m^{(s)}, \mathrm{cp}(\boldsymbol{\eta}_m^{(s)})) = \exp\left\{\mathbf{n}(\mathrm{ch}(\boldsymbol{\eta}_m^{(s)}), \mathrm{cp}(\boldsymbol{\eta}_m^{(s)}))^T \mathbf{t}(\boldsymbol{\eta}_m^{(s)}) - \lambda(\mathrm{ch}(\boldsymbol{\eta}_m^{(s)}), \mathrm{cp}(\boldsymbol{\eta}_m^{(s)}))\right\}. \tag{2.27}$$

Its prior distribution with respect to its parent variables $\mathrm{pa}(\boldsymbol{\eta}_m^{(s)})$ is in the exponential family, and it can be written as

$$p(\boldsymbol{\eta}_m^{(s)}|\mathrm{pa}(\boldsymbol{\eta}_m^{(s)})) = \exp\left\{\mathbf{n}(\mathrm{pa}(\boldsymbol{\eta}_m^{(s)}))^T \mathbf{t}(\boldsymbol{\eta}_m^{(s)}) - \lambda(\mathrm{pa}(\boldsymbol{\eta}_m^{(s)}))\right\}, \tag{2.28}$$

which indicates that the prior $p(\mathrm{ch}(\boldsymbol{\eta}_m^{(s)})|\boldsymbol{\eta}_m^{(s)}, \mathrm{cp}(\boldsymbol{\eta}_m^{(s)}))$ is conjugate to $p(\boldsymbol{\eta}_m^{(s)}|\mathrm{pa}(\boldsymbol{\eta}_m^{(s)}))$.

Condition 3. Any variable without a parent is a known quantity.

For models satisfying conditions 1–3, they all belong to the MPCEF model, and special cases can be found in various applications. Here we review two popular models.

Example 2.2 (*Relevance vector machine*) Relevance vector machine (RVM) [2] adopts a probabilistic model depicted in Fig. 2.6a. In the RVM model, the likelihood function is $p(\mathbf{y}|\mathbf{w}, \beta) = \mathcal{N}\left(\mathbf{y}|\mathbf{X}\mathbf{w}, \beta^{-1}\mathbf{I}_N\right)$, and the prior distributions of $\mathbf{w}$ and β are $p(\mathbf{w}|\{\gamma_l\}_{l=1}^L) = \prod_{l=1}^L \mathcal{N}\left(\mathbf{w}_l|0, \gamma_l^{-1}\right)$ and $p(\beta|\boldsymbol{\alpha}_\beta) = \mathrm{gamma}\left(\beta|\alpha_\beta, \alpha_\beta\right)$, respectively. Furthermore, a hyper-prior $p(\{\gamma_l\}_{l=1}^L|\boldsymbol{\lambda}_\gamma) = \prod_{l=1}^L \mathrm{gamma}\left(\gamma_l|\ \lambda_\gamma, \lambda_\gamma\right)$ is imposed on $\{\gamma_l\}_{l=1}^L$. In this model, $\mathbf{y} \in \mathbb{R}^{N\times 1}$ denotes the observation vector to be fitted, $\mathbf{X} \in \mathbb{R}^{N\times M}$ collects the feature vectors, $\mathbf{w} \in \mathbb{R}^{M\times 1}$ is the model parameter vector, and $\beta \in \mathbb{R}$ is the noise precision (i.e., inverse of variance).

The observed data are $\mathcal{D} = \{\mathbf{X}, \mathbf{y}\}$. The Layer 1 variables are $\boldsymbol{\eta}_1^{(1)} = \mathbf{w}$, $\boldsymbol{\eta}_2^{(1)} = \beta$. Starting from variable $\boldsymbol{\eta}_1^{(1)} = \mathbf{w}$, the Gaussian joint likelihood function $p(\mathbf{X}, \mathbf{y}|\mathbf{w}, \beta) = \mathcal{N}(\mathbf{y}|\mathbf{X}\mathbf{w}, \beta^{-1}\mathbf{I}_N)$ with respect to $\boldsymbol{\eta}_1^{(1)} = \mathbf{w}$ is in the form of (2.25),

$$p(\mathbf{X}, \mathbf{y}|\mathbf{w}, \beta) = \exp\left(\begin{bmatrix} \beta\mathbf{X}^T y \\ -\frac{1}{2}\text{vec}(\beta\mathbf{X}\mathbf{X}^T) \end{bmatrix}^T \begin{bmatrix} \mathbf{w} \\ \text{vec}(\mathbf{w}\mathbf{w}^T) \end{bmatrix} + \frac{1}{2}(N\log\beta - \beta\mathbf{y}^T\mathbf{y} - N\log 2\pi)\right), \tag{2.29}$$

with $\mathbf{n}(\mathbf{X}, \mathbf{y}, \beta) = \left[\beta\mathbf{X}^T y; -\frac{1}{2}\text{vec}(\beta\mathbf{X}\mathbf{X}^T)\right]$, $\mathbf{t}(\mathbf{w}) = \left[\mathbf{w}; \text{vec}(\mathbf{w}\mathbf{w}^T)\right]$, and $\lambda(\mathbf{X}, \mathbf{y}, \beta) = -\frac{1}{2}(N\log\beta - \beta\mathbf{y}^T\mathbf{y} - N\log 2\pi)$. Its prior distribution conditioned on its parent variables $\{\gamma_l\}_{l=1}^L$ takes the form of (2.26),

$$p(\mathbf{w}|\{\gamma_l\}_{l=1}^L) = \exp\left(\begin{bmatrix} \mathbf{0}_{M\times 1} \\ -\frac{1}{2}\text{vec}(\text{diag}\{\gamma_1, \ldots, \gamma_L\}) \end{bmatrix}^T \begin{bmatrix} \mathbf{w} \\ \text{vec}(\mathbf{w}\mathbf{w}^T) \end{bmatrix} + \frac{1}{2}\sum_{l=1}^L(\log\gamma - \log 2\pi)\right), \tag{2.30}$$

where $\mathbf{n}(\{\gamma_l\}_{l=1}^L) = \left[\mathbf{0}_{M\times 1}; -\frac{1}{2}\text{vec}(\text{diag}\{\gamma_1, \ldots, \gamma_L\})\right]$, $\mathbf{t}(\mathbf{w}) = \left[\mathbf{w}; \text{vec}(\mathbf{w}\mathbf{w}^T)\right]$, and $\lambda(\{\gamma_l\}_{l=1}^L) = -\frac{1}{2}\sum_{l=1}^L(\log\gamma - \log 2\pi)$. Therefore, Condition 1 of MPCEF is satisfied for Layer 1 variable $\mathbf{w}$.

Similarly, for Layer 1 variable $\boldsymbol{\eta}_2^{(1)} = \beta$, the Gaussian joint likelihood with respect to $\boldsymbol{\eta}_2^{(1)} = \beta$ takes the form of (2.25),

$$p(\mathbf{X}, \mathbf{y}|\mathbf{w}, \beta) = \exp\left(\begin{bmatrix} -\frac{1}{2}(\mathbf{y} - \mathbf{X}\mathbf{w})^T(\mathbf{y} - \mathbf{X}\mathbf{w}) \\ \frac{N}{2} \end{bmatrix}^T \begin{bmatrix} \beta \\ \log\beta \end{bmatrix} - \frac{N}{2}\log 2\pi\right). \tag{2.31}$$

Furthermore, its prior distribution is in the form of (2.26),

$$p(\beta|\boldsymbol{\alpha}_\beta) = \exp\left(\begin{bmatrix} -\alpha_\beta \\ \alpha_\beta - 1 \end{bmatrix}^T \begin{bmatrix} \beta \\ \log\beta \end{bmatrix} + \alpha_\beta \ln\alpha_\beta - \ln\Gamma(\alpha_\beta)\right), \tag{2.32}$$

which verifies Condition 1 of MPCEF for β. Since there are only two variables in Layer 1, Condition 1 holds.

For Layer 2 variable $\eta_l^{(2)} = \gamma_l$, its children distribution conditioned on its parent γ_l and co-parents $\{\gamma_j\}_{j=1,j\neq l}^L$ takes the form of (2.27),

$$p(\mathbf{w}|\gamma_l, \{\gamma_j\}_{j=1,j\neq l}^L) = \exp\left(\begin{bmatrix} -\frac{1}{2}\mathbf{w}_l^2 \\ \frac{1}{2} \end{bmatrix}^T \begin{bmatrix} \gamma_l \\ \log\gamma_l \end{bmatrix} - \sum_{j=1,j\neq l}^L \frac{1}{2}(\gamma_j\mathbf{w}_j^2 - \log\gamma_j) - \frac{L}{2}\log 2\pi\right), \tag{2.33}$$

where $\mathbf{n}(\mathbf{w}, \{\gamma_j\}_{j=1,j\neq l}^L) = \left[-\frac{1}{2}\mathbf{w}_l^2; \frac{1}{2}\right]$, $\mathbf{t}(\gamma_l) = \left[\gamma_l; \log\gamma_l\right]$, and $\lambda(\mathbf{w}, \{\gamma_j\}_{j=1,j\neq l}^L) = \sum_{j=1,j\neq l}^L \frac{1}{2}(\gamma_j \mathbf{w}_j^2 - \log\gamma_j) + \frac{L}{2}\log 2\pi$. The prior distribution of $\eta_l^{(2)} = \gamma_l$ admits the form of (2.28),

$$p(\gamma_l|\lambda_\gamma) = \exp\left(\begin{bmatrix} -\lambda_\gamma \\ \lambda_\gamma - 1 \end{bmatrix}^T \begin{bmatrix} \gamma_l \\ \log\gamma_l \end{bmatrix} + \lambda_\gamma \ln\lambda_\gamma - \ln\Gamma(\lambda_\gamma)\right). \tag{2.34}$$

Therefore, Condition 2 is satisfied. Finally, variables λ_γ and α_β are known quantities, and thus Condition 3 holds. To conclude, RVM belongs to MPCEF. ■

Example 2.3 (*Probabilistic matrix decomposition*) The probabilistic model is depicted in Fig. 2.6b [2]. In this model, the likelihood function is $p(\mathbf{Y}|\mathbf{A}, \mathbf{B}, \beta) = \prod_{n=1}^N \mathcal{N}\left(\mathbf{Y}_{:,n}|\mathbf{A}\mathbf{B}_{n,:}, \beta^{-1}\mathbf{I}_M\right)$, with prior distributions of $\mathbf{A}$, $\mathbf{B}$, and β given by $p(\mathbf{A}|\{\gamma_l\}_{l=1}^L) = \prod_{l=1}^L \mathcal{N}(\mathbf{A}_{:,l}|\mathbf{0}_{L\times 1}, \gamma_l^{-1}\mathbf{I}_L)$, $p(\mathbf{B}|\{\gamma_l\}_{l=1}^L) = \prod_{l=1}^L \mathcal{N}(\mathbf{B}_{:,l}|\mathbf{0}_{L\times 1}, \gamma_l^{-1}\mathbf{I}_L)$, and $p(\beta|\boldsymbol{\alpha}_\beta) = \text{gamma}\left(\beta|\boldsymbol{\alpha}_\beta, \boldsymbol{\alpha}_\beta\right)$, respectively. Furthermore, a hyper-prior $p(\{\gamma_l\}_{l=1}^L|\boldsymbol{\lambda}_\gamma) = \prod_{l=1}^L \text{gamma}\left(\gamma_l|\boldsymbol{\lambda}_\gamma, \boldsymbol{\lambda}_\gamma\right)$ is imposed on $\{\gamma_l\}_{l=1}^L$. In this model, $\mathbf{Y} \in \mathbb{R}^{M\times N}$ denotes the observed data matrix, while the random variables are factor matrices $\mathbf{A} \in \mathbb{R}^{M\times L}$ and $\mathbf{B} \in \mathbb{R}^{N\times L}$, and the noise precision $\beta \in \mathbb{R}$. With a similar proof for RVM, the likelihood function $p(\mathbf{Y}|\mathbf{A}, \mathbf{B}, \beta)$ with respect to $\mathbf{A}$ and β is in the form of (2.25). For variable $\mathbf{B}$, notice that the likelihood function can be re-expressed by factorizing over rows, i.e., $p(\mathbf{Y}|\mathbf{A}, \mathbf{B}, \beta) = \prod_{m=1}^M \mathcal{N}\left(\mathbf{Y}_{m,:}|\mathbf{B}\mathbf{A}_{m,:}, \beta^{-1}\mathbf{I}_N\right)$, and it can be seen that $p(\mathbf{Y}|\mathbf{A}, \mathbf{B}, \beta)$ with respect to $\mathbf{B}$ is also in the form of (2.25). The remaining conditions of MPCEF can be shown similarly to RVM. ■

While MPCEF is a general class of model, it has not been formally articulated in the literature, and the algorithms in previous works were derived independently. In the following, we show that VI derivations in the MPCEF model follow a regular pattern, and therefore the result of this chapter would facilitate the future development of applications obeying the MPCEF model.

2.5.4 Optimal Variational Pdfs for MPCEF Model

Referring to Fig. 2.5b, in the MPCEF model, the unknown variables $\boldsymbol{\theta} = \{\boldsymbol{\eta}_1^{(1)}, \ldots, \boldsymbol{\eta}_{L_1}^{(1)}, \boldsymbol{\eta}_1^{(2)}, \ldots, \boldsymbol{\eta}_{L_S}^{(S)}\}$. Correspondingly, the mean-field assumption is $Q(\boldsymbol{\theta}) = \prod_{m,s} Q(\boldsymbol{\eta}_m^{(s)}) \triangleq \prod_{k=1}^K Q(\boldsymbol{\theta}_k)$ where $K = \sum_{i=1}^S L_i$ with L_i being the number of variables in Layer i. As introduced in Sect. 2.5.1, the optimal variational pdfs $\{Q^*(\boldsymbol{\eta}_m^{(s)})\}$ can be obtained by carrying out the calculations in (2.10), and the results for the MPCEF model are summarized in the following theorem.

Theorem 2.1. *Given a dataset $\mathcal{D}$, and under the mean-field assumption $Q(\boldsymbol{\theta}) = \prod_{m,s} Q(\boldsymbol{\eta}_m^{(s)}) \triangleq \prod_{k=1}^{K} Q(\boldsymbol{\theta}_k)$, the optimal variational pdfs of the MPCEF model are given by the following:*
(a). For each variable $\boldsymbol{\eta}_l^{(1)}$ in Layer 1 with at least one parent, the optimal variational pdf $Q^*(\boldsymbol{\eta}_l^{(1)})$ takes the same functional form as (2.26) and its expression is derived as

$$Q^*(\boldsymbol{\eta}_l^{(1)}) \propto \exp\left(\mathbb{E}_{\prod_{\boldsymbol{\theta}_j \neq \boldsymbol{\eta}_l^{(1)}}}[\mathbf{n}(\mathcal{D}, \{\boldsymbol{\eta}_j^{(1)}\}_{j\neq l}) + \mathbf{n}(\mathrm{pa}(\boldsymbol{\eta}_l^{(1)}))]^T \mathbf{t}(\boldsymbol{\eta}_l^{(1)})\right). \quad (2.35)$$

(b). For each variable $\boldsymbol{\eta}_m^{(s)}$ in Layer $1 < s < S$ with at least one parent, the optimal variational pdf has the same functional form as (2.28) and its expression is

$$Q^*(\boldsymbol{\eta}_m^{(s)}) \propto \exp\left(\mathbb{E}_{\prod_{\boldsymbol{\theta}_j \neq \boldsymbol{\eta}_m^{(s)}}}[\mathbf{n}(\mathrm{ch}(\boldsymbol{\eta}_m^{(s)}), \mathrm{cp}(\boldsymbol{\eta}_m^{(s)})) + \mathbf{n}(\mathrm{pa}(\boldsymbol{\eta}_m^{(s)}))]^T \mathbf{t}(\boldsymbol{\eta}_m^{(s)})\right). \quad (2.36)$$

Proof
(a). Derivation for $Q^*(\boldsymbol{\eta}_l^{(1)})$

In the joint pdf $p(\mathcal{D}, \boldsymbol{\theta})$, after fixing variables other than $\boldsymbol{\eta}_l^{(1)}$, the term relevant to $\boldsymbol{\eta}_l^{(1)}$ is $p(\mathcal{D}|\boldsymbol{\eta}_l^{(1)}, \{\boldsymbol{\eta}_j^{(1)}\}_{j\neq l}) p(\boldsymbol{\eta}_l^{(1)}|\mathrm{pa}(\boldsymbol{\eta}_l^{(1)}))$, which is the product of (2.25) and (2.26). Substituting this term into (2.10), we have

$$\begin{aligned} Q^*(\boldsymbol{\eta}_l^{(1)}) &\propto \exp\left\{\mathbb{E}_{\prod_{\boldsymbol{\theta}_j \neq \boldsymbol{\eta}_l^{(1)}} q(\boldsymbol{\theta}_j)}\left[p(\mathcal{D}|\boldsymbol{\eta}_l^{(1)}, \{\boldsymbol{\eta}_j^{(1)}\}_{j\neq l}) p(\boldsymbol{\eta}_l^{(1)}|\mathrm{pa}(\boldsymbol{\eta}_l^{(1)}))\right]\right\} \\ &\propto \exp\left\{\mathbb{E}_{\prod_{\boldsymbol{\theta}_j \neq \boldsymbol{\eta}_l^{(1)}} q(\boldsymbol{\theta}_j)}\left[\mathbf{n}(\mathcal{D}, \{\boldsymbol{\eta}_j^{(1)}\}_{j\neq l})^T \mathbf{t}(\boldsymbol{\eta}_l^{(1)}) - \lambda(\mathcal{D}, \{\boldsymbol{\eta}_j^{(1)}\}_{j\neq l})\right.\right. \\ &\qquad \left.\left. + \mathbf{n}(\mathrm{pa}(\boldsymbol{\eta}_l^{(1)})^T \mathbf{t}(\boldsymbol{\eta}_l^{(1)}) - \lambda(\mathrm{pa}(\boldsymbol{\eta}_l^{(1)})\right]\right\}, \end{aligned} \quad (2.37)$$

which gives rise to (2.35).
(b). Derivation for $Q^*(\boldsymbol{\eta}_m^{(s)})$ where $1 < s < S$

Similar to (a), with other variables $\{\boldsymbol{\theta}_k, \boldsymbol{\theta}_k \neq \boldsymbol{\eta}_m^{(s)}\}$ being fixed, the term relevant to $\boldsymbol{\eta}_m^{(s)}$ in the joint pdf $p(\mathcal{D}, \boldsymbol{\theta})$ is the product of (2.27) and (2.28). By (2.10), the optimal variational pdf is derived as

$$
\begin{aligned}
Q^*(\boldsymbol{\eta}_m^{(s)}) &\propto \exp\left\{\mathbb{E}_{\prod_{\boldsymbol{\theta}_j \neq \boldsymbol{\eta}_m^{(s)}} q(\boldsymbol{\theta}_j)}\Big[p(\mathrm{ch}(\boldsymbol{\eta}_m^{(s)})|\boldsymbol{\eta}_m^{(s)}, \mathrm{cp}(\boldsymbol{\eta}_m^{(s)}))p(\boldsymbol{\eta}_m^{(s)}|\mathrm{pa}(\boldsymbol{\eta}_m^{(s)}))\Big]\right\} \\
&\propto \exp\left\{\mathbb{E}_{\prod_{\boldsymbol{\theta}_j \neq \boldsymbol{\eta}_m^{(s)}} q(\boldsymbol{\theta}_j)}\Big[\mathbf{n}(\mathrm{ch}(\boldsymbol{\eta}_m^{(s)}), \mathrm{cp}(\boldsymbol{\eta}_m^{(s)}))^T \mathbf{t}(\boldsymbol{\eta}_m^{(s)}) - \lambda(\mathrm{ch}(\boldsymbol{\eta}_m^{(s)}), \mathrm{cp}(\boldsymbol{\eta}_m^{(s)}))\right. \\
&\qquad \left. + \mathbf{n}(\mathrm{pa}(\boldsymbol{\eta}_m^{(s)}))^T \mathbf{t}(\boldsymbol{\eta}_m^{(s)}) - \lambda(\mathrm{pa}(\boldsymbol{\eta}_m^{(s)})))\Big]\right\},
\end{aligned} \tag{2.38}
$$

which verifies (2.36). □

Theorem 2.1 points out that if a model belongs to MPCEF, the optimal variational pdfs $\{Q^*(\boldsymbol{\theta}_k)\}_{k=1}^K$ are all in closed form. Since the computation of a variational pdf $Q(\boldsymbol{\theta}_k)$ relies on the statistics of $\{Q(\boldsymbol{\theta}_j)\}_{j\neq k}$, $\{Q(\boldsymbol{\theta}_k)\}_{k=1}^K$ are updated cyclically for $k = 1, 2, 3, \ldots, K$ using (2.35) or (2.36).

We next showcase the derivations of optimal variational pdfs given by **Theorem 2.1** for the RVM model (**Example 2.2**). For Layer 1 variable $\boldsymbol{\eta}_1^{(1)} = \mathbf{w}$, by comparing the likelihood function (2.29) to (2.25), we identify that $\mathbf{n}(\mathbf{X}, \mathbf{y}, \beta) = \left[\beta \mathbf{X}^T y; -\frac{1}{2}\mathrm{vec}(\beta \mathbf{X}\mathbf{X}^T)\right]$ and $\mathbf{t}(\mathbf{w}) = \left[\mathbf{w}; \mathrm{vec}(\mathbf{w}\mathbf{w}^T)\right]$. Furthermore, its prior distribution (2.30) takes the form of (2.26) with $\mathbf{n}(\{\gamma_l\}_{l=1}^L) = \left[\mathbf{0}_{M\times 1}; -\frac{1}{2}\mathrm{vec}(\mathrm{diag}\{\gamma_1, \ldots, \gamma_L\})\right]$, $\mathbf{t}(\mathbf{w}) = \left[\mathbf{w}; \mathrm{vec}(\mathbf{w}\mathbf{w}^T)\right]$. By substituting $\mathbf{n}(\mathbf{X}, \mathbf{y}, \beta)$, $\mathbf{n}(\{\gamma_l\}_{l=1}^L)$, and $\mathbf{t}(\mathbf{w})$ into (2.35), it is derived that

$$
\begin{aligned}
Q^*(\mathbf{w}) &\propto \exp\left(\mathbb{E}_{\prod_{\boldsymbol{\theta}_j \neq \boldsymbol{\eta}_l^{(1)}}}[\mathbf{n}(\mathbf{X}, \mathbf{y}, \beta) + \mathbf{n}(\{\gamma_l\}_{l=1}^L)]^T \mathbf{t}(\mathbf{w})\right) \\
&\propto \exp\left(\mathbb{E}\begin{bmatrix} \beta \mathbf{X}^T y \\ -\frac{1}{2}\mathrm{vec}(\beta \mathbf{X}\mathbf{X}^T) - \frac{1}{2}\mathrm{vec}(\mathrm{diag}\{\gamma_1, \ldots, \gamma_L\}) \end{bmatrix}^T \begin{bmatrix} \mathbf{w} \\ \mathrm{vec}(\mathbf{w}\mathbf{w}^T) \end{bmatrix}\right).
\end{aligned} \tag{2.39}
$$

From Table 2.2, it can be concluded that the optimal variational pdf is a Gaussian distribution, i.e., $Q^*(\mathbf{w}) = \mathcal{N}(\mathbf{w}|\boldsymbol{\mu}, \boldsymbol{\Sigma})$, where $\boldsymbol{\Sigma} = \left(\mathbb{E}[\beta]\mathbf{X}^T\mathbf{X} + \mathbb{E}\left[\mathrm{diag}\{\gamma_1, \ldots, \gamma_L\}\right]\right)^{-1}$ and $\boldsymbol{\mu} = \boldsymbol{\Sigma}\mathbb{E}[\beta]\mathbf{X}^T\mathbf{y}$. Similarly, for Layer 1 variable $\boldsymbol{\eta}_2^{(1)} = \beta$, the optimal variational pdf is a gamma distribution, $Q^*(\beta) = \mathrm{gamma}(\beta|a, b)$, where $a = \alpha_\beta + \frac{1}{2}N$ and $b = \alpha_\beta + \frac{1}{2}\mathbb{E}\left[(\mathbf{y} - \mathbf{X}\mathbf{w})^T(\mathbf{y} - \mathbf{X}\mathbf{w})\right]$.

For Layer 2 variable $\boldsymbol{\eta}_l^{(2)} = \gamma_l$, it is seen that (2.33) admits the form of (2.27), where $\mathbf{n}(\mathbf{w}, \{\gamma_j\}_{j=1, j\neq l}^L) = \left[-\frac{1}{2}\mathbf{w}_l^2; \frac{1}{2}\right]$ and $\mathbf{t}(\gamma_l) = \left[\gamma_l; \log \gamma_l\right]$. Furthermore, by matching (2.34) with (2.28), its natural parameter is $\mathbf{n}(\lambda_\gamma) = \left[-\lambda_\gamma; \lambda_\gamma - 1\right]$. By (2.36),

$$
\begin{aligned}
Q^*(\gamma_l) &\propto \exp\left(\mathbb{E}\left[\mathbf{n}(\mathbf{w}, \{\gamma_j\}_{j=1, j\neq l}^L) + \mathbf{n}(\lambda_\gamma))\right]^T \mathbf{t}(\gamma_l)\right) \\
&\propto \exp\left(\mathbb{E}\begin{bmatrix} -\lambda_\gamma - \frac{1}{2}\mathbf{w}_l^2 \\ \lambda_\gamma - \frac{1}{2} \end{bmatrix}^T \begin{bmatrix} \gamma_l \\ \log \gamma_l \end{bmatrix}\right).
\end{aligned} \tag{2.40}
$$

By comparing (2.40) with the gamma distribution in Table 2.2, the optimal variational pdf is found to be a gamma distribution $Q^*(\gamma_l) = \text{gamma}(\gamma_l | c_l, d_l)$ with $c_l = \lambda_\gamma + \frac{1}{2}$ and $d_l = \lambda_\gamma + \frac{1}{2}\mathbb{E}\left[\mathbf{w}_l^2\right]$. The derivations of the optimal variational pdfs for probabilistic matrix factorization model (**Example 2.3**) can be similarly achieved.

References

1. B. Efron, Bayes' theorem in the 21st century. Science **340**(6137), 1177–1178 (2013)
2. S. Theodoridis, *Machine Learning: A Bayesian and Optimization Perspective*, 2nd edn. (Academic, Cambridge, 2020)
3. K. Wang, F. Li, C.-M. Chen, M.M. Hassan, J. Long, N. Kumar, Interpreting adversarial examples and robustness for deep learning-based auto-driving systems. IEEE Trans. Intell. Transp. Syst. (2021)
4. I. Kononenko, Inductive and bayesian learning in medical diagnosis. Appl. Artif. Intell. Inter. J. **7**(4), 317–337 (1993)
5. M.E. Tipping, Bayesian inference: an introduction to principles and practice in machine learning, in *Summer School on Machine Learning* (Springer, Berlin, 2003), pp. 41–62
6. J.F. Claerbout, F. Muir, Robust modeling with erratic data. Geophysics **38**(5), 826–844 (1973)
7. Y.C. Eldar, G. Kutyniok, *Compressed Sensing: Theory and Applications* (Cambridge University Press, Cambridge, 2012)
8. M. Elad, *Sparse and Redundant Representations: From Theory to Applications in Signal and Image Processing* (Springer Science Business Media, Berlin, 2010)
9. K. Panousis, S. Chatzis, S. Theodoridis, Stochastic local winner-takes-all networks enable profound adversarial robustness, in *Proceedings of the Advances in Neural Information Processing Systems (NeurIPS)* (2022)
10. F. Yin, L. Pan, T. Chen, S. Theodoridis, Z.-Q. Luo, Linear multiple low-rank kernel based stationary Gaussian processes regression for time series. IEEE Trans. Signal Process. **68**, 5260–5275 (2020)
11. L. Cheng, Z. Chen, Q. Shi, Y.-C. Wu, S. Theodoridis, Towards flexible sparsity-aware modeling: automatic tensor rank learning using the generalized hyperbolic prior. IEEE Trans. Signal Process. **1**(1), 1–16 (2022)
12. D.F. Andrews, C.L. Mallows, Scale mixtures of normal distributions. J. R. Stat. Soc.: Ser. B (Methodological) **36**(1), 99–102 (1974)
13. C. Zhang, J. Bütepage, H. Kjellström, S. Mandt, Advances in variational inference. IEEE Trans. Pattern Anal. Mach. Intell. **41**(8), 2008–2026 (2018)
14. J. Winn, C.M. Bishop, T. Jaakkola, Variational message passing. J. Mach. Learn. Res. **6**(4) (2005)
15. M.J. Beal, *Variational Algorithms for Approximate Bayesian Inference* (University of London, University College London (United Kingdom), 2003)

Chapter 3
Bayesian Tensor CPD: Modeling and Inference

Abstract Having introduced the basic philosophies of Bayesian sparsity-aware learning in the last chapter, we formally start our Bayesian tensor decomposition journey in this chapter. For a pedagogical purpose, the first treatment is given on the most fundamental tensor decomposition format, namely CPD, which has been introduced in Chap. 1. As will be demonstrated in the following chapters, the key ideas developed for Bayesian CPD can be applied to other tensor decomposition models, including Tucker decomposition and tensor train decomposition. Therefore, this chapter serves as a stepping stone toward modern tensor machine learning and signal processing. Concretely, we will first show how the GSM family introduced in the last chapter can be adopted for the prior modeling of tensor CPD. Then, in this framework, we devise a Bayesian learning algorithm for CPD using the generalized hyperbolic (GH) prior, and introduce its widely adopted special case, namely Bayesian CPD using Gaussian-Gamma (GG) prior. At the end of this chapter, we introduce a different class of probabilistic modeling, namely non-parametric modeling, and present multiplicative gamma process (MGP) prior as an example.

3.1 A Unified Probabilistic Modeling Using GSM Prior

Before introducing unified probabilistic modeling, we first recap the definition of tensor CPD. Given an N dimensional (N-D) data tensor $\boldsymbol{\mathcal{Y}} \in \mathbb{R}^{I_1 \times \cdots \times I_N}$, a set of factor matrices $\{\mathbf{U}^{(n)} \in \mathbb{R}^{I_n \times R}\}$ are sought via solving the following problem:

$$\min_{\{\mathbf{U}^{(n)}\}_{n=1}^N} \|\boldsymbol{\mathcal{Y}} - \underbrace{\sum_{r=1}^{R} \mathbf{U}_{:,r}^{(1)} \circ \mathbf{U}_{:,r}^{(2)} \circ \cdots \circ \mathbf{U}_{:,r}^{(N)}}_{\boldsymbol{\mathcal{X}} \triangleq [\![\mathbf{U}^{(1)}, \mathbf{U}^{(2)}, \ldots, \mathbf{U}^{(N)}]\!]} \|_F^2, \tag{3.1}$$

where the symbol $\circ$ denotes the vector outer product, and the shorthand notation $[\![\cdots]\!]$ is termed as the Kruskal operator. As discussed in Chap. 1, the tensor CPD aims at decomposing an N-D tensor into a summation of R rank-1 tensors, with the rth component constructed as the vector outer product of the rth columns from all

L. Cheng et al., *Bayesian Tensor Decomposition for Signal Processing and Machine Learning*, https://doi.org/10.1007/978-3-031-22438-6_3

the factor matrices, i.e., $\{\mathbf{U}^{(n)}_{:,r}\}_{n=1}^N$. In problem (3.1), the number of columns R of each factor matrix, also known as tensor rank, determines the number of unknown model parameters and thus the model complexity. In practice, it needs to be carefully selected to achieve the best performance in both recovering the noise-free signals (e.g., in image denoising) and unveiling the underlying components (e.g., in social group clustering). In Bayesian perspective, (3.1) can be interpreted as the Gaussian likelihood function:

$$p\left(\mathcal{Y} \mid \{\boldsymbol{U}^{(n)}\}_{n=1}^N, \beta\right) \propto \exp\left(-\frac{\beta}{2} \parallel \mathcal{Y} - [\![\boldsymbol{U}^{(1)}, \boldsymbol{U}^{(2)}, \ldots, \boldsymbol{U}^{(N)}]\!] \parallel_F^2\right), \tag{3.2}$$

where β denotes the precision (i.e., inverse variance) of the observation error.

In the sequel, we show how the *low-rankness* is embedded into the CPD model using the GSM prior introduced in the last chapter. First, we employ an over-parameterized model for CPD by assuming an upper bound value L of tensor rank R with $L \gg R$. The low-rankness implies that many of the L rank-1 tensors should be zero. In other words, let all the lth columns of different factor matrices $\{\mathbf{U}^{(n)}\}_{n=1}^N$ be put into a vector $\mathfrak{q}_l \triangleq [\mathbf{U}^{(1)}_{:,l}; \mathbf{U}^{(2)}_{:,l}; \ldots; \mathbf{U}^{(N)}_{:,l}] \in \mathbb{R}^{\sum_{n=1}^N I_n}, \forall l$; the low-rankness indicates that a number of vectors in the set $\{\mathfrak{q}_l\}_{l=1}^L$ are zero vectors. To model such sparsity, we adopt the following multivariate extension of GSM prior from the last chapter:

$$\begin{aligned} p(\mathfrak{q}_l) &= \int \prod_{i=1}^{\sum_{p=1}^P I_p} \mathcal{N}([\mathfrak{q}_l]_i|0, \zeta_l) p(\zeta_l|\xi_n) d\zeta_l, \\ &= \int \mathcal{N}(\mathfrak{q}_l|0, \zeta_l \mathbf{I}) p(\zeta_l|\xi_n) d\zeta_l, \\ &= \int \prod_{n=1}^N \mathcal{N}(\mathbf{U}^{(n)}_{:,l}|0, \zeta_l \mathbf{I}) p(\zeta_l|\xi_n) d\zeta_l, \end{aligned} \tag{3.3}$$

where $[\mathfrak{q}_l]_i$ denotes the ith element of vector $\mathfrak{q}_l$. Since the elements in $\mathfrak{q}_l$ are assumed to be statistically independent, according to the definition of a multivariate Gaussian distribution, we have the second and third lines of (3.3) showing the equivalent prior modeling on the concatenated vector $\mathfrak{q}_l$ and the associated set of vectors $\{\mathbf{U}^{(n)}_{:,l}\}_{n=1}^N$, respectively. The mixing distribution $p(\zeta_l|\xi_n)$ can be any one listed in Table 2.2 (in Chap. 2). Note that in (3.3), the elements in vector $\mathfrak{q}_l$ are tied together via a common hyper-parameter ζ_l. Once the learning phase is over, if ζ_l approaches zero, the elements in $\mathfrak{q}_l$ will shrink to zero simultaneously, thus nulling a rank-1 tensor, as illustrated in Fig. 3.1. Since the prior distribution given in (3.3) favors zero-valued rank-1 tensors, it promotes the low-rankness of the CPD model.

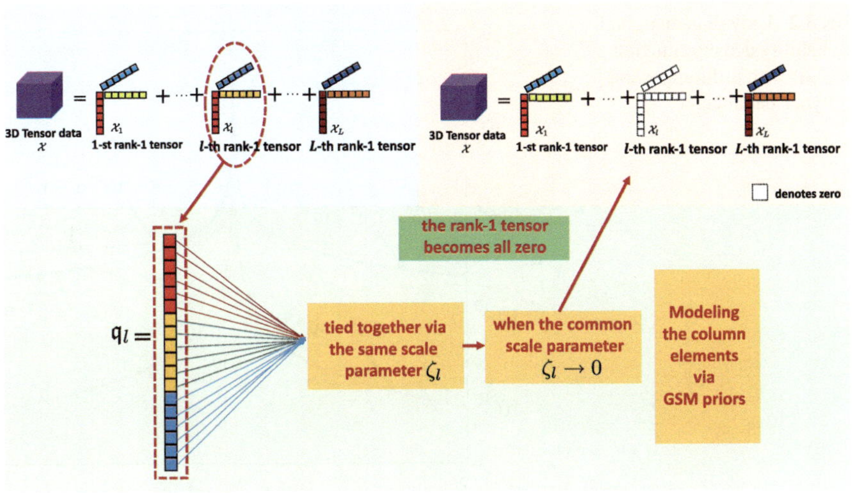

Fig. 3.1 Illustration of sparsity-aware modeling for rank-1 tensors using GSM priors

3.2 PCPD-GG: Probabilistic Modeling

In this section, we present the probabilistic CPD modeling using Gaussian-gamma prior, in which $p(\zeta_l|\xi_n)$ in (3.3) takes a gamma distribution. The modeling was first proposed in [1, 2] and widely adopted in follow-up works [3–9].

In tensor CPD, as shown in (3.1), the lth columns in all the factor matrices ($\{\boldsymbol{U}_{:,l}^{(n)}\}_{n=1}^{N}$) constitute the building block of the model. Given an upper bound value L of the tensor rank, for each factor matrix, since there are $L - R$ columns all being zero, sparsity-promoting priors should be imposed on the columns of each factor matrix to encode the information of over-parameterization.[1] In the pioneering works [1, 2], assuming statistical independence among the columns in $\{\boldsymbol{U}_{:,l}^{(n)}, \forall n, l\}$, a Gaussian-gamma prior was utilized to model them as

$$p(\{\boldsymbol{U}^{(n)}\}_{n=1}^{N}|\{\gamma_l\}_{l=1}^{L}) = \prod_{l=1}^{L} p(\{\boldsymbol{U}_{:,l}^{(n)}\}_{n=1}^{N}|\gamma_l) = \prod_{l=1}^{L}\prod_{n=1}^{N} \mathcal{N}(\boldsymbol{U}_{:,l}^{(n)}|\boldsymbol{0}_{I_n\times 1}, \gamma_l^{-1}\boldsymbol{I}_{I_n}), \tag{3.4}$$

$$p(\{\gamma_l\}_{l=1}^{L}|\{c_l^0, d_l^0\}_{l=1}^{L}) = \prod_{l=1}^{L} p(\gamma_l|c_l^0, d_l^0) = \prod_{l=1}^{L} \text{gamma}(\gamma_l|c_l^0, d_l^0), \tag{3.5}$$

where γ_l is the precision (i.e., the inverse of variance) of the lth columns $\{\boldsymbol{U}_{:,l}^{(n)}\}_{n=1}^{N}$, and $\{c_l^0, d_l^0\}$ are pre-determined hyper-parameters.

---

[1] The sparsity level of over-parameterized CPD model can be measured by $\frac{L-R}{L}$.

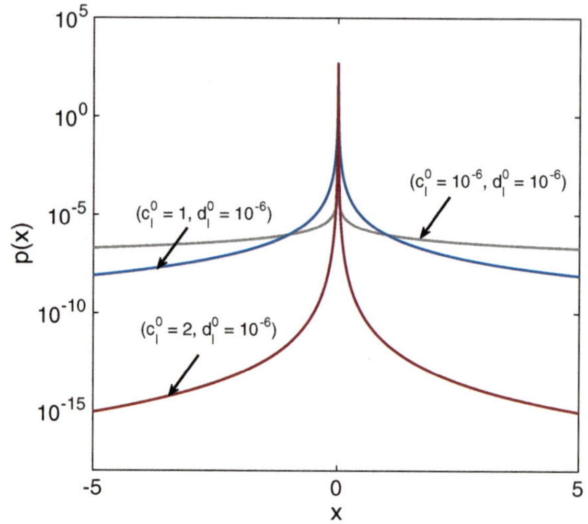

Fig. 3.2 Univariate marginal probability density function in (3.6) with different values of hyper-parameters

To see the sparsity-promoting property of the above Gaussian-gamma prior, we marginalize the precisions $\{\gamma_l\}_{l=1}^L$ to obtain the marginal probability density function (pdf) $p(\{\boldsymbol{U}^{(n)}\}_{n=1}^N)$ as follows:

$$
\begin{aligned}
& p\left(\{\boldsymbol{U}^{(n)}\}_{n=1}^N\right) \\
&= \prod_{l=1}^L p(\{\boldsymbol{U}_{:,l}^{(n)}\}_{n=1}^N) = \prod_{l=1}^L \int p(\{\boldsymbol{U}_{:,l}^{(n)}\}_{n=1}^N|\gamma_l) p(\gamma_l|c_l^0, d_l^0) d\gamma_l \\
&= \prod_{l=1}^L \left(\frac{1}{\pi}\right)^{\frac{\sum_{n=1}^N I_n}{2}} \frac{\Gamma(c_l^0 + \sum_{n=1}^N \frac{I_n}{2})}{2 d_l^{0^{-c_l^0}} \Gamma(c_l^0)} \left(2d_l^0 + \parallel \text{vec}\left(\{\boldsymbol{U}_{:,l}^{(n)}\}_{n=1}^N\right) \parallel_2^2\right)^{-c_l^0 - \sum_{n=1}^N \frac{I_n}{2}},
\end{aligned}
\tag{3.6}
$$

where $\Gamma(\cdot)$ denotes the gamma function and vec$(\cdot)$ denotes the vectorization[2] of its argument. Equation (3.6) characterizes a multivariate student's t distribution with hyper-parameters $\{c_l^0, d_l^0\}_{l=1}^L$. To get insights from this marginal distribution, we illustrate its univariate case in Fig. 3.2 with different hyper-parameters. It is clear that while different hyper-parameters lead to different tail behaviors, each student's t pdf is strongly peaked at zero. Therefore, with a proper setting of hyper-parameters (e.g., $c_l^0 = d_l^0 = 10^{-6}$), the Gaussian-gamma prior would be sparsity-promoting as the peak at zeros will inform the learning process to look for values around "zeros" while the heavy tails still allow the learning process to obtain components with

[2] The operation vec$\left(\{\boldsymbol{U}_{:,l}^{(n)}\}_{n=1}^N\right)$ simply stacks all these columns into a long vector, i.e., vec$\left(\{\boldsymbol{U}_{:,l}^{(n)}\}_{n=1}^N\right) = [\boldsymbol{U}_{:,l}^{(1)}; \boldsymbol{U}_{:,l}^{(2)}; \ldots; \boldsymbol{U}_{:,l}^{(N)}] \in \mathbb{R}^{Z \times 1}$, with $Z = \sum_{n=1}^N I_n$.

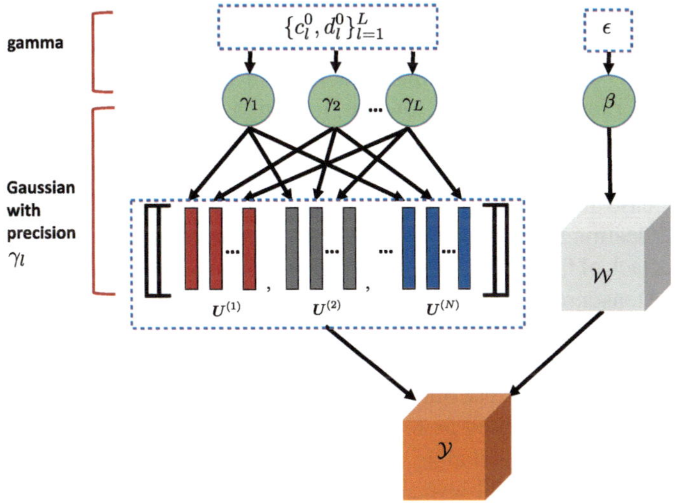

Fig. 3.3 Probabilistic CPD model with Gaussian-gamma prior

large values. Furthermore, as to be shown later, $\{c_l^0, d_l^0\}_{l=1}^L$ will be updated using the observed data during inference. This provides tailored sparsity probabilities for different rank-1 components.

The probabilistic CPD model is completed by specifying the likelihood function of $\mathcal{Y}$, given in (3.2). Equation (3.2) assumes that the signal tensor $[\![\boldsymbol{U}^{(1)}, \boldsymbol{U}^{(2)}, \ldots, \boldsymbol{U}^{(N)}]\!]$ is corrupted by additive white Gaussian noise (AWGN) tensor $\mathcal{W}$ with each element having power β^{-1}. This is consistent with the least-squares (LS) problem in (3.1). However, in Bayesian modeling, β is modeled as another random variable. Since we have no prior information about the noise power, a non-informative prior $p(\beta) = \text{gamma}(\beta|\epsilon, \epsilon)$ with a very small ϵ (e.g., 10^{-6}) is usually employed.

By using the introduced prior distributions and likelihood function, a probabilistic model for tensor CPD was constructed, as illustrated in Fig. 3.3. Based on this model, a VI-based algorithm was derived in [1] that can automatically drive most of the columns in each factor matrix to zero, by which the tensor rank is revealed. Inspired by the vanilla probabilistic CPD using the Gaussian-gamma prior, other structured and large-scale tensor CPDs with automatic tensor rank learning were further developed [3–9] in recent years.

3.3 PCPD-GH: Probabilistic Modeling

The success of the previous works on automatic tensor rank learning [1, 3–9] comes from the adoption of the sparsity-promoting Gaussian-gamma prior. However, their performances are also limited by the rigid central and tail behaviors of Gaussian-gamma prior in modeling different levels of sparsity. More specifically, when the tensor rank R is low, previous empirical results have shown that a relatively large upper bound L (e.g., set by the maximal value of tensor dimensions [1, 3]) gives accurate tensor rank estimation. However, for a high tensor rank R, the upper bound value L selected as in the low-rank case would be too small to render a sparsity pattern of the columns, and thus it leads to performance degradation. Even though we can increase the value of L to larger numbers, as it will be shown in later chapters, tensor rank learning accuracy using Gaussian-gamma prior is still not satisfactory, showing its lack of flexibility to adapt to different sparsity levels. Therefore, to further enhance the tensor rank learning capability, we explore the use of sparsity-promoting priors with more flexible central and tail behaviors.

In particular, we focus on the generalized hyperbolic (GH) prior (see Table 2.1 for its definition), since it not only includes the Gaussian-gamma prior as a special case, but also can be treated as the generalization of other widely used sparsity-promoting distributions including the Laplacian distribution, normal-inverse chi-squared distribution, normal-inverse gamma distribution, variance-gamma distribution, and Mckay's Bessel distribution. Therefore, it is expected that the functional flexibility of GH prior could lead to more adaptability in modeling different sparsity levels and thus more accurate learning of tensor rank.

Recall that the model building block in a CPD is the lth column group $\{\boldsymbol{U}^{(n)}_{:,l}\}_{n=1}^N$. With the GH prior on each column group, we have a new prior distribution for factor matrices:

$$
\begin{aligned}
&p(\{\boldsymbol{U}^{(n)}\}_{n=1}^N) \\
&= \prod_{l=1}^{L} \mathrm{GH}(\{\boldsymbol{U}^{(n)}_{:,l}\}_{n=1}^N | a_l^0, b_l^0, \lambda_l^0) \\
&= \prod_{l=1}^{L} \frac{(a_l^0)^{\frac{\sum_{n=1}^N I_n}{4}}}{(2\pi)^{\frac{\sum_{n=1}^N I_n}{2}}} \frac{(b_l^0)^{\frac{-\lambda_l^0}{2}}}{K_{\lambda_l^0}\left(\sqrt{a_l^0 b_l^0}\right)} \frac{K_{\lambda_l^0 - \frac{\sum_{n=1}^N I_n}{2}}\left(\sqrt{a_l^0\left(b_l^0 + \| \operatorname{vec}\left(\{\boldsymbol{U}^{(n)}_{:,l}\}_{n=1}^N\right) \|_2^2\right)}\right)}{\left(b_l^0 + \| \operatorname{vec}\left(\{\boldsymbol{U}^{(n)}_{:,l}\}_{n=1}^N\right) \|_2^2\right)},
\end{aligned}
\tag{3.7}
$$

where $K_{\cdot}(\cdot)$ is the modified Bessel function of the second kind, and $\mathrm{GH}(\{\boldsymbol{U}^{(n)}_{:,l}\}_{n=1}^N | a_l^0, b_l^0, \lambda_l^0)$ denotes the GH prior on the lth column group $\{\boldsymbol{U}^{(n)}_{:,l}\}_{n=1}^N$, in which the hyper-parameters $\{a_l^0, b_l^0, \lambda_l^0\}$ control the shape of the distribution. By setting $\{a_l^0, b_l^0, \lambda_l^0\}$ to specific values, the GH prior (3.7) reduces to other prevalent sparsity-promoting priors. For example:

(1) Student's t Distribution. When $a_l^0 \to 0$ and $\lambda_l^0 < 0$, it can be shown that the GH prior (3.7) reduces to [10, 11]

$$
\begin{aligned}
&p(\{\boldsymbol{U}^{(n)}\}_{n=1}^N) \\
&= \prod_{l=1}^{L} \mathrm{GH}(\{\boldsymbol{U}_{:,l}^{(n)}\}_{n=1}^N | a_l^0 \rightarrow 0, b_l^0, \lambda_l^0 < 0) \\
&= \prod_{l=1}^{L} \left(\frac{1}{\pi}\right)^{\frac{\sum_{n=1}^N J_n}{2}} \frac{\Gamma(\lambda_l^0 + \sum_{n=1}^N \frac{J_n}{2})}{b_l^{0\lambda_l^0} \Gamma(-\lambda_l^0)} \left(b_l^0 + \| \operatorname{vec}\left(\{\boldsymbol{U}_{:,l}^{(n)}\}_{n=1}^N\right) \|_2^2\right)^{\lambda_l^0 - \sum_{n=1}^N \frac{J_n}{2}}.
\end{aligned} \tag{3.8}
$$

By comparing the functional form of (3.8) to that of (3.6), it is clear that pdf (3.8) is a student's t distribution with hyper-parameters $\{-\lambda_l^0, \frac{b_l^0}{2}\}$.

(2) Laplacian Distribution. When $b_l^0 \rightarrow 0$ and $\lambda_l^0 > 0$, it can be shown that the GH prior (3.7) reduces to [10, 11]

$$
\begin{aligned}
p(\{\boldsymbol{U}^{(n)}\}_{n=1}^N) &= \prod_{l=1}^{L} \mathrm{GH}(\{\boldsymbol{U}_{:,l}^{(n)}\}_{n=1}^N | a_l^0, b_l^0 \rightarrow 0, \lambda_l^0 > 0) \\
&= \prod_{l=1}^{L} \frac{(a_l^0)^{\frac{\sum_{n=1}^N J_n}{4} + \frac{\lambda_l^0}{2}}}{\left(\pi^{\frac{\sum_{n=1}^N J_n}{2}}\right)\left(2^{\frac{\sum_{n=1}^N J_n}{2} + \lambda_l^0 - 1}\right)} \frac{\| \operatorname{vec}\left(\{\boldsymbol{U}_{:,l}^{(n)}\}_{n=1}^N\right) \|_2^{\lambda_l^0 - \frac{\sum_{n=1}^N J_n}{2}}}{2^{\lambda_l^0} \Gamma\left(\lambda_l^0\right)} \\
&\quad \times K_{\lambda_l^0 - \frac{\sum_{n=1}^N J_n}{2}} \left(\sqrt{a_l^0} \| \operatorname{vec}\left(\{\boldsymbol{U}_{:,l}^{(n)}\}_{n=1}^N\right) \|_2\right).
\end{aligned} \tag{3.9}
$$

The pdf (3.9) characterizes a generalized Laplacian distribution. By setting $\lambda_l^0 = \frac{\sum_{n=1}^N J_n}{2} + 1$, (3.9) reduces to a standard Laplacian pdf:

$$
\begin{aligned}
p(\{\boldsymbol{U}^{(n)}\}_{n=1}^N) &= \prod_{l=1}^{L} \mathrm{GH}\left(\{\boldsymbol{U}_{:,l}^{(n)}\}_{n=1}^N | a_l^0, b_l^0 \rightarrow 0, \lambda_l^0 = \frac{\sum_{n=1}^N J_n}{2} + 1\right) \\
&\propto \prod_{l=1}^{L} (a_l^0)^{\frac{\sum_{n=1}^N J_n}{2}} \exp\left(-\sqrt{a_l^0} \| \operatorname{vec}\left(\{\boldsymbol{U}_{:,l}^{(n)}\}_{n=1}^N\right) \|_2\right).
\end{aligned} \tag{3.10}
$$

To visualize the GH distribution and its special cases, the univariate GH pdfs with different hyper-parameters are illustrated in Fig. 3.4. It can be observed that the blue line is with a similar shape to those of the student's t distributions in Fig. 3.2, while the orange one resembles the shapes of Laplacian distributions. For other lines, they exhibit a wide range of the central and tail behaviors of the pdfs. Comparing Figs. 3.2 and 3.4 reveals that the GH prior is more flexible than the GG prior, and thus offers more modeling capability for different levels of sparsity.

On the other hand, the GH prior (3.7) can be expressed as a GSM formulation [10, 11]:

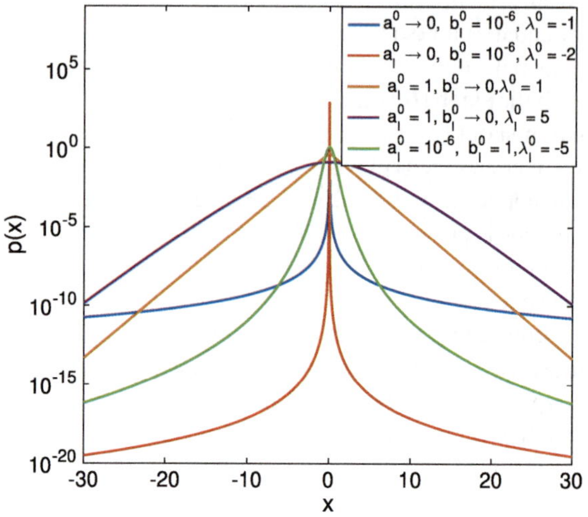

Fig. 3.4 Univariate marginal probability density function in (3.7) with different values of hyper-parameters

$$\begin{aligned} & p(\{\boldsymbol{U}^{(n)}\}_{n=1}^{N}) \\ &= \prod_{l=1}^{L} \text{GH}(\{\boldsymbol{U}_{:,l}^{(n)}\}_{n=1}^{N} | a_l^0, b_l^0, \lambda_l^0) \\ &= \prod_{l=1}^{L} \int \mathcal{N}\left(\text{vec}\left(\{\boldsymbol{U}_{:,l}^{(n)}\}_{n=1}^{N}\right) | \boldsymbol{0}_{\sum_{n=1}^{N} I_n \times 1}, z_l \boldsymbol{I}_{\sum_{n=1}^{N} I_n}\right) \text{GIG}(z_l | a_l^0, b_l^0, \lambda_l^0) dz_l, \end{aligned} \tag{3.11}$$

where z_l denotes the variance of the Gaussian distribution, and $\text{GIG}(z_l | a_l^0, b_l^0, \lambda_l^0)$ is the generalized inverse Gaussian (GIG) pdf:

$$\text{GIG}(z_l | a_l^0, b_l^0, \lambda_l^0) = \frac{\left(\frac{a_l^0}{b_l^0}\right)^{\frac{\lambda_l^0}{2}}}{2K_{\lambda_l^0}\left(\sqrt{a_l^0 b_l^0}\right)} z_l^{\lambda_l^0 - 1} \exp\left(-\frac{1}{2}\left(a_l^0 z_l + b_l^0 z_l^{-1}\right)\right). \tag{3.12}$$

This GSM formulation suggests that each GH distribution $\text{GH}(\{\boldsymbol{U}_{:,l}^{(n)}\}_{n=1}^{N} | a_l^0, b_l^0, \lambda_l^0)$ can be regarded as an infinite mixture of Gaussians with the mixing distribution being a GIG distribution. Besides connecting with the GSM framework, the formulation (3.11) allows a hierarchical construction of each GH prior by introducing the latent variable z_l, as illustrated in Fig. 3.5. Furthermore, it turns out that this hierarchical construction possesses the conjugacy property [10], which facilitates the derivation of the Bayesian inference algorithm later.

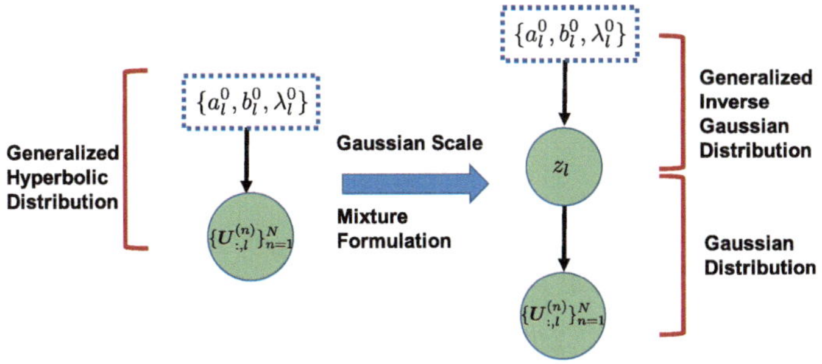

Fig. 3.5 Hierarchical construction of GH distribution

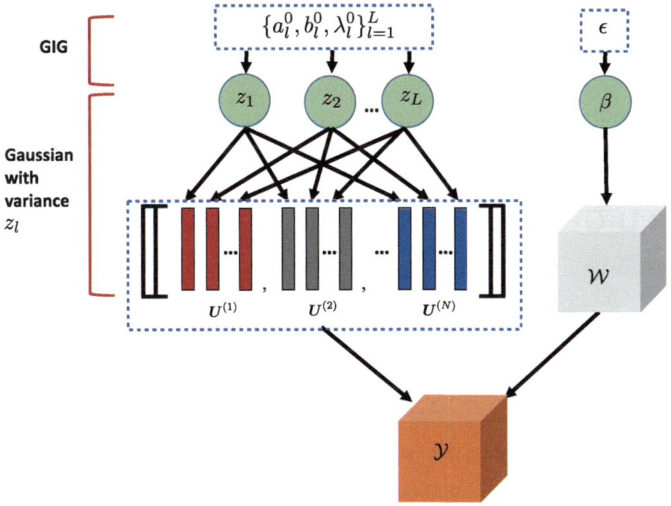

Fig. 3.6 The probabilistic tensor CPD model with GH prior

> **Property 3.1** The prior $p(z_l) = \mathrm{GIG}(z_l|a_l^0, b_l^0, \lambda_l^0)$ is conjugate to
>
> $$p\left(\{\boldsymbol{U}_{:,l}^{(n)}\}_{n=1}^N | z_l\right) = \mathcal{N}\left(\mathrm{vec}\left(\{\boldsymbol{U}_{:,l}^{(n)}\}_{n=1}^N\right) | \mathbf{0}_{\sum_{n=1}^N I_n \times 1}, z_l \boldsymbol{I}_{\sum_{n=1}^N I_n}\right). \quad (3.13)$$

Finally, together with the likelihood function in (3.2), the probabilistic model for tensor CPD using the hierarchical construction of the GH prior is shown in Fig. 3.6. Denoting the model parameter set $\boldsymbol{\Theta} = \{\{\boldsymbol{U}^{(n)}\}_{n=1}^N, \{z_l\}_{l=1}^L, \beta\}$, the GH-prior-based probabilistic tensor CPD model can be fully described by the joint pdf $p(\mathcal{Y}, \boldsymbol{\Theta})$ as

$$
\begin{aligned}
p(\mathcal{Y}, \boldsymbol{\Theta}) &= p\left(\mathcal{Y} \mid \{\boldsymbol{U}^{(n)}\}_{n=1}^{N}, \beta\right) p\left(\{\boldsymbol{U}^{(n)}\}_{n=1}^{N} | \{z_l\}_{l=1}^{L}\right) p\left(\{z_l\}_{l=1}^{L}\right) p(\beta) \\
&\propto \exp\Bigg\{ \frac{\prod_{n=1}^{N} I_n}{2} \ln \beta - \frac{\beta}{2} \parallel \mathcal{Y} - [\![\boldsymbol{U}^{(1)}, \boldsymbol{U}^{(2)}, \ldots, \boldsymbol{U}^{(N)}]\!] \parallel_F^2 \\
&\quad + \sum_{n=1}^{N} \left[\frac{I_n}{2} \sum_{l=1}^{L} \ln z_l^{-1} - \frac{1}{2} \mathrm{Tr}\left(\boldsymbol{U}^{(n)} \boldsymbol{Z}^{-1} \boldsymbol{U}^{(n)T} \right) \right] \\
&\quad + \sum_{l=1}^{L} \Bigg[\frac{\lambda_l^0}{2} \ln \frac{a_l^0}{b_l^0} - \ln \left[2 K_{\lambda_l^0} \left(\sqrt{a_l^0 b_l^0} \right) \right] + (\lambda_l^0 - 1) \ln z_l \\
&\quad - \frac{1}{2} \left(a_l^0 z_l + b_l^0 z_l^{-1} \right) \Bigg] + (\epsilon - 1) \ln \beta - \epsilon \beta \Bigg\},
\end{aligned}
\tag{3.14}
$$

where $\boldsymbol{Z} = \mathrm{diag}\{z_1, z_2, \ldots, z_L\}$.

3.4 PCPD-GH, PCPD-GG: Inference Algorithm

Since PCPD-GG is a special case of PCPD-GH, in this section, we first develop the variational inference algorithm for PCPD-GH, and then reduce the derived algorithm to cater to the PCPD-GG modeling.

The inference algorithm is derived based on the mean-field variational inference (MF-VI) framework introduced in the last chapter. Specifically, it is easy to show that the introduced PCPD-GH/PCPD-GG model falls into the MPCEF model, thus their optimal variational pdfs can be derived in closed form.

To make this chapter self-contained, we briefly review the MF-VI. Given the probabilistic model $p(\mathcal{Y}, \boldsymbol{\Theta})$, our goal is to learn the model parameters in $\boldsymbol{\Theta}$ from the tensor data $\mathcal{Y}$, in which the posterior distribution $p(\boldsymbol{\Theta}|\mathcal{Y})$ is to be sought. However, for such a complicated probabilistic model (3.14), the multiple integrations in computing the posterior distribution $p(\boldsymbol{\Theta}|\mathcal{Y})$ is not tractable. Rather than manipulating a huge number of samples from the probabilistic model, VI recasts the originally intractable multiple integration problem into the following functional optimization problem:

$$
\begin{aligned}
&\min_{Q(\boldsymbol{\Theta})} \mathrm{KL}\big(Q(\boldsymbol{\Theta}) \parallel p(\boldsymbol{\Theta} \mid \mathcal{Y})\big) \triangleq -\mathbb{E}_{Q(\boldsymbol{\Theta})} \left\{ \ln \frac{p(\boldsymbol{\Theta} \mid \mathcal{Y})}{Q(\boldsymbol{\Theta})} \right\} \\
&\text{s.t. } Q(\boldsymbol{\Theta}) \in \mathcal{F},
\end{aligned}
\tag{3.15}
$$

where $\mathrm{KL}(\cdot||\cdot)$ denotes the Kullback–Leibler (KL) divergence between two arguments, and $\mathcal{F}$ is a pre-selected family of pdfs. Its philosophy is to seek a tractable variational pdf $Q(\boldsymbol{\Theta})$ in $\mathcal{F}$ that is the closest to the true posterior distribution $p(\boldsymbol{\Theta}|\mathcal{Y})$

in the KL divergence sense. Therefore, the art is to determine the family $\mathcal{F}$ to balance the tractability of the algorithm and the accuracy of the posterior distribution learning.

Using the mean-field family, which restricts $Q(\boldsymbol{\Theta}) = \prod_{k=1}^{K} Q(\boldsymbol{\Theta}_k)$ where $\boldsymbol{\Theta}$ is partitioned into mutually disjoint non-empty subsets $\boldsymbol{\Theta}_k$ (i.e., $\boldsymbol{\Theta}_k$ is a part of $\boldsymbol{\Theta}$ with $\cup_{k=1}^{K} \boldsymbol{\Theta}_k = \boldsymbol{\Theta}$ and $\cap_{k=1}^{K} \boldsymbol{\Theta}_k = \emptyset$), the KL divergence minimization problem (3.15) becomes

$$\min_{\{Q(\boldsymbol{\Theta}_k)\}_{k=1}^{K}} -\mathbb{E}_{\{Q(\boldsymbol{\Theta}_k)\}_{k=1}^{K}} \left\{ \ln \frac{p(\boldsymbol{\Theta} \mid \mathcal{Y})}{\prod_{k=1}^{K} Q(\boldsymbol{\Theta}_k)} \right\}. \tag{3.16}$$

The factorable structure in (3.16) inspires the idea of block minimization from optimization theory. In particular, after fixing variational pdfs $\{Q(\boldsymbol{\Theta}_j)\}_{j \neq k}$ other than $Q(\boldsymbol{\Theta}_k)$, the remaining problem is

$$\min_{Q(\boldsymbol{\Theta}_k)} \int Q(\boldsymbol{\Theta}_k)(-\mathbb{E}_{\prod_{j\neq k} Q(\boldsymbol{\Theta}_j)} \left[\ln p(\boldsymbol{\Theta}, \mathcal{Y})\right] + \ln Q(\boldsymbol{\Theta}_k))d\boldsymbol{\Theta}_k, \tag{3.17}$$

and it has been shown that the optimal solution is

$$Q^*(\boldsymbol{\Theta}_k) = \frac{\exp\left(\mathbb{E}_{\prod_{j\neq k} Q(\boldsymbol{\Theta}_j)} \left[\ln p(\boldsymbol{\Theta}, \mathcal{Y})\right]\right)}{\int \exp\left(\mathbb{E}_{\prod_{j\neq k} Q(\boldsymbol{\Theta}_j)} \left[\ln p(\boldsymbol{\Theta}, \mathcal{Y})\right]\right) d\boldsymbol{\Theta}_k}. \tag{3.18}$$

3.4.1 Optimal Variational Pdfs

For the probabilistic CPD, the optimal variational pdfs $\{Q^*(\boldsymbol{\Theta}_k)\}_{k=1}^{K}$ can be obtained by substituting (3.14) into (3.18). Although straightforward as it may seem, the involvement of tensor algebras in (3.14) and the multiple integrations in the denominator of (3.18) make the derivation a challenge. On the other hand, since the probabilistic model employs the GH prior, and is different from previous works using Gaussian-gamma prior [1, 3–9], each optimal variational pdf $Q^*(\boldsymbol{\Theta}_k)$ needs to be derived from first principles. To keep this chapter concise, the lengthy derivations are omitted and the details can be found in [12]. In the following, we present only the inference results. For easy reference, the optimal variational pdfs are also summarized in Table 3.1.

In particular, the optimal variational pdf $Q^*(\boldsymbol{U}^{(k)})$ was derived to be a matrix normal distribution $\mathcal{MN}\left(\boldsymbol{U}^{(k)}|\boldsymbol{M}^{(k)}, \boldsymbol{I}_{I_n}, \boldsymbol{\Sigma}^{(k)}\right)$ with the covariance matrix

$$\boldsymbol{\Sigma}^{(k)} = \left[\mathbb{E}[\beta]\mathbb{E}\left[\left(\bigodot_{n=1,n\neq k}^{N} \boldsymbol{U}^{(n)}\right)^T \left(\bigodot_{n=1,n\neq k}^{N} \boldsymbol{U}^{(n)}\right)\right] + \mathbb{E}\left[\boldsymbol{Z}^{-1}\right]\right]^{-1}, \tag{3.19}$$

Table 3.1 Optimal variational density functions

Variational pdfs	Remarks
$Q^*\left(\boldsymbol{U}^{(k)}\right) = \mathcal{MN}\left(\boldsymbol{U}^{(k)} \| \boldsymbol{M}^{(k)}, \boldsymbol{I}_{I_n}, \boldsymbol{\Sigma}^{(k)}\right), \forall k$	Matrix normal distribution with mean $\boldsymbol{M}^{(k)}$ and covariance matrix $\boldsymbol{\Sigma}^{(k)}$ given in (3.19) and (3.20), respectively
$Q^*(z_l) = \mathrm{GIG}(z_l \| a_l, b_l, \lambda_l), \forall l$	Generalized inverse Gaussian distribution with parameters $\{a_l, b_l, \lambda_l\}$ given in (3.21)–(3.23)
$Q^*(\beta) = \mathrm{gamma}(\beta \| e, f)$	Gamma distribution with shape e and rate f given in (3.24), (3.25)

and mean matrix

$$\boldsymbol{M}^{(k)} = \mathbf{Y}^{(k)} \mathbb{E}[\beta] \left(\mathop{\odot}_{n=1, n\neq k}^{N} \mathbb{E}\left[\boldsymbol{U}^{(n)}\right] \right) \boldsymbol{\Sigma}^{(k)}. \tag{3.20}$$

In (3.19) and (3.20), $\mathbf{Y}^{(k)}$ is a matrix obtained by unfolding the tensor $\mathcal{Y}$ along its kth dimension, and the multiple Khatri–Rao products $\mathop{\odot}_{n=1,n\neq k}^{N} \boldsymbol{A}^{(n)} = \boldsymbol{A}^{(N)} \odot \boldsymbol{A}^{(N-1)} \odot \cdots \odot \boldsymbol{A}^{(k+1)} \odot \boldsymbol{A}^{(k-1)} \odot \cdots \odot \boldsymbol{A}^{(1)}$. The expectations are taken with respect to the corresponding variational pdfs of the arguments. For the optimal variational pdf $Q(z_l)$, by using the conjugacy result in **Property 3.1**, it can be derived to be a GIG distribution $\mathrm{GIG}(z_l | a_l, b_l, \lambda_l)$ with parameters

$$a_l = a_l^0, \tag{3.21}$$

$$b_l = b_l^0 + \sum_{n=1}^{N} \mathbb{E}\left[\left[\boldsymbol{U}_{:,l}^{(n)}\right]^T \boldsymbol{U}_{:,l}^{(n)} \right], \tag{3.22}$$

$$\lambda_l = \lambda_l^0 - \frac{1}{2} \sum_{n=1}^{N} I_n. \tag{3.23}$$

Finally, the optimal variational pdf $Q(\beta)$ was derived to be a gamma distribution $\mathrm{gamma}(\beta | e, f)$ with parameters

$$e = \epsilon + \frac{1}{2} \prod_{n=1}^{N} I_n, \tag{3.24}$$

$$f = \epsilon + \frac{1}{2} \mathbb{E}\left[\| \mathcal{Y} - [\![\boldsymbol{U}^{(1)}, \ldots, \boldsymbol{U}^{(N)}]\!] \|_F^2 \right]. \tag{3.25}$$

In (3.19)–(3.25), there are several expectations to be computed. They can be obtained either from the statistical literatures or similar results in related works [1, 3–9]. For easy reference, we listed the expected results needed for (3.19)–(3.25) in Table 3.2, where $\mathop{\circledast}_{n=1,n\neq k}^{N} \boldsymbol{A}^{(n)} = \boldsymbol{A}^{(N)} \circledast \boldsymbol{A}^{(N-1)} \circledast \cdots \circledast \boldsymbol{A}^{(k+1)} \circledast \boldsymbol{A}^{(k-1)} \circledast \cdots \circledast \boldsymbol{A}^{(1)}$ is the multiple Hadamard products.

Table 3.2 Computation results of expectations of PCDP-GH model

Expectations	Computation results
$\mathbb{E}\left[\boldsymbol{U}^{(k)}\right], \forall k$	$\boldsymbol{M}^{(k)}, \forall k$
$\mathbb{E}\left[z_l\right], \forall l$	$\left(\frac{b_l}{a_l}\right)^{\frac{1}{2}} \frac{K_{\lambda_l+1}(\sqrt{a_l b_l})}{K_{\lambda_l}(\sqrt{a_l b_l})}$
$\mathbb{E}\left[z_l^{-1}\right], \forall l$	$\left(\frac{b_l}{a_l}\right)^{-\frac{1}{2}} \frac{K_{\lambda_l-1}(\sqrt{a_l b_l})}{K_{\lambda_l}(\sqrt{a_l b_l})}$
$\mathbb{E}\left[\beta\right]$	$\frac{e}{f}$
$\mathbb{E}\left[\left[\boldsymbol{U}_{:,l}^{(n)}\right]^T \boldsymbol{U}_{:,l}^{(n)}\right]$	$\left[\boldsymbol{M}_{:,l}^{(n)}\right]^T \boldsymbol{M}_{:,l}^{(n)} + I_n \boldsymbol{\Sigma}_{l,l}^{(n)}$
$\mathbb{E}\left[\left(\bigodot_{n=1,n\neq k}^{N} \boldsymbol{U}^{(n)}\right)^T \left(\bigodot_{n=1,n\neq k}^{N} \boldsymbol{U}^{(n)}\right)\right]$	$\circledast_{n=1,n\neq k}^{N} \left[\left[\boldsymbol{M}^{(n)}\right]^T \boldsymbol{M}^{(n)} + I_n \boldsymbol{\Sigma}^{(n)}\right]$
$\mathbb{E}\left[\parallel \mathcal{Y} - [\![\boldsymbol{U}^{(1)}, \ldots, \boldsymbol{U}^{(N)}]\!] \parallel_F^2\right]$	$\parallel \mathcal{Y} \parallel_F^2 + \mathrm{Tr}\left(\circledast_{n=1}^{N} \left[\boldsymbol{M}^{(n)T} \boldsymbol{M}^{(n)} + I_n \boldsymbol{\Sigma}^{(n)}\right]\right)$ $-2\mathrm{Tr}\left(\mathbf{Y}^{(1)} \left(\bigodot_{n=2}^{N} \boldsymbol{M}^{(n)}\right) \boldsymbol{M}^{(1)T}\right)$

3.4.2 Setting the Hyper-parameters

From Table 3.1, it can be found that the shape of the variational pdf $Q(\boldsymbol{U}^{(k)})$ is affected by the variational pdf $\{Q(z_l)\}_{l=1}^{L}$. For each $Q(z_l)$, as seen in (3.21)–(3.23), its shape relies on the pre-selected hyper-parameters $\{a_l^0, b_l^0, \lambda_l^0\}$. In practice, we usually have no prior knowledge about the sparsity level before assessing the data, and a widely adopted approach is to make the prior non-informative.

In previous works using Gaussian-gamma prior [1, 3–9], hyper-parameters are set equal to very small values in order to approach a non-informative prior. Although nearly zero hyper-parameters lead to an improper prior, the derived variational pdf is still proper since these parameters are updated using information from observations. Therefore, in these works, the strategy of using a non-informative prior is valid. On the other hand, for the employed GH prior, non-informative prior requires $\{a_l^0, b_l^0, \lambda_l^0\}$ all go to zero, which however would lead to an improper variational pdf $Q(z_l)$, since its parameter $a_l = a_l^0$ is fixed (as seen in (3.21)). This makes the expectation computation $\mathbb{E}[z_l]$ in Table 3.2 problematic.

To tackle this issue, another viable approach is to optimize these hyper-parameters $\{a_l^0, b_l^0, \lambda_l^0\}$ so that they can be adapted during the procedure of model learning. However, as seen in (3.14), these three parameters are coupled together via the nonlinear modified Bessel function, and thus optimizing them jointly is prohibitively difficult. Therefore, [12] proposes to only optimize the most critical one, i.e., a_l^0, since it directly determines the shape of $Q(z_l)$ but will not be updated in the learning procedure. For the other two parameters $\{b_l^0, \lambda_l^0\}$, as seen in (3.22) and (3.23), since they are updated with model learning results or tensor dimension, according to the Bayesian theory, their effects on the posterior distribution would become negligible when the observation tensor is large enough. This justifies the optimization of a_l^0 while not $\{b_l^0, \lambda_l^0\}$.

For optimizing a_l^0, following related works [13], a conjugate hyper-prior $p(a_l^0) = \text{gamma}(a_l^0|\kappa_{a_1}, \kappa_{a_2})$ is introduced to ensure the positiveness of a_l^0 during the optimization. To bypass the nonlinearity from the modified Bessel function, we set $b_l^0 \to 0$ so that $K_{\lambda_l^0}\left(\sqrt{a_l^0 b_l^0}\right)$ becomes a constant. In the framework of VI, after fixing other variables, it has been derived in [12] that the hyper-parameter a_l^0 is updated via

$$a_l^0 = \frac{\kappa_{a_1} + \frac{\lambda_l^0}{2} - 1}{\kappa_{a_2} + \frac{\mathbb{E}[z_l]}{2}}. \tag{3.26}$$

Notice that it requires $\kappa_{a_1} > 1 - \lambda_l^0/2$ and $\kappa_{a_2} \geq 0$ to ensure the positiveness of a_l^0.

3.5 Algorithm Summary and Insights

From the equations above, it can be seen that the statistics of each variational pdf rely on other variational pdfs. Therefore, they need to be updated in an alternating fashion, giving rise to an iterative algorithm summarized in **Algorithm 4**. To gain more insights from the proposed algorithm, discussions on its convergence property, computational complexity, and automatic tensor rank learning are presented in the following.

3.5.1 Convergence Property

Notice that the algorithm is derived under the mean-field VI framework [13, 14]. In particular, in each iteration, after fixing other variational pdfs, the problem that optimizes a single variational pdf has been shown to be convex and has a unique solution [13, 14]. By treating each update step in mean-field VI as a block coordinate descent (BCD) step over the functional space, the limit point generated by the VI algorithm is at least a stationary point of the KL divergence [13, 14].

3.5.2 Automatic Tensor Rank Learning

During the iterations, the mean of parameter z_l^{-1} (denoted by $m[z_l^{-1}]$) will be learned using the updated parameters of other variational pdfs as seen in **Algorithm 4**. Due to the sparsity-promoting nature of the GH prior, some of $m[z_l^{-1}]$ will take very large values, e.g., in the order of 10^6. Since the inverse of $\{m[z_l^{-1}]\}_{l=1}^L$ contribute to the covariance matrix of each factor matrix $\boldsymbol{\Sigma}^{(k)}$ in (3.27), which scales the columns

Algorithm 4 PCPD-GH($\mathcal{Y}$, L)

Initializations: Choose $L > R$ and initial values $\{[\boldsymbol{M}^{(n)}]^0, [\boldsymbol{\Sigma}^{(n)}]^0\}_{n=1}^N$, $\{m[z_l^{-1}]^0, a_l^0, b_l^0, \lambda_l^0\}_{l=1}^L$, e^0, f^0. Choose $\kappa_{a_1} > -\lambda_l^0/2$ and $\kappa_{a_2} \geq 0$.

Iterations:

For the iteration $t+1$ $(t \geq 0)$,

For $k = 1, \ldots, N$, update the parameters of $Q(\boldsymbol{U}^{(k)})^{t+1}$:

$$\left[\boldsymbol{\Sigma}^{(k)}\right]^{t+1} = \left[\frac{e^t}{f^t} \underset{n=1,n\neq k}{\overset{N}{\circledast}} \left[\left(\left[\boldsymbol{M}^{(n)}\right]^s\right)^T \left[\boldsymbol{M}^{(n)}\right]^s + J_n \left[\boldsymbol{\Sigma}^{(n)}\right]^s\right]\right.$$
$$\left. + \operatorname{diag}\left\{m[z_1^{-1}]^t, m[z_2^{-1}]^t, \ldots, m[z_L^{-1}]^t\right\}\right]^{-1}, \tag{3.27}$$

$$\left[\boldsymbol{M}^{(k)}\right]^{t+1} = \mathbf{Y}^{(k)} \frac{e^t}{f^t} \left(\underset{n=1,n\neq k}{\overset{N}{\odot}} \left[\boldsymbol{M}^{(n)}\right]^s\right) \left[\boldsymbol{\Sigma}^{(n)}\right]^{t+1}, \tag{3.28}$$

where s denotes the most recent update index, i.e., $s = t+1$ when $n < k$, and $s = t$ otherwise.

Update the parameters of $Q(z_l)^{t+1}$:

$$a_l^{t+1} = [a_l^0]^t, \tag{3.29}$$

$$b_l^{t+1} = b_l^0 + \sum_{n=1}^{N} \left[\left(\left[\boldsymbol{M}_{:,l}^{(n)}\right]^{t+1}\right)^T \left[\boldsymbol{M}_{:,r}^{(n)}\right]^{t+1} + J_n \left[\boldsymbol{\Sigma}_{l,l}^{(n)}\right]^{t+1}\right], \tag{3.30}$$

$$[\lambda_l]^{t+1} = \lambda_l^0 - \frac{1}{2} \sum_{n=1}^{N} J_n, \tag{3.31}$$

$$m[z_l^{-1}]^{t+1} = \left(\frac{b_l^{t+1}}{a_l^{t+1}}\right)^{-\frac{1}{2}} \frac{K_{[\lambda_l]^{t+1}-1}\left(\sqrt{a_l^{t+1} b_l^{t+1}}\right)}{K_{[\lambda_l]^{t+1}}\left(\sqrt{a_l^{t+1} b_l^{t+1}}\right)}, \tag{3.32}$$

$$m[z_l]^{t+1} = \left(\frac{b_l^{t+1}}{a_l^{t+1}}\right)^{\frac{1}{2}} \frac{K_{[\lambda_l]^{t+1}+1}\left(\sqrt{a_l^{t+1} b_l^{t+1}}\right)}{K_{[\lambda_l]^{t+1}}\left(\sqrt{a_l^{t+1} b_l^{t+1}}\right)}. \tag{3.33}$$

Update the parameters of $Q(\beta)^{t+1}$:

$$e^{t+1} = \epsilon + \frac{\prod_{n=1}^{N} J_n}{2}, \tag{3.34}$$

$$f^{t+1} = \epsilon + \frac{\mathfrak{f}^{t+1}}{2}, \tag{3.35}$$

where $\mathfrak{f}^{t+1}$ is computed using the result in the last row of Table 3.2 with $\{\boldsymbol{M}^{(n)}, \boldsymbol{\Sigma}^{(n)}\}$ being replaced by $\{[\boldsymbol{M}^{(n)}]^{t+1}, [\boldsymbol{\Sigma}^{(n)}]^{t+1}\}$, $\forall n$.

Update the hyper-parameter $[a_l^0]^{t+1}$:

$$[a_l^0]^{t+1} = \frac{\kappa_{a_1} + \frac{\lambda_l^0}{2} - 1}{\kappa_{a_2} + \frac{m[z_l]^{t+1}}{2}}. \tag{3.36}$$

Until Convergence

in each factor matrix in (3.28), a very large $m[z_l^{-1}]$ will shrink the lth column of each factor matrix to all zero. Then, by enumerating how many non-zero columns are in each factor matrix, the tensor rank can be automatically learned.

In practice, to accelerate the learning algorithm, on-the-fly pruning is widely employed in Bayesian tensor research. In particular, in each iteration, if some of the columns in each factor matrix are found to be indistinguishable from all zeros, it indicates that these columns play no role in interpreting the data, and thus they can be safely pruned. This pruning procedure will not affect the convergence behavior of the algorithm, since each pruning is equivalent to restarting the algorithm for a reduced probabilistic model with the current variational pdfs acting as the initializations.

Note that the pruning would remove a column permanently. If the algorithm, fortunately, jumps out from one inferior local minima, the columns once deemed "irrelevance" might recover their importance. To address this, the birth process, which is opposite to the pruning process (also called the death process), can be adopted [13, 15]. Exploiting such schemes might further improve the tensor rank learning capability, especially in very low SNR and/or very high tensor rank regimes. However, from the extensive experiments, this issue does not frequently appear in a wide range of SNRs and tensor ranks.

3.5.3 Computational Complexity

For **Algorithm 4**, in each iteration, the computational complexity is dominated by updating the factor matrices, costing $O(N\prod_{n=1}^{N} J_n L^2 + L^3 \sum_{n=1}^{N} J_n)$. Therefore, the computational complexity of **Algorithm 4** is $O(q(N\prod_{n=1}^{N} J_n L^2 + L^3 \sum_{n=1}^{N} J_n))$ where q is the iteration number at convergence. The complexity is comparable to that of the inference algorithm using Gaussian-gamma prior [1].

3.5.4 Reducing to PCPD-GG

Now we show how the PCPD-GH algorithm (**Algorithm 4**) can be reduced to the PCPD-GG algorithm, which is obtained from the probabilistic tensor CPD model with GG prior (see Sect. 3.2 and Fig. 3.3).

When $a_l^0 \to 0$ and $\lambda_l^0 < 0$, as shown in Sect. 3.2, the GH prior reduces to the GG prior. Under this setting, there is no need to update a_l^0, and thus Eqs. (3.29) and (3.36) in **Algorithm 4** can be removed. With the value of a_l^0 goes to zero, other updating equations are simplified accordingly, resulting in the PCPD-GG algorithm, which is summarized in **Algorithm 5**.

Algorithm 5 PCPD-GG($\mathcal{Y}$, L)

Initializations: Choose $L > R$ and initial values $\{[\boldsymbol{M}^{(n)}]^0, [\boldsymbol{\Sigma}^{(n)}]^0\}_{n=1}^N$, $\{m[z_l^{-1}]^0, b_l^0, \lambda_l^0\}_{l=1}^L$, e^0, f^0.
Iterations:
For the iteration $t+1$ ($t \geq 0$),
For $k = 1, \ldots, N$, update the parameters of $Q(\boldsymbol{U}^{(k)})^{t+1}$:

$$\left[\boldsymbol{\Sigma}^{(k)}\right]^{t+1} = \left[\frac{e^t}{f^t} \underset{n=1, n\neq k}{\overset{N}{\circledast}} \left[\left(\left[\boldsymbol{M}^{(n)}\right]^s\right)^T \left[\boldsymbol{M}^{(n)}\right]^s + J_n \left[\boldsymbol{\Sigma}^{(n)}\right]^s\right] + \operatorname{diag}\left\{m[z_1^{-1}]^t, m[z_2^{-1}]^t, \ldots, m[z_L^{-1}]^t\right\}\right]^{-1}, \tag{3.37}$$

$$\left[\boldsymbol{M}^{(k)}\right]^{t+1} = \mathbf{Y}^{(k)} \frac{e^t}{f^t} \left(\underset{n=1, n\neq k}{\overset{N}{\odot}} \left[\boldsymbol{M}^{(n)}\right]^s\right) \left[\boldsymbol{\Sigma}^{(n)}\right]^{t+1}, \tag{3.38}$$

where s denotes the most recent update index, i.e., $s = t+1$ when $n < k$, and $s = t$ otherwise.
Update the parameters of $Q(z_l)^{t+1}$:

$$b_l^{t+1} = b_l^0 + \sum_{n=1}^{N} \left[\left(\left[\boldsymbol{M}_{:,l}^{(n)}\right]^{t+1}\right)^T \left[\boldsymbol{M}_{:,r}^{(n)}\right]^{t+1} + J_n \left[\boldsymbol{\Sigma}_{l,l}^{(n)}\right]^{t+1}\right], \tag{3.39}$$

$$[\lambda_l]^{t+1} = \lambda_l^0 - \frac{1}{2} \sum_{n=1}^{N} J_n, \tag{3.40}$$

$$m[z_l^{-1}]^{t+1} = \frac{-[\lambda_l]^{t+1}}{b_l^{t+1}/2} \tag{3.41}$$

Update the parameters of $Q(\beta)^{t+1}$:

$$e^{t+1} = \epsilon + \frac{\prod_{n=1}^{N} J_n}{2}, \tag{3.42}$$

$$f^{t+1} = \epsilon + \frac{\mathfrak{f}^{t+1}}{2}, \tag{3.43}$$

where $\mathfrak{f}^{t+1}$ is computed using the result in the last row of Table 3.2 with $\{\boldsymbol{M}^{(n)}, \boldsymbol{\Sigma}^{(n)}\}$ being replaced by $\{[\boldsymbol{M}^{(n)}]^{t+1}, [\boldsymbol{\Sigma}^{(n)}]^{t+1}\}$, $\forall n$.
Until Convergence

3.6 Non-parametric Modeling: PCPD-MGP

In this section, we introduce non-parametric modeling and exemplify it using the multiplicative gamma process (MGP). Before establishing the non-parametric modeling of CPD using MGP, we review the definition of CPD and introduce another equivalent formulation that will be utilized in this section. In particular, by explicitly introducing a coefficient λ_r of each rank-1 component in (3.1), we arrive at an equivalent formulation of tensor CPD:

$$\min_{\{\mathbf{U}^{(n)}\}_{n=1}^N, \{\lambda_r\}_{r=1}^R} \| \mathcal{Y} - \underbrace{\sum_{r=1}^{R} \lambda_r \mathbf{U}_{:,r}^{(1)} \circ \mathbf{U}_{:,r}^{(2)} \circ \cdots \circ \mathbf{U}_{:,r}^{(N)}}_{\mathcal{X} \triangleq [\![\mathbf{U}^{(1)}, \mathbf{U}^{(2)}, \ldots, \mathbf{U}^{(N)}; \boldsymbol{\lambda}]\!]} \|_F^2, \tag{3.44}$$

where the shorthand notation Kruskal operator now includes the rank-1 component coefficients $\boldsymbol{\lambda} = \left[\lambda_1, \ldots, \lambda_R\right]$ to represent this formulation [16].

From (3.44), it is seen that another viable approach to achieve tensor rank learning is to place sparsity-promoting prior on the rank-1 component coefficients $\{\lambda_r\}_{r=1}^R$, rather than on the columns of factor matrices as done in previous sections. In particular, we initialize with an over-parameterized model for CPD with $L \geq R$ rank-1 components. Due to the sparsity-promoting nature of the prior on the coefficients, the coefficients of the redundant rank-1 components will be driven to zero. Therefore, tensor rank learning is achieved by counting the rank-1 components with non-zero coefficients.

While the GSM prior introduced earlier can also be utilized for modeling sparsity on rank-1 component coefficients $\{\lambda_l\}_{l=1}^L$, we introduce in this section another class of sparsity-promoting modeling, namely non-parametric modeling, by using MGP [17] as an example. Specifically, MGP is placed on rank-1 component coefficients λ_l,

$$p(\lambda_l | \{\delta_k\}_{k=1}^l) = \mathcal{N}\left(\lambda_l | 0, \left(\prod_{k=1}^{l} \delta_k\right)^{-1}\right), \tag{3.45}$$

$$p(\delta_l) = \text{gamma}(\delta_l | a_c, 1), \tag{3.46}$$

where $a_c > 1$ is a pre-determined hyper-parameter. The MGP prior presented in (3.45) and (3.46) incorporates the prior belief that the precision of λ_l is more likely to shrink to zero as the index l increases, since the precision is a product of increasing numbers of gamma-distributed random variables.

Given that (3.44) is a least-squares cost function, the corresponding Gaussian likelihood function is expressed as

$$p(\mathcal{Y} | \{\mathbf{U}^{(n)}\}_{n=1}^N, \boldsymbol{\lambda}, \beta) \propto \exp\left(-\frac{\beta}{2} \left\| \mathcal{Y} - [\![\mathbf{U}^{(1)}, \mathbf{U}^{(2)}, \ldots, \mathbf{U}^{(N)}; \boldsymbol{\lambda}]\!] \right\|_F^2\right). \tag{3.47}$$

To complete the Bayesian modeling, the prior for factor matrices $\{\mathbf{U}^{(n)}\}_{n=1}^N$ is specified as independent columns with each having zero-mean unit-variance Gaussian distribution,

$$p(\{\mathbf{U}^{(n)}\}_{n=1}^N) = \prod_{n=1}^{N} \prod_{l=1}^{L} \mathcal{N}(\mathbf{U}_{:,l}^{(n)} | \mathbf{0}_{I_n \times 1}, \mathbf{I}_{I_n}), \tag{3.48}$$

and the precision β is assigned with a Gamma prior $\mathrm{Ga}(\beta|\epsilon, \epsilon)$, where ϵ is chosen as a very small number (e.g., 10^{-6}). To summarize, the PCPD-MGP model is represented by the joint probability density function of tensor data $\boldsymbol{\mathcal{Y}}$ and random variables $\boldsymbol{\Theta} = \{\{\mathbf{U}^{(n)}\}_{n=1}^{N}, \lambda, \{\delta_l\}_{l=1}^{L}, \beta\}$:

$$p(\boldsymbol{\mathcal{Y}}, \boldsymbol{\Theta}) = p(\boldsymbol{\mathcal{Y}}|\{\mathbf{U}^{(n)}\}_{n=1}^{N}, \boldsymbol{\lambda}, \beta)p(\{\mathbf{U}^{(n)}\}_{n=1}^{N}) \prod_{l=1}^{L} p(\lambda_l|\{\delta_k\}_{k=1}^{l}) \prod_{l=1}^{L} p(\delta_l)p(\beta). \tag{3.49}$$

3.7 PCPD-MGP: Inference Algorithm

The exact inference of random variables $\boldsymbol{\Theta} = \{\{\mathbf{U}^{(n)}\}_{n=1}^{N}, \lambda, \{\delta_l\}_{l=1}^{L}, \beta\}$ from tensor data $\boldsymbol{\mathcal{Y}}$ requires the computation of the posterior distribution $p(\boldsymbol{\Theta}|\boldsymbol{\mathcal{Y}})$, which is intractable due to the multiple integrations. Different from the Gibbs sampling method adopted in [17], we employ here the mean-field variational inference (MF-VI) to minimize the Kullback–Leibler divergence between the posterior and the variational pdf $Q(\boldsymbol{\Theta})$, where $Q(\boldsymbol{\Theta})$ lies in the mean-field family, i.e., $Q(\boldsymbol{\Theta}) = \prod_{k=1}^{K} Q(\boldsymbol{\Theta}_k)$. As derived in Chap. 2, the general formula of the optimal variational pdf is

$$Q^*(\boldsymbol{\Theta}_k) = \frac{\exp\left(\mathbb{E}_{\prod_{j\neq k} Q(\boldsymbol{\Theta}_k)}\left[\ln p\left(\boldsymbol{\mathcal{Y}}, \boldsymbol{\theta}\right)\right]\right)}{\int \exp\left(\mathbb{E}_{\prod_{j\neq k} Q(\boldsymbol{\Theta}_j)}\left[\ln p\left(\boldsymbol{\mathcal{Y}}, \boldsymbol{\Theta}, \right)\right]\right) d\boldsymbol{\Theta}_k}. \tag{3.50}$$

For PCPD-MGP model, the mean-field family is assumed as $Q(\boldsymbol{\Theta}) = \prod_{n=1}^{N} Q(\mathbf{U}^{(n)}) \times \prod_{l=1}^{L} Q(\lambda_l) \prod_{l=1}^{L} Q(\delta_l)Q(\beta)$. By substituting (3.49) into (3.50), the optimal variational pdfs for various variables can be derived. In particular, the optimal variational pdf $Q^*(\mathbf{U}^{(n)})$ is derived to be a matrix normal distribution $\mathcal{MN}\left(\mathbf{U}^{(n)}|\mathbf{M}^{(n)}, \mathbf{I}_{I_n}, \boldsymbol{\Sigma}^{(n)}\right)$ with the covariance matrix

$$\begin{aligned}\boldsymbol{\Sigma}^{(n)} = \Bigg[&\mathbb{E}[\beta]\,\mathbb{E}[\boldsymbol{\Lambda}]\,\mathbb{E}\Bigg[\left(\mathop{\odot}_{k=1,k\neq n}^{N} \mathbf{U}^{(k)}\right)^T \left(\mathop{\odot}_{k=1,k\neq n}^{N} \mathbf{U}^{(k)}\right)\Bigg]\mathbb{E}[\boldsymbol{\Lambda}] \\ &\mathbb{E}[\beta]\,\mathbb{E}\left[\boldsymbol{\Lambda}^2\right] \mathrm{diag}\left\{\mathbb{E}\Bigg[\left(\mathop{\odot}_{k=1,k\neq n}^{N} \mathbf{U}^{(k)}\right)^T \left(\mathop{\odot}_{k=1,k\neq n}^{N} \mathbf{U}^{(k)}\right)\Bigg]\right\} + \mathbf{I}_{I_n}\Bigg]^{-1},\end{aligned} \tag{3.51}$$

and mean matrix

$$\mathbf{M}^{(n)} = \mathbf{Y}^{(n)}\mathbb{E}[\beta]\left(\bigodot_{k=1,k\neq n}^{N} \mathbb{E}\left[\mathbf{U}^{(k)}\right]\right)\mathbb{E}[\mathbf{\Lambda}]\,\mathbf{\Sigma}^{(n)}. \tag{3.52}$$

In (3.51) and (3.52), $\mathbf{Y}^{(n)}$ is a matrix obtained by unfolding the tensor $\mathcal{Y}$ along its nth dimension, the multiple Khatri–Rao products $\bigodot_{k=1,k\neq n}^{N} \mathbf{A}^{(k)} = \mathbf{A}^{(N)} \odot \mathbf{A}^{(N-1)} \odot \cdots \odot \mathbf{A}^{(n+1)} \odot \mathbf{A}^{(n-1)} \odot \cdots \odot \mathbf{A}^{(1)}$, and $\mathbf{\Lambda} = \mathrm{diag}(\boldsymbol{\lambda})$.

On the other hand, the functional form of the optimal variational pdf $Q^*(\lambda_l)$ coincides with the normal distribution $\mathcal{N}(\lambda_l|m_l, s_l)$ with the variance

$$s_l = \left(\prod_{k=1}^{l} \mathbb{E}[\delta_k] + \mathbb{E}[\beta]\,\mathbb{E}\left[< \mathcal{X}_l, \mathcal{X}_l >\right]\right)^{-1}, \tag{3.53}$$

and mean

$$m_l = s_l\mathbb{E}[\beta]\left(< \mathcal{Y}, \mathbb{E}\left[\mathcal{X}_l\right] > - \sum_{k=1,k\neq l}^{L} \mathbb{E}[\lambda_k]\,\mathbb{E}\left[< \mathcal{X}_k, \mathcal{X}_l >\right]\right), \tag{3.54}$$

where $\mathcal{X}_l$ denotes the lth rank-1 component $\mathcal{X}_l = \mathbf{U}^{(1)}_{:,l} \circ \mathbf{U}^{(2)}_{:,l} \circ \cdots \circ \mathbf{U}^{(N)}_{:,l}$, and the inner product between two tensors $< X, Y >$ has been defined in (1.6).

Finally, the optimal variational pdf $Q^*(\delta_l)$ was derived to be a Gamma distribution gamma $(\delta_l|c_l, d_l)$ with the parameters

$$c_l = a_c + \frac{1}{2}(L - l + 1), \tag{3.55}$$

$$d_l = 1 + \frac{1}{2}\sum_{k=l}^{L} \mathbb{E}\left[\lambda_k^2\right] \prod_{j=1,j\neq l}^{k} \mathbb{E}\left[\delta_j\right]. \tag{3.56}$$

Furthermore, the optimal variational pdf $Q^*(\beta)$ is also a Gamma distribution gamma $(\beta|a, b)$ with the parameters

$$a = \epsilon + \frac{1}{2}\prod_{n=1}^{N} I_n, \tag{3.57}$$

$$b = \epsilon + \frac{1}{2}\mathbb{E}\left[\left\| \mathcal{Y} - [\![\mathbf{U}^{(1)}, \mathbf{U}^{(2)}, \ldots, \mathbf{U}^{(N)}; \boldsymbol{\lambda}]\!]\right\|_F^2\right]. \tag{3.58}$$

Table 3.3 Computation results of expectations of PCDP-MGP model

Expectations	Computation results
$\mathbb{E}\left[\mathbf{U}^{(n)}\right], \forall n$	$\mathbf{M}^{(n)}$
$\mathbb{E}\left[\lambda_l\right], \forall l$	m_l
$\mathbb{E}\left[\lambda_l^2\right], \forall l$	$m_l^2 + s_l$
$\mathbb{E}\left[\boldsymbol{\Lambda}\right]$	$\operatorname{diag}\{m_1, \ldots, m_L\} \triangleq \mathbf{M}_{\boldsymbol{\Lambda}}$
$\mathbb{E}\left[\boldsymbol{\Lambda}^2\right]$	$\operatorname{diag}\{m_1^2 + s_1, \ldots, m_L^2 + s_L\} \triangleq \mathbf{S}_{\boldsymbol{\Lambda}}$
$\mathbb{E}\left[\delta_l\right], \forall l$	$\frac{c_l}{d_l}$
$\mathbb{E}\left[\mathcal{X}_l\right], \forall l$	$\frac{a}{b}$
$\mathbb{E}\left[< \mathcal{X}_k, \mathcal{X}_l >\right], \forall k, l$	$\prod_{n=1}^{N} \mathbf{M}_{:,k}^{(n)T} \mathbf{M}_{:,l}^{(n)} + I_n \boldsymbol{\Sigma}_{k,l}^{(n)}$
$\mathbb{E}\left[\left(\bigodot_{k=1,k\neq n}^{N} \mathbf{U}^{(k)}\right)^T \left(\bigodot_{k=1,k\neq n}^{N} \mathbf{U}^{(k)}\right)\right]$	$\underset{k=1,k\neq n}{\overset{N}{\circledast}} \left[\left[\mathbf{M}^{(k)}\right]^T \mathbf{M}^{(k)} + I_k \boldsymbol{\Sigma}^{(k)}\right]$
$\mathbb{E}\left[\left\Vert \mathcal{Y} - [\![\mathbf{U}^{(1)}, \mathbf{U}^{(2)}, \ldots, \mathbf{U}^{(N)}; \boldsymbol{\lambda}]\!] \right\Vert_F^2\right]$	$\Vert\mathcal{Y}\Vert_F^2 + \operatorname{Tr}\Big(\underset{n=2}{\overset{N}{\circledast}} \left[\mathbf{M}^{(n)T}\mathbf{M}^{(n)} + I_n \boldsymbol{\Sigma}^{(n)}\right]$ $\circledast \Big[\mathbf{M}_{\boldsymbol{\Lambda}} \left(\mathbf{M}^{(1)T}\mathbf{M}^{(1)} + I_1 \boldsymbol{\Sigma}^{(1)}\right) \mathbf{M}_{\boldsymbol{\Lambda}}$ $+\mathbf{S}_{\boldsymbol{\Lambda}} \operatorname{diag}\left(\mathbf{M}^{(1)T}\mathbf{M}^{(1)} + I_1 \boldsymbol{\Sigma}^{(1)}\right)\Big]\Big)$ $-2 < \mathcal{Y}, [\![\mathbf{M}^{(1)}, \ldots, \mathbf{M}^{(N)}; \left[m_1, \ldots, m_L\right]]\!] >$

In (3.51)–(3.58), there are several expectations to be computed, and they are summarized in Table 3.3. As discussed in previous chapters, the MF-VI algorithm estimates the variational pdfs in an alternating fashion, since the update of each variational pdf depends on the statistics from other variational pdfs. This gives rise to **Algorithm 6**.

Algorithm 6 PCPD-MGP($\mathcal{Y}$, L)

Initializations: Choose $L > R$, $a_c > 1$, and initial values $\{[\mathbf{M}^{(n)}]^0, [\Sigma^{(n)}]^0\}_{n=1}^N$, $\{m_l^0, s_l^0\}_{l=1}^L$, a^0, b^0. Set $\mathbf{M}_\Lambda^0 = \{m_1^0, \ldots, m_L^0\}$, $\mathbf{S}_\Lambda^0 = \{(m_1^0)^2 + s_1^0, \ldots, (m_L^0)^2 + s_L^0\}$.
Iterations:
For the iteration $t+1$ ($t \geq 0$),
For $n = 1, \ldots, N$, update the parameters of $Q(\mathbf{U}^{(n)})^{t+1}$:

$$\left[\Sigma^{(n)}\right]^{t+1} = \left[\frac{a^t}{b^t}\mathbf{M}_\Lambda^t \mathop{\circledast}_{k=1,k\neq n}^{N}\left[\left(\left[\mathbf{M}^{(k)}\right]^s\right)^T\left[\mathbf{M}^{(k)}\right]^s + I_k\left[\Sigma^{(k)}\right]^s\right]\mathbf{M}_\Lambda^t\right.$$
$$\left.\times\frac{a^t}{b^t}\mathbf{S}_\Lambda^t \operatorname{diag}\left\{\left[\left(\left[\mathbf{M}^{(k)}\right]^s\right)^T\left[\mathbf{M}^{(k)}\right]^s + I_k\left[\Sigma^{(k)}\right]^s\right]\right\} + \mathbf{I}_n\right]^{-1}, \tag{3.59}$$

$$\left[\mathbf{M}^{(n)}\right]^{t+1} = \mathbf{Y}^{(n)}\frac{a^t}{b^t}\left(\mathop{\odot}_{k=1,k\neq n}^{N}\left[\mathbf{M}^{(k)}\right]^s\right)\mathbf{M}_\Lambda^t\left[\Sigma^{(k)}\right]^{t+1}, \tag{3.60}$$

where s denotes the most recent update index, i.e., $s = t+1$ when $n < k$, and $s = t$ otherwise.
Update the parameters of $Q(\lambda_l)^{t+1}$:

$$s_l^{t+1} = \left(\prod_{k=1}^{l}\frac{c_l^t}{d_l^t} + \frac{a^t}{b^t}\prod_{n=1}^{N}\left(\left[\mathbf{M}_{:,l}^{(n)}\right]^s\right)^T\left[\mathbf{M}_{:,l}^{(n)}\right]^s + I_n\Sigma_{l,l}^{(n)}\right)^{-1} \tag{3.61}$$

$$m_l^{t+1} = s_l^{t+1}\frac{a^t}{b^t}\Big(< \mathcal{Y}, [\![\left[\mathbf{M}_{:,l}^{(1)}\right]^s, \ldots, \left[\mathbf{M}_{:,l}^{(N)}\right]^s]\!] >$$
$$- \sum_{k=1,k\neq l}^{L} m_k^t \prod_{n=1}^{N}\left(\left[\mathbf{M}_{:,k}^{(n)}\right]^s\right)^T\left[\mathbf{M}_{:,l}^{(n)}\right]^s + I_n\Sigma_{k,l}^{(n)}\Big). \tag{3.62}$$

Set $\mathbf{M}_\Lambda^{t+1} = \{m_1^{t+1}, \ldots, m_L^{t+1}\}$, $\mathbf{S}_\Lambda^{t+1} = \{(m_1^{t+1})^2 + s_1^{t+1}, \ldots, (m_L^{t+1})^2 + s_L^{t+1}\}$.
Update the parameters of $Q(\delta_l)^{t+1}$:

$$c_l^{t+1} = a_c + \frac{1}{2}(L - l + 1) \tag{3.63}$$

$$d_l^{t+1} = 1 + \frac{1}{2}\sum_{k=l}^{L}\left(\left(m_l^{t+1}\right)^2 + s_l^{t+1}\right)\prod_{j=1,j\neq l}^{k}\frac{c_j^s}{d_j^s}, \tag{3.64}$$

where s denotes the most recent update index, i.e., $s = t+1$ when $j < l$, and $s = t$ otherwise.
Update the parameters of $Q(\beta)^{t+1}$:

$$a^{t+1} = \epsilon + \frac{\prod_{n=1}^N I_n}{2}, \tag{3.65}$$

$$b^{t+1} = \epsilon + \frac{\mathfrak{b}^{t+1}}{2}, \tag{3.66}$$

where $\mathfrak{b}^{t+1}$ is computed using the result in the last row of Table 3.3 with $\{\mathbf{M}^{(n)}, \Sigma^{(n)}\}$, $\{m_l\}$, $\mathbf{M}_\Lambda$, $\mathbf{S}_\Lambda$ being replaced by $\{[\mathbf{M}^{(n)}]^{t+1}, [\Sigma^{(n)}]^{t+1}\}, \forall n$, $\{m_l^{t+1}\}, \forall l$, $\mathbf{M}_\Lambda^{t+1}$, $\mathbf{S}_\Lambda^{t+1}$.
Until Convergence

References

1. Q. Zhao, L. Zhang, A. Cichocki, Bayesian cp factorization of incomplete tensors with automatic rank determination. IEEE Trans. Pattern Anal. Mach. Intell. **37**(9), 1751–1763 (2015)
2. M. Mørup, L.K. Hansen, Automatic relevance determination for multi-way models. J. Chemom.: J. Chemom. Soc. **23**(7–8), 352–363 (2009)
3. Q. Zhao, G. Zhou, L. Zhang, A. Cichocki, S.-I. Amari, Bayesian robust tensor factorization for incomplete multiway data. IEEE Trans. Neural Netw. Learn. Syst. **27**(4), 736–748 (2015)
4. L. Cheng, Y.-C. Wu, H.V. Poor, Probabilistic tensor canonical polyadic decomposition with orthogonal factors. IEEE Trans. Signal Process. **65**(3), 663–676 (2016)
5. L. Cheng, Y.-C. Wu, H.V. Poor, Scaling probabilistic tensor canonical polyadic decomposition to massive data. IEEE Trans. Signal Process. **66**(21), 5534–5548 (2018)
6. L. Cheng, X. Tong, S. Wang, Y.-C. Wu, H.V. Poor, Learning nonnegative factors from tensor data: probabilistic modeling and inference algorithm. IEEE Trans. Signal Process. **68**, 1792–1806 (2020)
7. Z. Zhang, C. Hawkins, Variational bayesian inference for robust streaming tensor factorization and completion, in *2018 IEEE International Conference on Data Mining (ICDM)* (IEEE, 2018), pp. 1446–1451
8. J. Luan, Z. Zhang, Prediction of multidimensional spatial variation data via bayesian tensor completion. IEEE Trans. Comput.-Aided Design Integr. Circuits Syst. **39**(2), 547–551 (2019)
9. Q. Zhao, L. Zhang, A. Cichocki, Bayesian sparse tucker models for dimension reduction and tensor completion (2015). arXiv:1505.02343
10. L. Thabane, M. Safiul Haq, On the matrix-variate generalized hyperbolic distribution and its bayesian applications. Statistics **38**(6), 511–526 (2004)
11. S.D. Babacan, S. Nakajima, M.N. Do, Bayesian group-sparse modeling and variational inference. IEEE Trans. Signal Process. **62**(11), 2906–2921 (2014)
12. L. Cheng, Z. Chen, Q. Shi, Y.-C. Wu, S. Theodoridis, Towards flexible sparsity-aware modeling: Automatic tensor rank learning using the generalized hyperbolic prior. IEEE Trans. Signal Process. (2022)
13. M.J. Beal, *Variational Algorithms for Approximate Bayesian Inference* (University of London, University College London (United Kingdom), 2003)
14. C. Zhang, J. Bütepage, H. Kjellström, S. Mandt, Advances in variational inference. IEEE Trans. Pattern Anal. Mach. Intell. **41**(8), 2008–2026 (2018)
15. P.J. Green, Reversible jump markov chain monte carlo computation and bayesian model determination. Biometrika **82**(4), 711–732 (1995)
16. T.G. Kolda, B.W. Bader, Tensor decompositions and applications. SIAM Rev. **51**(3), 455–500 (2009)
17. P. Rai, Y. Wang, S. Guo, G. Chen, D. Dunson, L. Carin, Scalable bayesian low-rank decomposition of incomplete multiway tensors, in *International Conference on Machine Learning* (PMLR, 2014), pp. 1800–1808

Chapter 4
Bayesian Tensor CPD: Performance and Real-World Applications

Abstract In this chapter, extensive numerical results using synthetic datasets and real-world datasets are presented to reveal the insights and performance of the introduced algorithms in the previous chapter. Since the GH prior provides a more flexible sparsity-aware modeling than the Gaussian-gamma prior, it has the potential to act as a better regularizer against noise corruption and to adapt to a wider range of sparsity levels. Numerical studies have confirmed the improved performance of the PCPD-GH method over the PCPD-GG in terms of tensor rank learning and factor matrix recovery, especially in the challenging high-rank and/or low-SNR regimes. Note that the principle followed in PCDP-GG and PCPD-GH is a parametric way to seek flexible sparsity-aware modeling. In parallel to this path, the PCPD-MGP is a non-parametric Bayesian CPD modeling. Due to the decaying effects of the length scales (i.e., the variance of the rank-1 component coefficient) learned through MGP, the inference algorithm is capable of learning low tensor rank, but it has the tendency to underestimate the tensor rank when the ground-truth rank is high, making it not very flexible in the high-rank regime. Numerical results will be presented in this chapter to demonstrate this phenomenon.

4.1 Numerical Results on Synthetic Data

In this section, extensive numerical results are presented to compare the performance of the algorithms using synthetic data. All experiments were conducted in Matlab R2015b with an Intel Core i7 CPU at 2.2 GHz.

4.1.1 Simulation Setup

We consider 3D tensors $\mathcal{X} = [\![\boldsymbol{A}^{(1)}, \boldsymbol{A}^{(2)}, \boldsymbol{A}^{(3)}]\!] \in \mathbb{R}^{30\times 30\times 30}$ with different tensor ranks. Each element in the factor matrices $\{\boldsymbol{A}^{(n)}\}_{n=1}^{3}$ is independently drawn from a zero-mean Gaussian distribution with unit power. The observation model is $\mathcal{Y} = \mathcal{X} + \mathcal{W}$, where each element of the noise tensor $\mathcal{W}$ is independently drawn from

L. Cheng et al., *Bayesian Tensor Decomposition for Signal Processing and Machine Learning*, https://doi.org/10.1007/978-3-031-22438-6_4

a zero-mean Gaussian distribution with variance σ_w^2. This data generation process follows that of [1]. The SNR is defined as $10\log_{10}\left(\text{var}\left(\mathcal{X}\right)/\sigma_w^2\right)$ [1], where $\text{var}\left(\mathcal{X}\right)$ is the variance[1] of the noise-free tensor $\mathcal{X}$. All simulation results in this section are obtained by averaging 100 Monte Carlo runs unless stated otherwise.

4.1.2 PCPD-GH Versus PCPD-GG

We first compare the probabilistic CPD using GH prior (labeled as PCPD-GH) [2] with the benchmarking algorithm using GG prior [1] (labeled as PCPD-GG).

4.1.2.1 Tensor Rank Learning

The performance of tensor rank learning is firstly evaluated. We regard the tensors as low-rank tensors when their ranks are smaller than or equal to half of the maximal tensor dimension, i.e., $R \leq \max\{J_n\}_{n=1}^N/2$. Similarly, high-rank tensors are those with $R > \max\{J_n\}_{n=1}^N/2$. In particular, in Fig. 4.1, we assess the tensor rank learning performances of the two algorithms for low-rank tensors with $R = \{3, 6, 9, 12, 15\}$ and high-rank tensors with $R = \{18, 21, 24, 27\}$ under SNR = 10 dB.

In Fig. 4.1a, the two algorithms are both with the tensor rank upper bound $L = \max\{J_n\}_{n=1}^N$. It can be seen that the PCPD-GH algorithm and the PCPD-GG algorithm achieve comparable performances in learning low tensor ranks. More specifically, the PCPD-GH algorithm achieves higher learning accuracies when $R = \{3, 6\}$ while the PCPD-GG method performs better when $R = \{9, 15\}$. However, when tackling high-rank tensors with $R > 15$, as seen in Fig. 4.1a, both algorithms with tensor rank upper bound $L = \max\{J_n\}_{n=1}^N$ fail to work properly. The reason is that the upper bound value $L = \max\{J_n\}_{n=1}^N$ results in too small sparsity level $(L - R)/L$ to leverage the power of the sparsity-promoting priors in tensor rank learning. Therefore, the upper bound value should be set larger in case that the tensor rank is high.

An immediate choice is $L = f \times \max\{J_n\}_{n=1}^N$ where $f = 1, 2, 3, \ldots$. In Fig. 4.1b and c, we assess the performances of tensor rank learning for the two methods using the upper bound $L = 2\max\{J_n\}_{n=1}^N$ and $L = 5\max\{J_n\}_{n=1}^N$, respectively. It can be seen that the PCPD-GG algorithm is very sensitive to the rank upper bound value, in the sense that its performance deteriorates significantly for low-rank tensors after employing the larger upper bounds. While PCPD-GG has improved performance for high-rank tensors after adopting a larger upper bound, the chance of getting the correct rank is still very low. In contrast, the performance of the PCPD-GH algorithm is stable for all cases and it achieves nearly 100% accuracies of tensor rank learning in a wide range of scenarios, showing its flexibility in adapting to different levels of sparsity. In Appendix L of [2], further numerical results on the

[1] It means the empirical variance computed by treating all entries of the tensor as independent realizations of a same scalar random variable.

Fig. 4.1 Performance of tensor rank learning when the rank upper bound is **a** $\max\{J_n\}_{n=1}^N$, **b** $2\max\{J_n\}_{n=1}^N$, and **c** $5\max\{J_n\}_{n=1}^N$

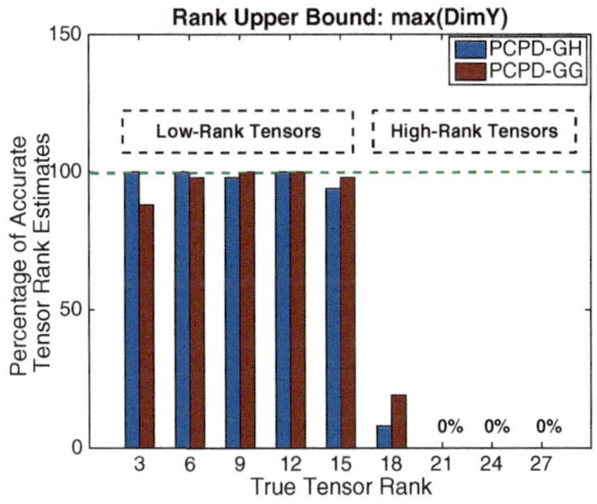

(a)

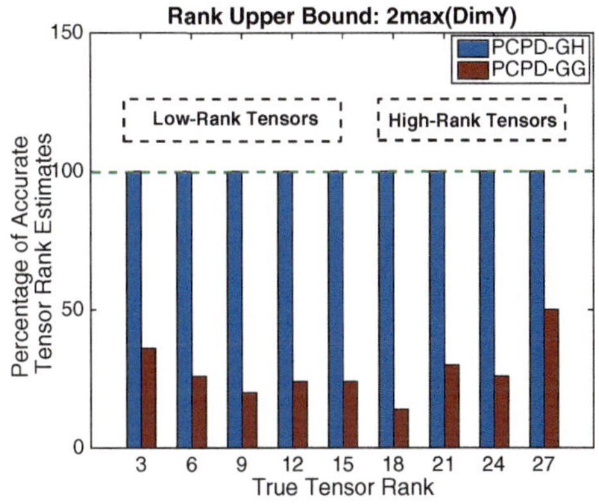

(b)

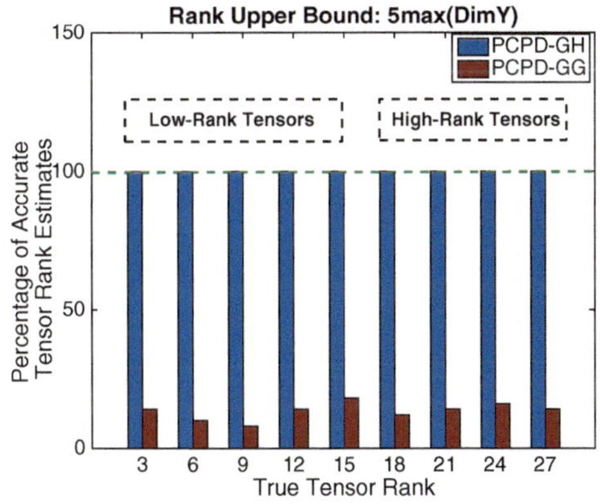

(c)

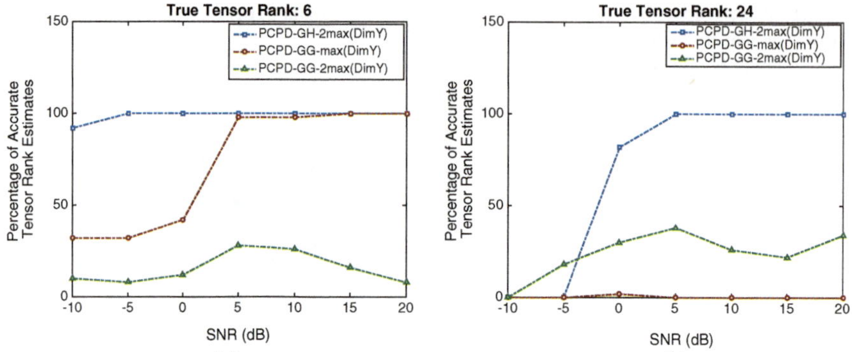

(a) (b)

Fig. 4.2 Performance of tensor rank learning versus different SNRs: **a** low-rank tensors and **b** high-rank tensors

tensor rank learning accuracies versus different sparsity levels are presented, which show the better performance of the PCPD-GH algorithm in a wide range of sparsity levels.

To assess the tensor rank learning performance under different SNRs, in Fig. 4.2, the percentages of accurate tensor rank learning from the two methods are presented. We consider two scenarios: (1) low-rank tensor with $R = 6$ shown in Fig. 4.2a and (2) high-rank tensor with $R = 24$ shown in Fig. 4.2b. For the PCPD-GH algorithm, due to its robustness to different rank upper bounds, $2\max\{J_n\}_{n=1}^N$ is adopted as the upper bound value (labeled as PCPD-GH-2max(DimY)). For the PCPD-GG algorithm, both the upper bound value $\max\{\{J_n\}_{n=1}^N\}$ and $2\max\{J_n\}_{n=1}^N$ are considered (labeled as PCPD-GG-max(DimY) and PCPD-GH-2max(DimY), respectively). From Fig. 4.2, it is clear that the performance of the PCPD-GG method, for all cases, highly relies on the choice of the rank upper bound value. In particular, when adopting $2\max\{J_n\}_{n=1}^N$, its performance in tensor rank learning is not good (i.e., accuracy below 50%) for both the low-rank tensor and the high-rank tensor cases. In contrast, when adopting $\max\{\{J_n\}_{n=1}^N\}$, its performance becomes much better for the low-rank cases. In Fig. 4.2a, when SNR is larger than 5 dB, the PCPD-GG with upper bound value $\max\{\{J_n\}_{n=1}^N\}$ achieves nearly 100% accuracy, which is very close to the accuracies of the PCPD-GH method. However, when the SNR is smaller than 5 dB, the PCPD-GH method still achieves nearly 100% accuracies in tensor rank learning, but the accuracies of the PCPD-GG method fall below 50%. For the high-rank case, as seen in Fig. 4.2b, both the PCPD-GH and the PCPD-GG methods fail to recover the true tensor rank when SNR is smaller than 0 dB. However, when the SNR is larger than 0 dB, the accuracies of the PCPD-GH method are near 100% while those of the PCPD-GG at most achieve about 50% accuracy. Consequently, it can be concluded from Fig. 4.2 that the PCPD-GH method achieves more stable and accurate tensor rank learning.

In summary, Figs. 4.1 and 4.2 show that PCPD-GH method finds the correct tensor rank even if the initial tensor rank is *exceedingly overestimated*, which is a *practically useful* feature since the rank is unknown in real-life cases.

4.1.2.2 Insights from Learned Length Scales

To clearly show the substantial difference between the GG and GH prior, we compare the two algorithms (PCPD-GG and PCPD-GH) in terms of their learned length scales. The length scale powers of GG and GH prior are denoted by $\{\gamma_l^{-1}\}_{l=1}^{L}$ and $\{z_l\}_{l=1}^{L}$, respectively. To assess the patterns of learned length scales, we turn off the pruning procedure and let the two algorithms directly output $\{\gamma_l^{-1}\}_{l=1}^{L}$ and $\{z_l\}_{l=1}^{L}$ after convergence. Since the learned length scale powers are possibly of different sparsity patterns in different Monte Carlo trials, averaging them over Monte Carlo trials is not informative. Instead, we present the learned values of $\{\gamma_l^{-1}\}_{l=1}^{L}$ and $\{z_l\}_{l=1}^{L}$ in a single trial.

In particular, Fig. 4.3 shows the result for a typical low-SNR and low-rank case (SNR $= -5$ dB, $R = 6$) with rank upper bound being 60. From this figure, it can be seen that the learned length scales of the two algorithms substantially *differ* from each other, in the sense that the number of learned length scales (and the associated components) *with non-negligible magnitudes* are different. For example, in Fig. 4.3, PCPD-GG recovers 7 components with non-negligible magnitudes[2] (the smallest one has value $4.3 \gg 0$), while PCPD-GH recovers 6 components. Note that the ground-truth rank is 6, and PCPD-GG produces a "ghost" component *with magnitude much larger than zero*. Additional simulation runs, and results of other simulation settings (e.g., high-SNR and high-rank case: SNR $= 5$ dB, $R = 21$) are included in [2], from which similar conclusions can be drawn.

4.1.2.3 Insights on Noise Precision Learning

The learning of the noise precision β is crucial for reliable inference, since incorrect estimates will cause over-/under-regularization. To examine how the speed of noise learning affects the tensor rank (sparsity) learning when SNR is low (SNR $= -5$ dB), we turn on the pruning and present the rank learning results over iterations in three cases: (1) Case I: update β every iteration; (2) Case II: update β every 10th iteration; (3) Case III: update β every 20th iteration. In Fig. 4.4, the rank estimates are averaged over 100 Monte Carlo runs. It can be seen that updating the noise precision β at earlier iterations will help the learning process to unveil the sparsity pattern more quickly.

Then, we investigate under which scenario slowing the noise precision learning will be helpful. We consider a very low-SNR case, that is, SNR $= -10$ dB, and

[2] The magnitude of the lth component is defined as $\left(\sum_{n=1}^{3}\left[\boldsymbol{A}_{:,l}^{(n)}\right]^{T}\boldsymbol{A}_{:,l}^{(n)}\right)^{\frac{1}{2}}$ [1].

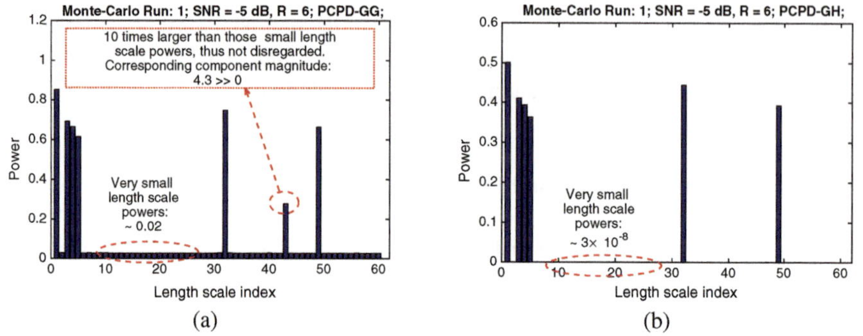

Fig. 4.3 **a** The powers of learned length scales (i.e., $\left\{\gamma_l^{-1}\right\}_{l=1}^{L}$) for PCPD-GG; **b** The powers of learned length scales (i.e., $\{z_l\}_{l=1}^{L}$) for PCPD-GH. It can be seen that PCPD-GG recovers 7 components with non-negligible magnitudes, while PCPD-GH recovers 6 components. The two algorithms are with the same upper bound value: 60

Fig. 4.4 Tensor rank estimates of PCPD-GH versus iteration number (averaged over 100 Monte Carlo runs) with different noise precision learning speeds

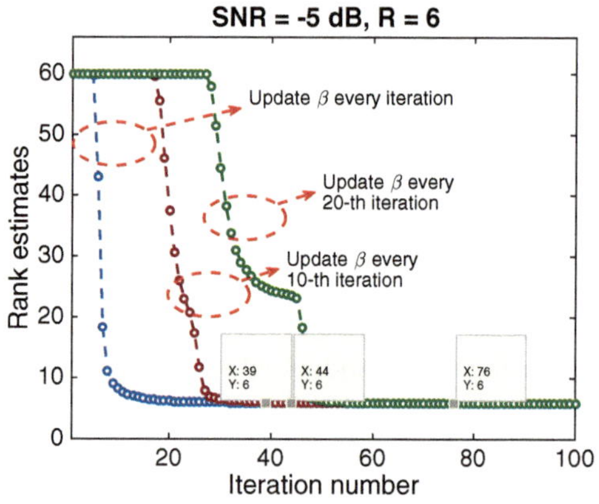

then evaluate the percentages of accurate rank learning over 100 Monte Carlo runs. The results are (1) Case I: 76%; (2) Case II: 100%; (3) Case III: 100%. In other words, when the noise power is very large (e.g., SNR $= -10$ dB), slowing the noise precision learning will make the algorithm more robust to the noises.

Finally, from further experiments in [2], it is found that if we fix the noise precision β and do not allow its update, PCPD-GH fails to identify the underlying sparsity pattern (tensor rank). Particularly, a small value of β (e.g., 0.01) leads to over-regularization, thus causing underestimation of non-zero components, while a large value of β (e.g., 100) causes under-regularization, thus inducing overestimation of non-zero components. This shows the importance of modeling and updating of noise precision.

4.1.2.4 Other Performance Metrics

Additional results and discussions on the run time, tensor recovery root mean square error (RMSE), algorithm performance under factor matrix correlation, convergence behavior of the PCPD-GH algorithm in terms of evidence lower bound (ELBO), and hyper-parameter learning of PCPD-GG are also included in [2]. The *key messages* of these simulation results are given as follows: (1) PCPD-GH generally costs more run time than PCPD-GG; (2) Incorrect estimation of tensor rank degrades the tensor signal recovery; (3) PCPD-GH performs well under factor matrix correlation; (4) PCPD-GH monotonically increases the ELBO; (5) Further update of hyper-parameter of PCPD-GH does not help too much in improving rank estimation in the low-SNR regime. Interested readers can refer to the details in [2].

4.1.3 Comparisons with Non-parametric PCPD-MGP

After comparing to the parametric PCPD-GG, further comparisons are performed with the non-parametric Bayesian tensor CPD using MGP prior (labeled as PCPD-MGP).[3] The initializations and hyper-parameters settings follow those used in [3].

4.1.3.1 Tensor Rank Learning

We first assess the performance of tensor rank learning of the three algorithms (PCPD-MGP, PCPD-GG, and PCPD-GH). The simulation settings follow those of Figs. 4.1 and 4.2.

In Table 4.1, the percentages of accurate tensor rank estimates under different rank upper bound values are presented. From the table, we can draw the following conclusions. (1) When the tensor rank is low (e.g., $R = 3, 6, 9$) and the SNR is high (e.g., SNR = 10 dB), PCPD-MGP correctly learns the tensor rank over 100 Monte Carlo trials under different rank upper bound values. Its performance is insensitive to the selection of the rank upper bound values, due to the decaying effects of the learned length scales [3]. Therefore, in the low-rank and high-SNR scenario, the rank learning performance of PCPD-MGP is comparable to that of PCPD-GH and much better than PCPD-GG. (2) When the tensor rank is high (e.g., $R = 18, 21, 24, 27$), PCPD-MGP fails to learn the correct tensor rank under different rank upper bound values and even perform worse than PCPD-GG, since the decaying effects of learned length scales tend to *underestimate* the tensor rank. On the contrary, the PCPD-GH method always accurately estimates the high tensor ranks.

In Table 4.2, we present the rank estimation performance under different SNRs, with the same settings as those of Fig. 4.2. It can be seen that for low tensor rank case (e.g., $R = 6$), PCPD-MGP shows good performance when SNR is larger than

[3] We appreciate Prof. Piyush Rai for sharing the code and data with us.

Table 4.1 Performance of tensor rank learning under different rank upper bound values. Algorithm: PCPD-GH, PCPD-MGP, PCPD-GG; SNR = 10 dB. The simulation settings are the same as those of Fig. 4.1

Rank upper bound	True tensor rank R														
	3			6			9			12			15		
	GH (%)	MGP (%)	GG (%)	GH (%)	MGP (%)	GG (%)	GH (%)	MGP (%)	GG (%)	GH (%)	MGP (%)	GG (%)	GH (%)	MGP (%)	GG (%)
max(DimY)	**100**	**100**	88	**100**	**100**	98	98	**100**	**100**	**100**	98	**100**	94	90	**98**
2max(DimY)	**100**	**100**	36	**100**	**100**	26	**100**	**100**	20	**100**	94	24	**100**	64	24
5max(DimY)	**100**	**100**	14	**100**	**100**	10	**100**	**100**	8	**100**	26	14	**100**	24	18

Rank upper bound	True tensor rank R											
	18			21			24			27		
	GH (%)	MGP (%)	GG (%)	GH (%)	MGP (%)	GG (%)	GH (%)	MGP (%)	GG (%)	GH (%)	MGP (%)	GG (%)
max(DimY)	8	0	**19**	0	0	0	0	0	0	0	0	0
2max(DimY)	**100**	0	14	**100**	0	30	**100**	0	26	**100**	0	50
5max(DimY)	**100**	0	12	**100**	0	14	**100**	0	16	**100**	0	14

Table 4.2 Performance of tensor rank learning under different SNRs. Algorithm: PCPD-GG, PCPD-MGP; Upper bound value: 2max(DimY). The simulation settings are the same as those of Fig. 4.2

True tensor rank	SNR (dB)											
	−10			−5			0			5		
	GH (%)	MGP (%)	GG (%)	GH (%)	MGP (%)	GG (%)	GH (%)	MGP (%)	GG (%)	GH (%)	MGP (%)	GG (%)
6	**92**	2	10	**100**	78	8	**100**	90	12	**100**	98	28
24	0	0	0	0	0	**18**	**82**	0	30	**100**	0	38
True tensor rank	SNR (dB)											
	10			15			20					
	GH (%)	MGP (%)	GG (%)	GH (%)	MGP (%)	GG (%)	GH (%)	MGP (%)	GG (%)			
6	**100**	**100**	26	**100**	**100**	16	**100**	**100**	8			
24	**100**	00	26	**100**	0	22	**100**	0	34			

−5 dB. It outperforms the PCPD-GG method, but it is still not as good as PCPD-GH. At SNR $= -10$ dB, PCPD-MGP fails to correctly estimate the tensor rank, while PCPD-GH shows good performance in a wider range of SNRs (from −10 to 20 dB). Furthermore, when the tensor rank becomes large (e.g., $R = 24$), PCPD-MGP fails to learn the underlying true tensor rank, making it inferior to PCPD-GH (and even PCPD-GG) in the high-rank regime.

4.1.3.2 Insights from Learned Length Scales

To reveal more insights, we present the learned length scales (without pruning) of PCPD-MPG. Following the notations in [3], the learned length scale powers of PCPD-MGP are denoted by $\left\{\tau_l^{-1} = \left(\prod_{k=1}^{l} \delta_k\right)^{-1}\right\}_{l=1}^{L}$.

Two typical cases are considered: (1) Case I: SNR $= 5$ dB, $R = 21$, corresponding to high rank and high SNR; (2) Case II: SNR $= -10$ dB, $R = 6$, corresponding to very low SNR and low rank. For each case, we present the learned length scales in a single trial in Figs. 4.5 and 4.6, respectively. From these figures, we have the following observations. (1) Due to the *decaying effect* of the MGP prior, the learned length scale power τ_l^{-1} quickly decreases as l becomes larger. This drives PCPD-MGP to *fail* to recover the sparsity pattern of the high-rank CPD (e.g., $R = 21$), in which a large number of length scale powers should be much larger than zero; see Fig. 4.5a. In contrast, without the decaying effect, the PCPD-GH successfully identifies the 21 non-negligible components, as seen in Fig. 4.5b. (2) When the SNR is very low (e.g., SNR $= -10$ dB) and the rank is low (e.g., $R = 6$), while the rank estimation given by PCPD-MGP is close to the ground truth, its sparsity pattern of the learned length scales is *not as accurate as* that of PCPD-GH (see Fig. 4.6). This

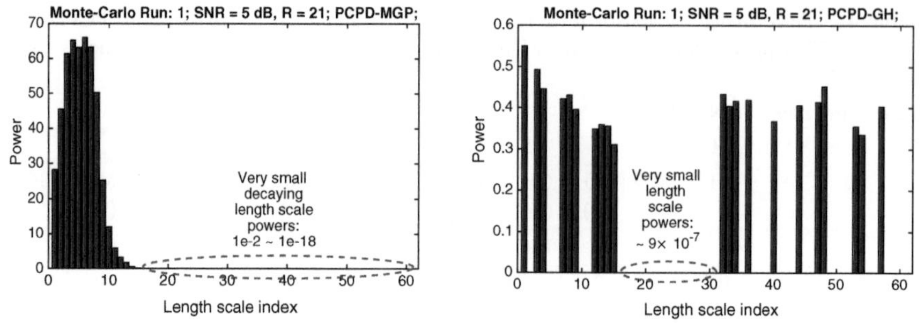

(a) (b)

Fig. 4.5 **a** The powers of learned length scales (i.e., $\{\tau_l^{-1}\}_{l=1}^{L}$) for PCPD-MGP; **b** The powers of learned length scales (i.e., $\{z_l\}_{l=1}^{L}$) for PCPD-GH. It can be seen that PCPD-MGP recovers 15 components with non-negligible magnitudes, while PCPD-GH recovers 21 components. The two algorithms are with the same upper bound value $L = 60$. Simulation setting: SNR $= 5$ dB, $R = 21$

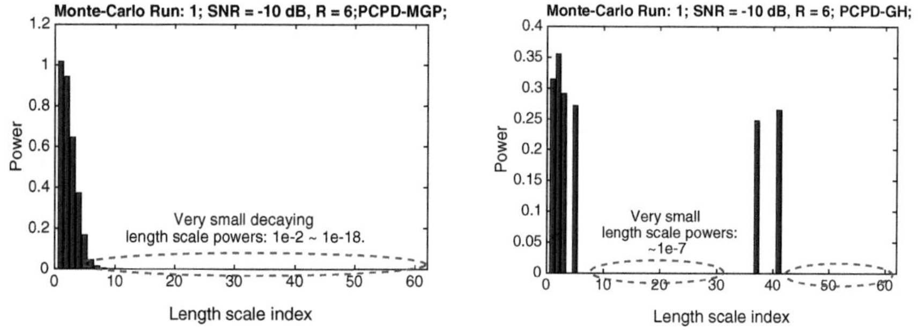

(a) (b)

Fig. 4.6 **a** The powers of learned length scales (i.e., $\{\tau_l^{-1}\}_{l=1}^{L}$) for PCPD-MGP; **b** The powers of learned length scales (i.e., $\{z_l\}_{l=1}^{L}$) for PCPD-GH. It can be seen that PCPD-MGP recovers 5 components with non-negligible magnitudes, while PCPD-GH recovers 6 components. The two algorithms are with the same upper bound value $L = 60$. Simulation setting: SNR $= -10$ dB, $R = 6$

is due to the mismatch in the decaying component amplitude assumption in the MGP model.

Additional simulation runs and results of more simulation settings are included in [2], from which similar conclusions can be drawn.

4.2 Real-World Applications

In this section, two real-world applications of CPD are presented, with a focus on performance comparisons between PCPD-GG and PCPD-GH.

4.2.1 *Fluorescence Data Analytics*

Fluorescence spectroscopy is a fast, simple, and inexpensive method to determine the concentration of any solubilized sample based on its fluorescent properties (see Fig. 4.7) and is widely used in chemical, pharmaceutical, and biomedical fields [4]. In fluorescence spectroscopy, an excitation beam with a certain wavelength λ_i passes through a solution in a cuvette. The excited chemical species in the sample will change their electronic states and then emit a beam of light, of which its spectrum is measured at the detector. Mathematically, let the concentration of the rth species in the sample be c_r, and the excitation value at wavelength λ_i be $a_r(\lambda_i)$. Then, the noise-free measured spectrum intensity at the wavelength λ_j is $a_r(\lambda_i)b_r(\lambda_j)c_r$, where $b_r(\lambda_j)$ is the emission value of the rth species at the wavelength λ_j. If there are R different species in the sample, the noise-free fluorescence excitation–emission measured (EEM) data at λ_j is

$$x_{i,j} = \sum_{r=1}^{R} a_r(\lambda_i)b_r(\lambda_j)c_r. \tag{4.1}$$

Assume the excitation beam contains I wavelengths, and the noise-free EEM data is collected at J different wavelengths, then an $I \times J$ data matrix is obtained as

$$\boldsymbol{X} = \sum_{r=1}^{R} \boldsymbol{A}_{:,r} \circ \boldsymbol{B}_{:,r} c_r, \tag{4.2}$$

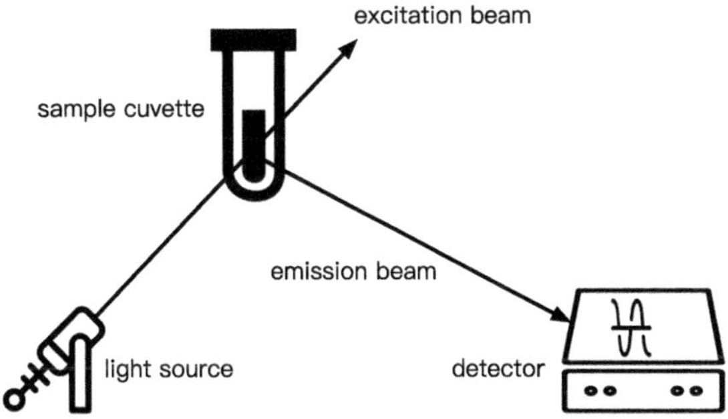

Fig. 4.7 An application example of CPD: fluorescence excitation–emission measurements

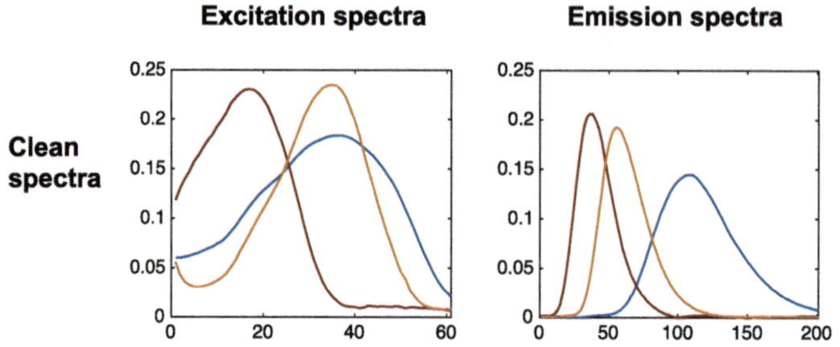

Fig. 4.8 The clean spectra recovered from the noise-free fluorescence tensor data assuming the knowledge of tensor rank

where symbol $\circ$ denotes vector outer product, $\boldsymbol{A}_{:,r} \in \mathbb{R}^{I\times 1}$ is a vector with the ith element being $a_r(\lambda_i)$, and $\boldsymbol{B}_{:,r} \in \mathbb{R}^{J\times 1}$ is a vector with the jth element being $b_r(\lambda_j)$. Assume $K > 1$ samples with the same chemical species but with different concentrations of each species are measured. Let the concentration of the rth species in the kth sample be $c_{k,r}$, then after stacking the noise-free EEM data for each sample along a third dimension, a three-dimensional (3D) tensor data $\mathcal{X} \in \mathbb{R}^{I\times J\times K}$ can be obtained as

$$\mathcal{X} = \sum_{r=1}^{R} \boldsymbol{A}_{:,r} \circ \boldsymbol{B}_{:,r} \circ \boldsymbol{C}_{:,r} \triangleq [\![\boldsymbol{A}, \boldsymbol{B}, \boldsymbol{C}]\!], \tag{4.3}$$

where $\boldsymbol{C}_{:,r} \in \mathbb{R}^{K\times 1}$ is a vector with the kth element being $c_{k,r}$; matrices $\boldsymbol{A} \in \mathbb{R}^{I\times R}$, $\boldsymbol{B} \in \mathbb{R}^{J\times R}$, and $\boldsymbol{C} \in \mathbb{R}^{K\times R}$ are matrices with their rth columns being $\boldsymbol{A}_{:,r}$, $\boldsymbol{B}_{:,r}$, and $\boldsymbol{C}_{:,r}$, respectively. It is easy to see that the noise-free data model in (4.3) yields exactly the tensor CPD model, and that is why CPD algorithms work very well for EEM data analysis.

To showcase various CPD algorithms for the fluorescence data analysis application, we consider the popular amino acids fluorescence data[4] $\mathcal{X}$ with size $5 \times 201 \times 61$ [5]. It consists of five laboratory-made samples, with each sample containing different amounts of tyrosine, tryptophan, and phenylalanine dissolved in phosphate buffered water. Since there are three different types of amino acids, when adopting the CPD model, the optimal tensor rank should be 3. In particular, with the optimal tensor rank 3, the clean spectra for the three types of amino acids, which are recovered by the alternative least-squares (ALS) algorithm (**Algorithm 1** of Chap. 1), are presented in Fig. 4.8 as the benchmark.

In practice, it is impossible to know how many components are present in the data in advance, and this calls for automatic tensor rank learning. Therefore, we assess both the rank learning performance and the noise mitigation performance for the two algorithms (i.e., PCPD-GH and PCPD-GG) under different levels of noise sources.

[4] http://www.models.life.ku.dk.

Table 4.3 Fit values and estimated tensor ranks of fluorescence data under different SNRs

SNR (dB)	Rank upper bound = max(DimX)				Rank upper bound = 2max(DimX)			
	PCPD-GG		PCPD-GH		PCPD-GG		PCPD-GH	
	Fit value	Estimated tensor rank	Fit value	Estimated tensor rank	Fit value	Estimated tensor rank	Fit value	Estimated tensor rank
−10	71.8109	4	**72.6401**	**3**	71.8197	4	**72.6401**	**3**
−5	83.9269	4	**84.3424**	**3**	83.5101	4	**84.3424**	**3**
0	90.6007	4	**90.8433**	**3**	90.3030	5	**90.8433**	**3**
5	94.2554	4	**94.3554**	**3**	94.0928	5	**94.3555**	**3**
10	96.0907	**3**	96.0951	**3**	96.0369	4	**96.0955**	**3**
15	96.8412	**3**	96.8431	**3**	96.8412	**3**	96.8432	**3**
20	97.1197	**3**	97.1204	**3**	97.1197	**3**	97.1204	**3**

In particular, the Fit value [6], which is defined as $(1 - \frac{||\hat{\mathcal{X}}-\mathcal{X}||_F}{||\mathcal{X}||_F}) \times 100\%$, is adopted, where $\hat{\mathcal{X}}$ represents the reconstructed fluorescence tensor data from the algorithm.

In Table 4.3, the performances of the two algorithms are presented assuming different upper bound values of tensor rank. It can be observed that with different upper bound values, the PCPD-GH algorithm always gives the correct tensor rank estimates, even when the SNR is smaller than 0 dB. On the other hand, the PCPD-GG method is quite sensitive to the choice of the upper bound value. Its performance with upper bound $2\max\{J_n\}_{n=1}^N$ becomes much worse than that with $\max\{J_n\}_{n=1}^N$ in tensor rank learning. Even with the upper bound being equal to $\max\{J_n\}_{n=1}^N$, PCD-GG fails to recover the optimal tensor rank 3 in the low-SNR region (i.e., SNR $\leq$ 5 dB). With the overestimated tensor rank, the reconstructed fluorescence tensor data $\hat{\mathcal{X}}$ will be overfitted to the noise sources, leading to lower Fit values. As a result, the Fit values of the PCPD-GH method are generally higher than those of the PCPD-GG method under different SNRs.

In this application, since the tensor rank represents the number of underlying components inside the data, its incorrect estimation will not only lead to overfitting to the noise, but also will cause "ghost" components that cannot be interpreted. This is illustrated in Fig. 4.9 where ghost component appears in SNR $= -10$ dB and 0 dB for PCPD-GG. In addition, in Fig. 4.10, we present the learned length scale powers from PCPD-GG and PCPD-GH. It can be seen that PCPD-GG recovers four components with non-negligible powers. The smallest component's power is 8 times larger than those of negligible components. In data analysis, disregarding such a large learned latent component is not reasonable. On the other hand, the PCPD-GH gives very clean 3 components with non-negligible powers.

Since the "ghost" component in PCPD-GG has a relatively large magnitude, it degrades the performance of the tensor signal recovery. For example, from Table 4.3, when the SNR is low (e.g., -10 dB), the Fit value of PCPD-GH is 0.8 higher than

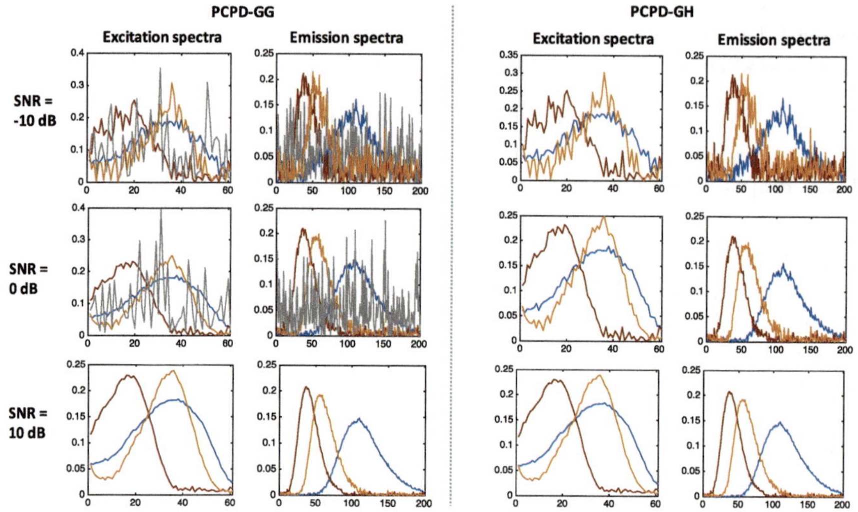

Fig. 4.9 The recovered spectra of fluorescence data under different SNRs

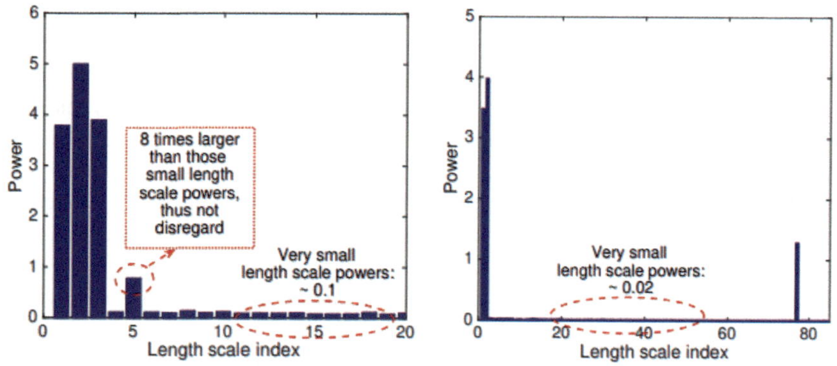

(a) PCPD-GG (b) PCPD-GH

Fig. 4.10 Amino acids fluorescence data analysis. **a** The powers of learned length scales (i.e., $\{\gamma_l^{-1}\}_{l=1}^L$) for PCPD-GG; **b** The powers of learned length scales (i.e., $\{z_l\}_{l=1}^L$) for PCPD-GH. It can be seen that PCPD-GG recovers 4 components with non-negligible magnitudes, while PCPD-GH recovers 3 components. The two algorithms are with the same upper bound value: 201. Simulation setting: SNR $= -5$ dB. Since the x-axis is too long (containing 201 points), we only present partial results that include non-zero components. Those values not shown in the figures are all very close to zero

that of PCPD-GG. This is in great contrast in the high-SNR regime (no "ghost" component) where their Fit value difference is about 0.001. Therefore, the "ghost" component significantly degrades the tensor signal recovery performance.

4.2.2 Hyperspectral Images Denoising

Hyperspectral image (HSI) data are naturally three-dimensional (two spatial dimensions and one spectral dimension), in which tensor CPD is a suitable tool to analyze such data. However, due to the radiometric noise, photon effects, and calibration errors, it is crucial to mitigate these corruptions before putting the HSI data into use. Since each HSI is rich in details, previous works using searching-based methods [7, 8] revealed that the tensor rank in HSI data is usually larger than half of the maximal tensor dimension. This corresponds to the *high tensor rank* scenario defined in Sect. 4.1.

In this application, we consider two real-world datasets: the Salinas-A HSI and the Indian Pines HSI, where different bands of HSIs were corrupted by different levels of noises. Some of the HSIs are quite clean while some of them are quite noisy. For such types of real-world data, since no ground truth is available, a no-reference quality assessment score is usually adopted [7, 8]. In particular, following [7], the SNR output $10 \log_{10} ||\hat{\mathcal{X}}||_F^2 / ||\mathcal{X} - \hat{\mathcal{X}}||_F^2$ is utilized as the denoising performance measure, where $\hat{\mathcal{X}}$ is the restored tensor data and $\mathcal{X}$ is the original HSI data. In Table 4.4, the SNR outputs of the two methods using different rank upper bound values are presented, from which it can be seen that the PCPD-GH method gives higher SNR outputs than PCPD-GG.

Samples of denoised HSIs are depicted in Fig. 4.11. On the left side of Fig. 4.11, the relatively clean Salinas-A HSI in band 190 is presented to serve as a reference, from which it can be observed that the landscape exhibits "stripe" pattern. For the noisy HSI in band 1, the denoising results from the two methods using the rank upper bound $\max\{J_n\}_{n=1}^N$ are presented. It is clear that the PCPD-GH method recovers better "stripe" pattern than the PCPD-GG method. Similarly, the results from the Indian Pines dataset are presented on the right side of Fig. 4.11. For noisy HSI in band 1, with the relatively clean image in band 10 serving as the reference, it can be observed that the PCPD-GH method recovers more details than the PCPD-GG method, when both using rank upper bound $2\max\{J_n\}_{n=1}^N$.

Since the HSIs in band 1 are quite noisy, inspecting the performance difference between the two methods requires a closer look. In Fig. 4.11, we have used

Table 4.4 SNR outputs and estimated tensor ranks of HSI data under different rank upper bounds

Algorithm		PCPD-GG		PCPD-GH	
Dataset	Rank upper bound	SNR output (dB)	Estimated tensor rank	SNR output (dB)	Estimated tensor rank
Salinas-A	$\max\{J_n\}_{n=1}^N$	43.7374	137	**44.0519**	143
	$2\max\{J_n\}_{n=1}^N$	46.7221	257	**46.7846**	260
Indian pines	$\max\{J_n\}_{n=1}^N$	30.4207	169	**30.5541**	178
	$2\max\{J_n\}_{n=1}^N$	31.9047	317	**32.0612**	335

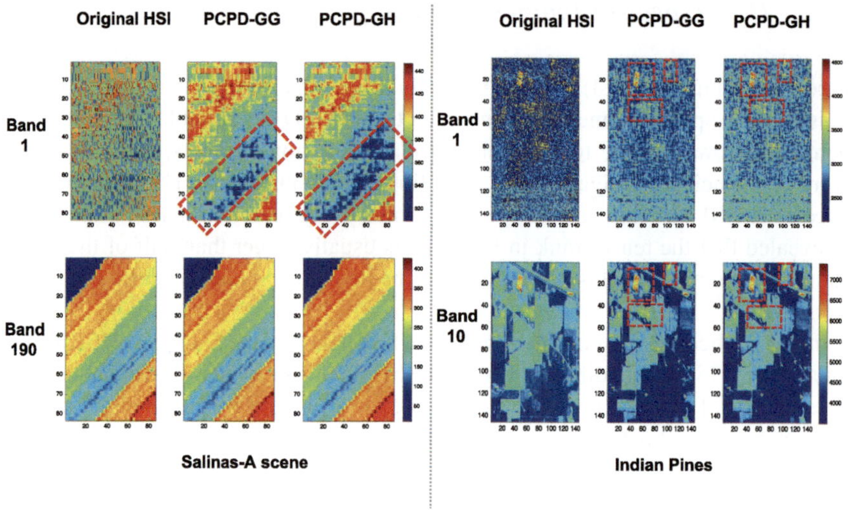

Fig. 4.11 The hyper-spectral image denoising results

red boxes to highlight those differences. Note that although HSIs in different frequency bands have different pixel intensities (different color bars), they share the same "clustering" structure. The goal of HSI denoising is to reconstruct the "clustering" structure in each band in order to facilitate the downstream segmentation task [7, 8]. Therefore, the assessment is based on whether the recovered HSI exhibits correct "clustering" patterns. Specifically, for Salinas-A scene data, the recovered images are supposed to render explicit "stripe" patterns, in each of which the intensities (colors) are almost the same. As indicated by the red boxes, it can be observed that PCPD-GH recovers better "stripe" pattern than PCPD-GG, since much more pixels in the red box of PCPD-GH have the same blue color. Similarly, for the Indian Pines dataset, as indicated by each red box, the area supposed to be identified as a cluster (with warmer colors than nearby areas) is more accurately captured by PCPD-GH.

References

1. Q. Zhao, L. Zhang, A. Cichocki, Bayesian cp factorization of incomplete tensors with automatic rank determination. IEEE Trans. Pattern Anal. Mach. Intell. **37**(9), 1751–1763 (2015)
2. L. Cheng, Z. Chen, Q. Shi, Y.-C. Wu, S. Theodoridis, Towards flexible sparsity-aware modeling: automatic tensor rank learning using the generalized hyperbolic prior. IEEE Trans. Signal Process. **70**, 1834–1849 (2022)
3. P. Rai, Y. Wang, S. Guo, G. Chen, D. Dunson, L. Carin, Scalable bayesian low-rank decomposition of incomplete multiway tensors, in *International Conference on Machine Learning* (PMLR, 2014), pp. 1800–1808
4. J.R. Albani, *Principles and Applications of Fluorescence Spectroscopy* (Wiley, New York, 2008)

5. R. Bro, Multi-way analysis in the food industry, in *Models, Algorithms, and Applications* (Academish proefschrift, Dinamarca, 1998)
6. A. Cichocki, R. Zdunek, A.H. Phan, S.-I. Amari, *Nonnegative Matrix and Tensor Factorizations: Applications to Exploratory Multi-way Data Analysis and Blind Source Separation* (Wiley, New York, 2009)
7. X. Liu, S. Bourennane, C. Fossati, Denoising of hyperspectral images using the parafac model and statistical performance analysis. IEEE Trans. Geosci. Remote Sens. **50**(10), 3717–3724 (2012)
8. B. Rasti, P. Scheunders, P. Ghamisi, G. Licciardi, J. Chanussot, Noise reduction in hyperspectral imagery: overview and application. Remote Sens. **10**(3), 482 (2018)

Chapter 5
When Stochastic Optimization Meets VI: Scaling Bayesian CPD to Massive Data

Abstract In previous chapters, Bayesian tensor CPD algorithms are derived for batch-mode operation, meaning that it needs to process the whole dataset at the same time. Obviously, this is no longer suitable for large datasets. To enable Bayesian tensor CPD in the Big Data era, the idea of stochastic optimization can be incorporated, rendering a scalable algorithm that only processes a mini-batch data at a time. In this chapter, we develop a scalable algorithm for Bayesian tensor CPD with automatic rank determination. Numerical examples in synthetic and real-world data demonstrate the excellent performance of the algorithm, both in terms of computation time and accuracy.

5.1 CPD Problem Reformulation

According to (1.8), tensor CPD assumes that a $P+1$ dimensional data tensor $\dot{\mathcal{Y}} \in \mathbb{R}^{I_1 \times I_2 \times \cdots \times I_P \times I_{P+1}}$ obeys the following model:

$$\begin{aligned}\dot{\mathcal{Y}} &= \sum_{r=1}^{R} \Xi^{(1)}_{:,r} \circ \Xi^{(2)}_{:,r} \circ \cdots \circ \Xi^{(P)}_{:,r} \circ \Xi^{(P+1)}_{:,r} + \dot{\mathcal{W}} \\ &\triangleq [\![\Xi^{(1)}, \Xi^{(1)}, \ldots, \Xi^{(P)}, \Xi^{(P+1)}]\!] + \dot{\mathcal{W}},\end{aligned} \tag{5.1}$$

where $\dot{\mathcal{W}}$ represents an additive noise tensor. The vector $\Xi^{(p)}_{:,r} \in \mathbb{R}^{I_p}$ is the rth column of the factor matrix $\Xi^{(p)} \in \mathbb{R}^{I_p \times R}$, and $\circ$ denotes the vector outer product. The number of rank-1 components R is defined as the tensor rank.

The core problem of tensor CPD is to find the factor matrices $\{\Xi^{(p)}\}_{p=1}^{P+1}$ from $\dot{\mathcal{Y}}$ under the unknown tensor rank R. The problem can be stated as

$$\min_{\{\Xi^{(p)}\}_{p=1}^{P+1}} \frac{\beta}{2} \parallel \dot{\mathcal{Y}} - [\![\Xi^{(1)}, \Xi^{(2)}, \ldots, \Xi^{(P+1)}]\!] \parallel_F^2 + \sum_{l=1}^{L} \frac{\gamma_l}{2} \Big(\sum_{p=1}^{P+1} \Xi^{(p)T}_{:,l} \Xi^{(p)}_{:,l} \Big), \tag{5.2}$$

where L is the maximum possible value of tensor rank R. Notice that the tensor rank acquisition is generally NP-hard, due to its discrete nature. As an effective heuristic

L. Cheng et al., *Bayesian Tensor Decomposition for Signal Processing and Machine Learning*, https://doi.org/10.1007/978-3-031-22438-6_5

approach, a regularization term $\sum_{l=1}^{L} \frac{\gamma_l}{2} \left(\sum_{p=1}^{P+1} \Xi_{:,l}^{(p)T} \Xi_{:,l}^{(p)} \right)$ is added in order to control the complexity of the model and avoid overfitting of noise.

It is easy to show that problem (5.2) is non-convex, since all the factor matrices $\{\Xi^{(p)}\}_{p=1}^{P+1}$ are coupled via Khatri–Rao products. To solve problem (5.2), the widely used ALS optimization framework iteratively updates each factor matrix with other factor matrices being fixed. However, in each update, the computation requires analyzing the whole data tensor $\dot{\mathcal{Y}}$, and thus is not computationally efficient. To achieve highly scalable algorithms, a recent work [1] rewrites the objective function into an equivalent summation form so that stochastic optimization can be readily applied. In particular, using the definition of tensor CPD, it is easy to show that problem (5.2) is equivalent to

$$\min_{\{\Xi^{(p)}\}_{p=1}^{P}} \frac{\beta}{2} \sum_{n=1}^{I_{P+1}} \| \mathcal{Y}(n) - [\![\Xi^{(1)}, \Xi^{(2)}, \ldots, \Xi^{(P)}, \boldsymbol{\xi}(n)]\!] \|_F^2 + \sum_{l=1}^{L} \frac{\gamma_l}{2} \left(\sum_{p=1}^{P} \Xi_{:,l}^{(p)T} \Xi_{:,l}^{(p)} + \sum_{n=1}^{I_{P+1}} \boldsymbol{\xi}(n)_l^2 \right) \tag{5.3}$$

where the data slices $\{\mathcal{Y}(n) \in \mathbb{R}^{I_1 \times I_2 \times \cdots \times I_P}\}_{n=1}^{I_{P+1}}$ are obtained by slicing the data tensor $\dot{\mathcal{Y}}$ along the last dimension.[1] In this expression, factor matrices $\{\Xi^{(p)} \in \mathbb{R}^{I_p \times L}\}_{p=1}^{P}$ span the tensor subspace where data tensor $\mathcal{Y}(n)$ lies, and vector $\boldsymbol{\xi}(n) = \Xi_{n,:}^{(P+1)}$ is the feature vector associated with the tensor data slice $\mathcal{Y}(n)$. From (5.3), it is obvious that the gradient of the objective function is a summation of I_{P+1} components, with each component depending only on a slice of tensor data $\mathcal{Y}(n)$. This motivates the use of stochastic optimization to improve the scalability.

5.1.1 Probabilistic Model and Inference for the Reformulated Problem

The probabilistic model for the reformulated problem (5.3) could be established by interpreting various terms in (5.3) via probability density functions (pdfs). Firstly, the squared error term in problem (5.3) can be interpreted as the negative log of a Gaussian likelihood function:

[1] The CPD definition implies that the summation form holds for the data tensor $\dot{\mathcal{Y}}$ being sliced along any dimension. Without loss of generality, the discussions here assume slicing along the last dimension.

$$
\begin{aligned}
& p\left(\{\mathcal{Y}(n)\}_{n=1}^{I_{P+1}} \mid \{\boldsymbol{\Xi}^{(p)}\}_{p=1}^{P}, \{\boldsymbol{\xi}(n)\}_{n=1}^{I_{P+1}}, \beta^{-1}\right) \\
& = \prod_{n=1}^{I_{P+1}} \left(\frac{\beta}{2\pi}\right)^{\frac{I_1 I_2 \cdots I_P}{2}} \exp\left(-\frac{\beta}{2} \parallel \mathcal{Y}(n) - [\![\boldsymbol{\Xi}^{(1)}, \boldsymbol{\Xi}^{(2)}, \ldots, \boldsymbol{\Xi}^{(P)}, \boldsymbol{\xi}(n)]\!] \parallel_F^2\right).
\end{aligned} \tag{5.4}
$$

Secondly, the regularization term in problem (5.3) can be interpreted as arising from a Gaussian prior distribution over the columns of the factor matrices and feature vectors, i.e.,

$$
p(\{\boldsymbol{\Xi}^{(p)}\}_{p=1}^{P}|\{\gamma_l\}_{l=1}^{L}) = \prod_{p=1}^{P}\prod_{l=1}^{L} \mathcal{N}\left(\boldsymbol{\Xi}_{:,l}^{(p)}|\mathbf{0}_{I_p\times 1}, \gamma_l^{-1}\boldsymbol{I}_L\right) = \prod_{p=1}^{P}\prod_{\tau=1}^{I_p} \mathcal{N}\left(\boldsymbol{\Xi}_{\tau,:}^{(p)}|\mathbf{0}_{1\times L}, \boldsymbol{\Gamma}^{-1}\right), \tag{5.5}
$$

$$
p(\{\boldsymbol{\xi}(n)\}_{n=1}^{I_{P+1}}|\{\gamma_l\}_{l=1}^{L}) = \prod_{n=1}^{I_{P+1}}\prod_{l=1}^{L} \mathcal{N}\left(\boldsymbol{\xi}(n)_l|0, \gamma_l^{-1}\right) = \prod_{n=1}^{I_{P+1}} \mathcal{N}\left(\boldsymbol{\xi}(n)|\mathbf{0}_{1\times L}, \boldsymbol{\Gamma}^{-1}\right), \tag{5.6}
$$

where $\boldsymbol{\Gamma} = \mathrm{diag}\{\gamma_1, \gamma_2, \ldots, \gamma_L\}$ and L is an upper bound of R. The inverse of the regularization parameter γ_l^{-1} has a physical interpretation of the power of the lth column of various factor matrices. When power γ_l^{-1} goes to zero, it indicates the corresponding columns in various factor matrices play no role, and can be pruned out. For the parameters β and $\{\gamma_l\}_{l=1}^{L}$, since we have no information about their distributions, non-informative gamma priors [4] are imposed on them, i.e., $p(\beta|\boldsymbol{\alpha}_\beta) = \mathrm{gamma}(\beta|10^{-6}, 10^{-6})$ and $p(\{\gamma_l\}_{l=1}^{L}|\boldsymbol{\lambda}_\gamma) = \prod_{l=1}^{L} \mathrm{gamma}(\gamma_l|10^{-6}, 10^{-6})$. This corresponds to the Gaussian-gamma model in (3.4) and (3.5).

The complete probabilistic model is shown in Fig. 5.1. Let $\boldsymbol{\Theta}$ be a set containing the unknown factor matrices $\{\boldsymbol{\Xi}^{(p)}\}_{p=1}^{P}$, feature vectors $\{\boldsymbol{\xi}(n)\}_{n=1}^{I_{P+1}}$, and other variables β, $\{\gamma_l\}_{l=1}^{L}$. From the probabilistic model established above, the goal of Bayesian inference is to calculate the posterior distribution $p(\boldsymbol{\Theta}|\{\mathcal{Y}(n)\}_{n=1}^{I_{P+1}}) = p(\boldsymbol{\Theta}, \{\mathcal{Y}(n)\}_{n=1}^{I_{P+1}})/\int p(\boldsymbol{\Theta}, \{\mathcal{Y}(n)\}_{n=1}^{I_{P+1}})d\boldsymbol{\Theta}$, which however is analytically intractable due to the multiple integrations involved. Therefore, we make use of variational inference in Sect. 3.4. In particular, under the mean-field approximation $Q(\boldsymbol{\Theta}) = \prod_{k=1}^{K} Q(\boldsymbol{\Theta}_k)$, where $\boldsymbol{\Theta}_k$ is part of $\boldsymbol{\Theta}$ with $\cup_{k=1}^{K}\boldsymbol{\Theta}_k = \boldsymbol{\Theta}$ and $\cap_{k=1}^{K}\boldsymbol{\Theta}_k = \emptyset$, the optimal variational pdf $Q^*(\boldsymbol{\Theta}_k)$ is obtained by solving the following problem with other $\{Q(\boldsymbol{\Theta}_j)\}_{j\neq k}$ fixed [5] (also see (3.17)):

$$
\begin{aligned}
& \min_{Q(\boldsymbol{\Theta}_k)} \int Q\left(\boldsymbol{\Theta}_k\right)\left(-\mathbb{E}_{\prod_{j\neq k} Q(\boldsymbol{\Theta}_j)}\left[\ln p\left(\boldsymbol{\Theta}, \{\mathcal{Y}(n)\}_{n=1}^{I_{P+1}}\right)\right] + \ln Q\left(\boldsymbol{\Theta}_k\right)\right)d\boldsymbol{\Theta}_k \\
& \text{s.t.} \int Q(\boldsymbol{\Theta}_k)d\boldsymbol{\Theta}_k = 1\,,\; Q(\boldsymbol{\Theta}_k) \geq 0.
\end{aligned} \tag{5.7}
$$

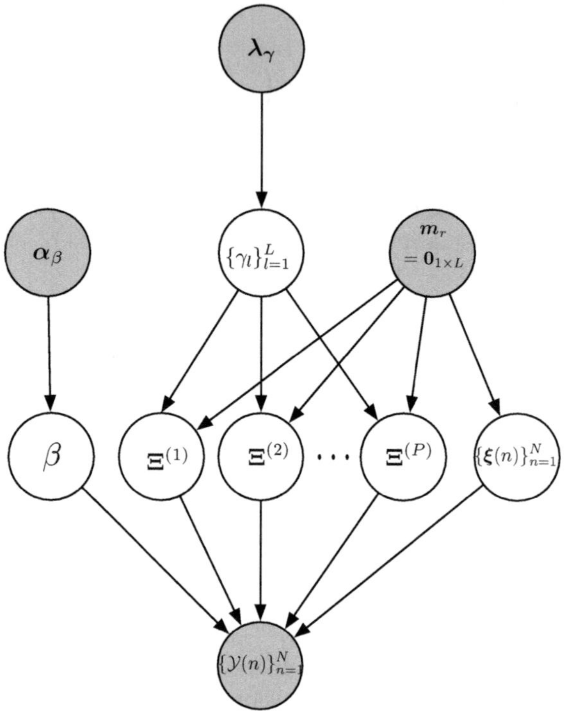

Fig. 5.1 Probabilistic model for reformulated tensor CPD (© [2018] IEEE. Reprinted, with permission, from [L. Cheng, Y.-C. Wu, and H. V. Poor, Scaling Probabilistic Tensor Canonical Polyadic Decomposition to Massive Data, IEEE Transactions on Signal Processing, Nov 2018]. It applies to all figures and tables in this chapter)

For this convex problem, the Karush–Kuhn–Tucker (KKT) condition gives the optimal variational pdf $Q^*(\boldsymbol{\Theta}_k)$ as [5]

$$Q^*(\boldsymbol{\Theta}_k) \propto \exp\left(\mathbb{E}_{\prod_{j\neq k} Q(\boldsymbol{\Theta}_j)}\left[\ln p\left(\boldsymbol{\Theta}, \{\mathcal{Y}(n)\}_{n=1}^{I_{P+1}}\right)\right]\right). \tag{5.8}$$

If we compute various $Q^*(\boldsymbol{\Theta})$ directly, we would end up **Algorithm 5** in Chap. 3, with the slight variation that the last factor matrix $\boldsymbol{\Xi}^{(P+1)}$ is learned row-by-row via $\boldsymbol{\xi}(n)$.

5.2 Interpreting VI Update from Natural Gradient Descent Perspective

While it may seem that the reformulation in Sect. 5.1 is just a trivial variation of the original Bayesian CPD problem and inference, it turns out this reformulation is the first step in the process of revealing a connection between VI and natural gradient descent, which is the gradient descent for distributions rather than parameters. If such a connection is established, we can make use of an idea from stochastic optimization, which has recently enabled many large-scale machine learning tasks [6–8].

5.2.1 Optimal Variational Pdfs in Exponential Family Form

In a previous work [3], it has been shown that for some special probabilistic models, such as the two-layer model specified in [3], the VI update due to (5.8) can be interpreted from a natural gradient descent perspective, which paves the way to integrate stochastic optimization in the VI framework. However, the probabilistic model of tensor CPD does not belong to the model family specified in [3], as evidenced by Fig. 5.1 that the considered model contains three layers. To bridge the framework of [3] to tensor CPD, we notice from **Algorithm 5** of Chap. 3 (also [15]) that the optimal variational pdfs in batch-mode Gaussian-gamma-based CPD are of Gaussian or gamma distribution, and thus can be characterized by their natural parameters (see **Definition 2.1**). In particular, from (3.38) and (3.39) in **Algorithm 5**, we can see that the optimal $\{Q^*(\Xi^{(p)})\}_{p=1}^P$ are Gaussian distributions. Tailoring to the model in Fig. 5.1, $Q^*(\Xi^{(p)})$ can be written in the exponential family form (for details, see [18]) as

$$Q^*(\Xi^{(p)}) = p(\mathrm{vec}(\Xi^{(p)})|\boldsymbol{\alpha}_p) \propto \exp\left\{\boldsymbol{\alpha}_p^T\left[\mathrm{vec}\left(\Xi^{(p)}\right)^T, -\mathrm{vec}\left(\Xi^{(p)T}\Xi^{(p)}\right)^T\right]^T\right\},$$

with the natural parameter being

$$\boldsymbol{\alpha}_p = \mathbb{E}_{\prod_{\Theta_j \neq \Xi^{(p)}} Q(\Theta_j)}\begin{bmatrix} \sum_{n=1}^{I_{P+1}} \boldsymbol{t}\left(\mathcal{Y}(n), \{\boldsymbol{\eta}_j, \boldsymbol{\eta}_j \neq \Xi^{(p)}\}\right) \\ \mathrm{vec}\left(\frac{1}{2}\boldsymbol{\Gamma}\right) + \sum_{n=1}^{I_{P+1}} \boldsymbol{l}_n\left(\{\boldsymbol{\eta}_j, \boldsymbol{\eta}_j \neq \Xi^{(p)}\}\right)\end{bmatrix}, \tag{5.9}$$

where[2]

$$\boldsymbol{t}\left(\mathcal{Y}(n), \left\{\boldsymbol{\eta}_j, \boldsymbol{\eta}_j \neq \Xi^{(p)}\right\}\right) = \mathrm{vec}\Big(\beta\left[\mathcal{Y}(n)\right]^{(p)}\left(\boldsymbol{\xi}(n) \diamond \big(\mathop{\diamond}_{m=1, m\neq p}^{P} \Xi^{(m)}\big)\right)\Big), \tag{5.10}$$

$$\boldsymbol{l}_n\left(\left\{\boldsymbol{\eta}_j, \boldsymbol{\eta}_j \neq \Xi^{(p)}\right\}\right) = \mathrm{vec}\Big(\frac{\beta}{2}\big(\boldsymbol{\xi}(n) \diamond \big(\mathop{\diamond}_{m=1, m\neq p}^{P} \Xi^{(m)}\big)\big)^T\big(\boldsymbol{\xi}(n) \diamond \big(\mathop{\diamond}_{m=1, m\neq p}^{P} \Xi^{(m)}\big)\big)\Big). \tag{5.11}$$

In (5.9), $\boldsymbol{\eta} = \{\{\Xi^{(p)}\}_{p=1}^P, \beta, \{\boldsymbol{\xi}(n)\}_{n=1}^{I_{P+1}}\}$ denotes the unknown variables of layer 1 in Fig. 5.1. In (5.10), the operation $[\mathcal{A}]^{(p)}$ unfolds the tensor $\mathcal{A}$ along the pth dimension, and $\mathop{\diamond}_{n=1, n\neq k}^{N} \boldsymbol{A}^{(n)} = \boldsymbol{A}^{(N)} \diamond \cdots \diamond \boldsymbol{A}^{(k+1)} \diamond \boldsymbol{A}^{(k-1)} \diamond \cdots \diamond \boldsymbol{A}^{(1)}$ in (5.11) is the multiple Khatri–Rao products.

[2] We use overloaded notation for the functions $\boldsymbol{t}(\cdot)$ and $\boldsymbol{l}(\cdot)$ for brevity. Their detailed functional forms depend on their arguments.

Similarly, the optimal variational pdf $Q^*(\boldsymbol{\xi}(n)) = p(\boldsymbol{\xi}(n)|\boldsymbol{\alpha}_{\boldsymbol{\xi}(n)})$ can be shown to be a Gaussian distribution, with the natural parameter being

$$\boldsymbol{\alpha}_{\boldsymbol{\xi}(n)} = \mathbb{E}_{\prod_{\boldsymbol{\Theta}_j \neq \boldsymbol{\xi}(n)} Q(\boldsymbol{\Theta}_j)} \begin{bmatrix} \boldsymbol{t}\left(\mathcal{Y}(n), \left\{\boldsymbol{\eta}_j, \boldsymbol{\eta}_j \neq \boldsymbol{\xi}(n)\right\}\right) \\ \mathrm{vec}\left(\frac{1}{2}\boldsymbol{\Gamma}\right) + \boldsymbol{l}\left(\left\{\boldsymbol{\eta}_j, \boldsymbol{\eta}_j \neq \boldsymbol{\xi}(n)\right\}\right) \end{bmatrix}, \tag{5.12}$$

where

$$\boldsymbol{t}\left(\mathcal{Y}(n), \left\{\boldsymbol{\eta}_j, \boldsymbol{\eta}_j \neq \boldsymbol{\xi}(n)\right\}\right) = \beta\Big(\mathop{\diamond}_{p=1}^{P} \boldsymbol{\Xi}^{(p)}\Big)^T \mathrm{vec}\left(\mathcal{Y}(n)\right), \tag{5.13}$$

$$\boldsymbol{l}\left(\left\{\boldsymbol{\eta}_j, \boldsymbol{\eta}_j \neq \boldsymbol{\xi}(n)\right\}\right) = \mathrm{vec}\Big(\frac{\beta}{2}\big(\mathop{\diamond}_{p=1}^{P} \boldsymbol{\Xi}^{(p)}\big)^T\big(\mathop{\diamond}_{p=1}^{P} \boldsymbol{\Xi}^{(p)}\big)\Big). \tag{5.14}$$

On the other hand, the optimal variational pdf $Q^*(\beta)$ is a gamma distribution [15]. Writing the gamma distribution in exponential family form, we have $Q^*(\beta) = p(\beta|\boldsymbol{\alpha}_\beta) \propto \exp\left\{\boldsymbol{\alpha}_\beta^T [\beta, \ln\beta]^T\right\}$, with the natural parameter being

$$\boldsymbol{\alpha}_\beta = \mathbb{E}_{\prod_{\boldsymbol{\Theta}_j \neq \beta} Q(\boldsymbol{\Theta}_j)} \begin{bmatrix} -10^{-6} + \sum_{n=1}^{I_{P+1}} \boldsymbol{t}\left(\mathcal{Y}(n), \left\{\boldsymbol{\eta}_j, \boldsymbol{\eta}_j \neq \beta\right\}\right) \\ 10^{-6} - 1 + \sum_{n=1}^{I_{P+1}} \boldsymbol{l}_n\left(\left\{\boldsymbol{\eta}_j, \boldsymbol{\eta}_j \neq \beta\right\}\right) \end{bmatrix}, \tag{5.15}$$

where

$$\boldsymbol{t}\left(\mathcal{Y}(n), \left\{\boldsymbol{\eta}_j, \boldsymbol{\eta}_j \neq \beta\right\}\right) = -\frac{1}{2} \parallel \mathcal{Y}(n) - [\![\boldsymbol{\Xi}^{(1)}, \boldsymbol{\Xi}^{(2)}, \ldots, \boldsymbol{\Xi}^{(P)}, \boldsymbol{\xi}(n)]\!] \parallel_F^2, \tag{5.16}$$

$$\boldsymbol{l}_n\left(\left\{\boldsymbol{\eta}_j, \boldsymbol{\eta}_j \neq \beta\right\}\right) = \frac{I_1 I_2 \cdots I_P}{2}. \tag{5.17}$$

Finally, the optimal $Q^*(\boldsymbol{\gamma}) = p(\boldsymbol{\gamma}|\boldsymbol{\lambda}_\gamma)$ is a gamma distribution [15] with the natural parameter being

$$\boldsymbol{\lambda}_\gamma = \mathbb{E}_{\prod_{\boldsymbol{\Theta}_j \neq \gamma} Q(\boldsymbol{\Theta}_j)} \begin{bmatrix} -\boldsymbol{d} + \boldsymbol{u}(\{\boldsymbol{\Xi}^{(p)}\}_{p=1}^P) + \sum_{n=1}^{I_{P+1}} \boldsymbol{u}(\boldsymbol{\xi}(n)) \\ \boldsymbol{c} - \boldsymbol{1} + \boldsymbol{w} \end{bmatrix}, \tag{5.18}$$

where

$$\boldsymbol{u}(\{\boldsymbol{\Xi}^{(p)}\}_{p=1}^P) = \left[\sum_{p=1}^{P} -\frac{1}{2}\boldsymbol{\Xi}_{:,1}^{(p)T}\boldsymbol{\Xi}_{:,1}^{(p)}, \ldots, \sum_{p=1}^{P} -\frac{1}{2}\boldsymbol{\Xi}_{:,L}^{(p)T}\boldsymbol{\Xi}_{:,L}^{(p)}\right]^T, \tag{5.19}$$

$$\boldsymbol{u}(\boldsymbol{\xi}(n)) = \left[-\frac{1}{2}\boldsymbol{\xi}(n)_1^2, \ldots, -\frac{1}{2}\boldsymbol{\xi}(n)_L^2\right]^T, \tag{5.20}$$

$$\boldsymbol{w} = \left[\frac{1}{2} \sum_{p=1}^{P} I_p + \frac{1}{2} I_{P+1}, \ldots, \frac{1}{2} \sum_{p=1}^{P} I_p + \frac{1}{2} I_{P+1} \right]^T, \tag{5.21}$$

vector $\mathbf{1}$ is the all-ones vector with length L, and $\boldsymbol{c} = \boldsymbol{d} = 10^{-6}\mathbf{1}$. From (5.9), (5.12), (5.15), and (5.18), since the update in each equation depends on the parameters in other equations, the VI algorithm consists of cyclical update among these equations.

Remark: If we compute the expectations in (5.9), (5.12), (5.15), and (5.18), after tedious derivations, we would obtain the batch-mode VI algorithm for tensor CPD [13, 15]. However, computing the expectations at this stage would obscure the interpretation of update Eqs. (5.9), (5.12), (5.15), and (5.18) as natural gradient descent steps in Riemannian space, an insight critical in developing the stochastic optimization for probabilistic tensor CPD algorithm.

5.2.2 VI Updates as Natural Gradient Descent

In order to reveal the connection between the optimal variational parametric updates (5.9), (5.12), (5.15), (5.18), and gradient descent steps, we substitute them back into problem (5.7), transforming the functional optimization problem (5.7) into a parametric optimization problem.

For example, using the knowledge that $Q^*(\boldsymbol{\Xi}^{(p)}) = p(\mathrm{vec}(\boldsymbol{\Xi}^{(p)})|\boldsymbol{\alpha}_p)$ is a Gaussian distribution, the constraints in problem (5.7) are automatically satisfied. Since $Q(\boldsymbol{\Xi}^{(p)})$ is parametrized by $\boldsymbol{\alpha}_p$, the problem (5.7) becomes

$$\min_{\boldsymbol{\alpha}_p} \int p(\mathrm{vec}(\boldsymbol{\Xi}^{(p)})|\boldsymbol{\alpha}_p)\Big(- \mathbb{E}_{\prod_{\boldsymbol{\Theta}_j \neq \boldsymbol{\Xi}^{(p)}} Q(\boldsymbol{\Theta}_j)} \left[\ln p\left(\boldsymbol{\Theta}, \mathcal{Y}\right)\right] + \ln p(\mathrm{vec}(\boldsymbol{\Xi}^{(p)})|\boldsymbol{\alpha}_p)\Big) d\boldsymbol{\alpha}_p. \tag{5.22}$$

After performing the integration and discarding the terms irrelevant to $\boldsymbol{\Xi}^{(p)}$, it can be shown that problem (5.22) becomes problem (5.23):

$$\begin{aligned} \min_{\boldsymbol{\alpha}_p} \quad & - \mathbb{E}_{\prod_{\boldsymbol{\Theta}_j \neq \boldsymbol{\Xi}^{(p)}} Q(\boldsymbol{\Theta}_j)} \begin{bmatrix} \sum_{n=1}^{I_{P+1}} \boldsymbol{t}\left(\mathcal{Y}(n), \{\boldsymbol{\eta}_j, \boldsymbol{\eta}_j \neq \boldsymbol{\Xi}^{(p)}\}\right) \\ \mathrm{vec}\left(\frac{1}{2}\boldsymbol{\Gamma}\right) + \sum_{n=1}^{I_{P+1}} \boldsymbol{l}_n\left(\{\boldsymbol{\eta}_j, \boldsymbol{\eta}_j \neq \boldsymbol{\Xi}^{(p)}\}\right) \end{bmatrix}^T \mathbb{E}_{p(\mathrm{vec}(\boldsymbol{\Xi}^{(p)})|\boldsymbol{\alpha}_p)} \left[\boldsymbol{t}(\boldsymbol{\Xi}^{(p)})\right] \\ & + {\boldsymbol{\alpha}_p}^T \mathbb{E}_{p(\mathrm{vec}(\boldsymbol{\Xi}^{(p)})|\boldsymbol{\alpha}_p)} \left[\boldsymbol{t}(\boldsymbol{\Xi}^{(p)})\right] - \mathfrak{L}(\boldsymbol{\Xi}^{(p)}), \end{aligned} \tag{5.23}$$

where $\boldsymbol{t}(\boldsymbol{\Xi}^{(p)}) = \left[\mathrm{vec}\left(\boldsymbol{\Xi}^{(p)}\right), -\mathrm{vec}\left(\boldsymbol{\Xi}^{(p)T}\boldsymbol{\Xi}^{(p)}\right)\right]^T$ is the sufficient statistic of the prior distribution (5.5), and $\mathfrak{L}(\boldsymbol{\Xi}^{(p)}) = I_p \sum_{l=1}^{L} \ln \gamma_l$ is the log normalizer of the prior distribution (5.5).

Given that (5.23) is a parametric optimization problem, denoting the objective function in (5.23) as $f(\boldsymbol{\alpha}_p)$, its derivative is shown as follows:

$$
\begin{aligned}
&\nabla_{\boldsymbol{\alpha}_p} f(\boldsymbol{\alpha}_p) \\
&= -\nabla_{\boldsymbol{\alpha}_p} \mathbb{E}_{p(\text{vec}(\Xi^{(p)})|\boldsymbol{\alpha}_p)}[\boldsymbol{t}(\Xi^{(p)})]\mathbb{E}_{\prod_{\Theta_j \neq \Xi^{(p)}} Q(\Theta_j)} \begin{bmatrix} \sum_{n=1}^{I_{P+1}} \boldsymbol{t}\left(\mathcal{Y}(n), \{\boldsymbol{\eta}_j, \boldsymbol{\eta}_j \neq \Xi^{(p)}\}\right) \\ \text{vec}\left(\frac{1}{2}\boldsymbol{\Gamma}\right) + \sum_{n=1}^{I_{P+1}} \boldsymbol{l}_n\left(\{\boldsymbol{\eta}_j, \boldsymbol{\eta}_j \neq \Xi^{(p)}\}\right) \end{bmatrix} \\
&+ \nabla_{\boldsymbol{\alpha}_p} \mathbb{E}_{p(\text{vec}(\Xi^{(p)})|\boldsymbol{\alpha}_p)}[\boldsymbol{t}(\Xi^{(p)})]\boldsymbol{\alpha}_p + \mathbb{E}_{p(\text{vec}(\Xi^{(p)})|\boldsymbol{\alpha}_p)}[\boldsymbol{t}(\Xi^{(p)})] - \nabla_{\boldsymbol{\alpha}_p}\left[\mathfrak{L}(\Xi^{(p)})\right].
\end{aligned}
\tag{5.24}
$$

Since $p(\text{vec}(\Xi^{(p)})|\boldsymbol{\alpha}_p)$ is in the exponential family, with $\boldsymbol{t}(\Xi^{(p)})$ and $\mathfrak{L}(\Xi^{(p)})$ being the sufficient statistic and the log normalizer, respectively, we have the property that $\mathbb{E}_{p(\text{vec}(\Xi^{(p)})|\boldsymbol{\alpha}_p)}[\boldsymbol{t}(\Xi^{(p)})] = \nabla_{\boldsymbol{\alpha}_p}\left[\mathfrak{L}(\Xi^{(p)})\right]$ [4]. Using this result, the last two terms of (5.24) cancel each other, and the remaining terms can be organized as

$$
\begin{aligned}
&\nabla_{\boldsymbol{\alpha}_p} f(\boldsymbol{\alpha}_p) \\
&= \nabla^2_{\boldsymbol{\alpha}_p}\left[\mathfrak{L}(\Xi^{(p)})\right]\left(- \mathbb{E}_{\prod_{\Theta_j \neq \Xi^{(p)}} Q(\Theta_j)} \begin{bmatrix} \sum_{n=1}^{I_{P+1}} \boldsymbol{t}\left(\mathcal{Y}(n), \{\boldsymbol{\eta}_j, \boldsymbol{\eta}_j \neq \Xi^{(p)}\}\right) \\ \text{vec}\left(\frac{1}{2}\boldsymbol{\Gamma}\right) + \sum_{n=1}^{I_{P+1}} \boldsymbol{l}_n\left(\{\boldsymbol{\eta}_j, \boldsymbol{\eta}_j \neq \Xi^{(p)}\}\right) \end{bmatrix} + \boldsymbol{\alpha}_p \right).
\end{aligned}
\tag{5.25}
$$

This gradient, which is commonly used in optimization, is defined in Euclidean distance metric, i.e., $\nabla_{\boldsymbol{\alpha}_p} f(\boldsymbol{\alpha}_p) = \arg\max_{d\boldsymbol{\alpha}_p} f(\boldsymbol{\alpha}_p + d\boldsymbol{\alpha}_p)$ subject to $d\boldsymbol{\alpha}_p^T d\boldsymbol{\alpha}_p < \epsilon$ for sufficiently small ϵ. However, since the parameter $\boldsymbol{\alpha}_p$ has its physical meaning as the natural parameter in exponential family distribution $p(\text{vec}(\Xi^{(p)})|\boldsymbol{\alpha}_p)$, the Euclidean distance between $\boldsymbol{\alpha}_p$ and $\boldsymbol{\alpha}'_p$ does not account for the information geometry of its parameter space, and thus is not a good distance measure between $p(\text{vec}(\Xi^{(p)})|\boldsymbol{\alpha}_p)$ and $p(\text{vec}(\Xi^{(p)})|\boldsymbol{\alpha}'_p)$.

Instead, the symmetric KL divergence, a metric in the Riemannian space, is found to be a natural distance measure for parameters $\boldsymbol{\alpha}_p$, and it is defined as [9]

$$
\begin{aligned}
&\boldsymbol{KL}^{sym}(\boldsymbol{\alpha}_p, \boldsymbol{\alpha}'_p) \\
&= \mathbb{E}_{p(\text{vec}(\Xi^{(p)})|\boldsymbol{\alpha}_p)}\left[\ln \frac{p(\text{vec}(\Xi^{(p)})|\boldsymbol{\alpha}_p)}{p(\text{vec}(\Xi^{(p)})|\boldsymbol{\alpha}'_p)}\right] + \mathbb{E}_{p(\text{vec}(\Xi^{(p)})|\boldsymbol{\alpha}'_p)}\left[\ln \frac{p(\text{vec}(\Xi^{(p)})|\boldsymbol{\alpha}'_p)}{p(\text{vec}(\Xi^{(p)})|\boldsymbol{\alpha}_p)}\right].
\end{aligned}
$$

Using this metric, the natural gradient that points in the direction of the steepest ascent in the Riemannian space is defined as $\nabla^{\text{n}}_{\boldsymbol{\alpha}_p} f(\boldsymbol{\alpha}_p) = \arg\max_{d\boldsymbol{\alpha}_p} f(\boldsymbol{\alpha}_p + d\boldsymbol{\alpha}_p)$ subject to $\boldsymbol{KL}^{sym}(\boldsymbol{\alpha}_p, \boldsymbol{\alpha}_p + d\boldsymbol{\alpha}_p) < \epsilon$ for sufficiently small ϵ [9]. Since this natural gradient utilizes the implicit information of parameter space, it is expected to give faster convergence than the traditional Euclidean metric-based gradient in stochastic gradient descent, and this has been demonstrated in the maximum-likelihood estimation problem [10].

Computing the natural gradient of $f(\boldsymbol{\alpha}_p)$ under the KL divergence of $p(\text{vec}(\Xi^{(p)})|\boldsymbol{\alpha}_p)$ and $p(\text{vec}(\Xi^{(p)})|\boldsymbol{\alpha}'_p)$ is easy, since $p(\text{vec}(\Xi^{(p)})|\boldsymbol{\alpha}_p)$ lies in the exponential family. It is shown in [10] that the natural gradient $\nabla^{\text{n}}_{\boldsymbol{\alpha}_p} f(\boldsymbol{\alpha}_p)$ can be calculated by premultiplying the traditional gradient $\nabla_{\boldsymbol{\alpha}_p} f(\boldsymbol{\alpha}_p)$ with the inverse of the Fisher information matrix of $p(\text{vec}(\boldsymbol{\Xi}^{(p)})|\boldsymbol{\alpha}_p)$, i.e., $\nabla^{\text{n}}_{\boldsymbol{\alpha}_p} f(\boldsymbol{\alpha}_p) = \boldsymbol{G}(\boldsymbol{\alpha}_p)^{-1} \nabla_{\boldsymbol{\alpha}_p} f(\boldsymbol{\alpha}_p)$ where $\boldsymbol{G}(\boldsymbol{\alpha}_p) \triangleq \mathbb{E}_{p(\text{vec}(\boldsymbol{\Xi}^{(p)})|\boldsymbol{\alpha}_p)}\big[\big(\nabla_{\boldsymbol{\alpha}_p} \ln p(\text{vec}(\boldsymbol{\Xi}^{(p)})|\boldsymbol{\alpha}_p)\big)\big(\nabla_{\boldsymbol{\alpha}_p} \ln p(\text{vec}(\boldsymbol{\Xi}^{(p)})|\boldsymbol{\alpha}_p)\big)^T\big]$. Substituting the functional form of $p(\text{vec}(\boldsymbol{\Xi}^{(p)})|\boldsymbol{\alpha}_p)$ into the definition of Fisher information matrix, it can be shown that $\boldsymbol{G}(\boldsymbol{\alpha}_p)$ is just the second derivative of the log normalizer [10], i.e., $\boldsymbol{G}(\boldsymbol{\alpha}_p) = \nabla^2_{\boldsymbol{\alpha}_p}\big[\mathfrak{L}(\boldsymbol{\Xi}^{(p)})\big]$. Using this result and (5.25), the natural gradient $\nabla^{\text{n}}_{\boldsymbol{\alpha}_p} f(\boldsymbol{\alpha}_p)$ is

$$\nabla^{\text{n}}{}_{\boldsymbol{\alpha}_p} f(\boldsymbol{\alpha}_p) = -\mathbb{E}_{\prod_{\boldsymbol{\Theta}_j \neq \boldsymbol{\Xi}^{(p)}} Q(\boldsymbol{\Theta}_j)} \begin{bmatrix} \sum_{n=1}^{I_{P+1}} \boldsymbol{t}\left(\mathcal{Y}(n), \left\{\boldsymbol{\eta}_j, \boldsymbol{\eta}_j \neq \boldsymbol{\Xi}^{(p)}\right\}\right) \\ \text{vec}\left(\frac{1}{2}\boldsymbol{\Gamma}\right) + \sum_{n=1}^{I_{P+1}} \boldsymbol{l}_n\left(\left\{\boldsymbol{\eta}_j, \boldsymbol{\eta}_j \neq \boldsymbol{\Xi}^{(p)}\right\}\right) \end{bmatrix} + \boldsymbol{\alpha}_p. \tag{5.26}$$

Similar derivations can be performed for other variables $\{\boldsymbol{\xi}(n)\}_{n=1}^{I_{P+1}}$, β, and $\boldsymbol{\gamma}$. In particular, after substituting $Q^*(\boldsymbol{\xi}(n)) = p(\boldsymbol{\xi}(n)|\boldsymbol{\alpha}_{\boldsymbol{\xi}(n)})$, $Q^*(\beta) = p(\beta|\boldsymbol{\alpha}_\beta)$ and $Q^*(\boldsymbol{\gamma}) = p(\boldsymbol{\gamma}|\boldsymbol{\lambda}_\gamma)$ into problem (5.7), and using the definition of natural gradient, we obtain

$$\nabla^{\text{n}}{}_{\boldsymbol{\alpha}_{\boldsymbol{\xi}(n)}} f(\boldsymbol{\alpha}_{\boldsymbol{\xi}(n)}) = -\mathbb{E}_{\prod_{\boldsymbol{\Theta}_j \neq \boldsymbol{\xi}(n)} Q(\boldsymbol{\Theta}_j)} \begin{bmatrix} \boldsymbol{t}\left(\mathcal{Y}(n), \left\{\boldsymbol{\eta}_j, \boldsymbol{\eta}_j \neq \boldsymbol{\xi}(n)\right\}\right) \\ \text{vec}\left(\frac{1}{2}\boldsymbol{\Gamma}\right) + \boldsymbol{l}\left(\left\{\boldsymbol{\eta}_j, \boldsymbol{\eta}_j \neq \boldsymbol{\xi}(n)\right\}\right) \end{bmatrix} + \boldsymbol{\alpha}_{\boldsymbol{\xi}(n)} \tag{5.27}$$

$$\nabla^{\text{n}}{}_{\boldsymbol{\alpha}_\beta} f(\boldsymbol{\alpha}_\beta) = -\mathbb{E}_{\prod_{\boldsymbol{\Theta}_j \neq \beta} Q(\boldsymbol{\Theta}_j)} \begin{bmatrix} -10^{-6} + \sum_{n=1}^{I_{P+1}} \boldsymbol{t}\left(\mathcal{Y}(n), \left\{\boldsymbol{\eta}_j, \boldsymbol{\eta}_j \neq \beta\right\}\right) \\ 10^{-6} - 1 + \sum_{n=1}^{I_{P+1}} \boldsymbol{l}_n\left(\left\{\boldsymbol{\eta}_j, \boldsymbol{\eta}_j \neq \beta\right\}\right) \end{bmatrix} + \boldsymbol{\alpha}_\beta \tag{5.28}$$

$$\nabla^{\text{n}}{}_{\boldsymbol{\lambda}_\gamma} f(\boldsymbol{\lambda}_\gamma) = -\mathbb{E}_{\prod_{\boldsymbol{\Theta}_j \neq \gamma} Q(\boldsymbol{\Theta}_j)} \begin{bmatrix} -\boldsymbol{d} + \boldsymbol{u}(\{\boldsymbol{\Xi}^{(p)}\}_{p=1}^P) + \sum_{n=1}^{I_{P+1}} \boldsymbol{u}(\boldsymbol{\xi}(n)) \\ \boldsymbol{c} - \boldsymbol{1} + \boldsymbol{w} \end{bmatrix} + \boldsymbol{\lambda}_\gamma. \tag{5.29}$$

From the expressions (5.26)–(5.29), it is easy to see that the cyclical updates of (5.9), (5.12), (5.15), and (5.18) are equivalent to the natural gradient descents with step size 1 as follows:

$$\boldsymbol{\alpha}_p^{t+1} = \boldsymbol{\alpha}_p^t - \nabla^{\text{n}}{}_{\boldsymbol{\alpha}_p} f(\boldsymbol{\alpha}_p^t) \tag{5.30}$$

$$\boldsymbol{\alpha}_{\boldsymbol{\xi}(n)}^{t+1} = \boldsymbol{\alpha}_{\boldsymbol{\xi}(n)}^t - \nabla^{\text{n}}{}_{\boldsymbol{\alpha}_{\boldsymbol{\xi}(n)}} f(\boldsymbol{\alpha}_{\boldsymbol{\xi}(n)}^t) \tag{5.31}$$

$$\boldsymbol{\alpha}_\beta^{t+1} = \boldsymbol{\alpha}_\beta^t - \nabla^{\text{n}}{}_{\boldsymbol{\alpha}_\beta} f(\boldsymbol{\alpha}_\beta^t) \tag{5.32}$$

$$\boldsymbol{\lambda}_\gamma^{t+1} = \boldsymbol{\lambda}_\gamma^t - \nabla^{\text{n}}{}_{\boldsymbol{\lambda}_\gamma} f(\boldsymbol{\lambda}_\gamma^t). \tag{5.33}$$

5.3 Scalable VI Algorithm for Tensor CPD

In update (5.31), computing the natural gradient is inexpensive, as the computation in (5.27) only involves one data slice $\mathcal{Y}(n)$. However, computing natural gradients for updates (5.30), (5.32), and (5.33) are much more expensive, as the whole dataset $\{\mathcal{Y}(n)\}_{n=1}^{I_{P+1}}$ is involved. To scale the inference algorithm to a massive data paradigm, one promising way is to leverage stochastic optimization, where noisy, unbiased, and cheap-to-compute gradients are used to substitute the original expensive gradients in (5.30), (5.32), and (5.33) as shown in (5.34)–(5.36) as follows:

$$\boldsymbol{\alpha}_p^{t+1} = \boldsymbol{\alpha}_p^t - \rho_t \underbrace{\left(-\frac{1}{S}\sum_{i=1}^{S} \mathbb{E}_{\prod_{\boldsymbol{\Theta}_j \neq \boldsymbol{\Xi}^{(p)}} Q(\boldsymbol{\Theta}_j)} \begin{bmatrix} I_{P+1}\boldsymbol{t}\left(\mathcal{Y}(n_i), \{\boldsymbol{\eta}_j, \boldsymbol{\eta}_j \neq \boldsymbol{\Xi}^{(p)}\}\right) \\ \mathrm{vec}\left(\frac{1}{2}\boldsymbol{\Gamma}\right) + I_{P+1}\boldsymbol{l}_{n_i}\left(\{\boldsymbol{\eta}_j, \boldsymbol{\eta}_j \neq \boldsymbol{\Xi}^{(p)}\}\right) \end{bmatrix} + \boldsymbol{\alpha}_p^t \right)}_{\triangleq \hat{\nabla}^{\mathrm{n}}_{\boldsymbol{\alpha}_p} f(\boldsymbol{\alpha}_p^t)} \tag{5.34}$$

$$\boldsymbol{\alpha}_\beta^{t+1} = \boldsymbol{\alpha}_\beta^t - \rho_t \underbrace{\left(-\frac{1}{S}\sum_{i=1}^{S} \mathbb{E}_{\prod_{\boldsymbol{\Theta}_j \neq \beta} Q(\boldsymbol{\Theta}_j)} \begin{bmatrix} -10^{-6} + I_{P+1}\boldsymbol{t}\left(\mathcal{Y}(n_i), \{\boldsymbol{\eta}_j, \boldsymbol{\eta}_j \neq \beta\}\right) \\ 10^{-6} - 1 + I_{P+1}\boldsymbol{l}_{n_i}\left(\{\boldsymbol{\eta}_j, \boldsymbol{\eta}_j \neq \beta\}\right) \end{bmatrix} + \boldsymbol{\alpha}_\beta^t \right)}_{\triangleq \hat{\nabla}^{\mathrm{n}}_{\boldsymbol{\alpha}_\beta} f(\boldsymbol{\alpha}_\beta^t)} \tag{5.35}$$

$$\boldsymbol{\lambda}_\gamma^{t+1} = \boldsymbol{\lambda}_\gamma^t - \rho_t \underbrace{\left(\frac{1}{S}\sum_{i=1}^{S} \mathbb{E}_{\prod_{\boldsymbol{\Theta}_j \neq \gamma} Q(\boldsymbol{\Theta}_j)} \begin{bmatrix} -\boldsymbol{d} + \boldsymbol{u}(\{\boldsymbol{\Xi}^{(p)}\}_{p=1}^{P}) + I_{P+1}\boldsymbol{u}(\boldsymbol{\xi}(n_i)) \\ \boldsymbol{c} - \boldsymbol{1} + \boldsymbol{w} \end{bmatrix} - \boldsymbol{\lambda}_\gamma^t \right)}_{\triangleq \hat{\nabla}^{\mathrm{n}}_{\boldsymbol{\lambda}_\gamma} f(\boldsymbol{\lambda}_\gamma^t)}. \tag{5.36}$$

In (5.34)–(5.36), unbiased noisy gradients $\hat{\nabla}^{\mathrm{n}}_{\boldsymbol{\alpha}_p} f(\boldsymbol{\alpha}_p)$, $\hat{\nabla}^{\mathrm{n}}_{\boldsymbol{\alpha}_\beta} f(\boldsymbol{\alpha}_\beta)$, and $\hat{\nabla}^{\mathrm{n}}_{\boldsymbol{\lambda}_\gamma} f(\boldsymbol{\lambda}_\gamma)$ are computed from uniformly sampled mini-batch data $\{\mathcal{Y}(n_1), \mathcal{Y}(n_2), \ldots, \mathcal{Y}(n_S)\}$ with index $n_i \sim \mathrm{Uniform}(1, 2, 3, \ldots, I_{P+1})$ for $i = 1, 2, \ldots, S$ [3]. Obviously, their computations only involve S data slices, and thus can be very economical when $S \ll I_{P+1}$. To guarantee convergence, the step size ρ_t for updates with noisy gradients must follow the Robbins and Monro conditions: $\sum_t \rho_t = \infty$ and $\sum_t \rho_t^2 < \infty$ [2].

With these stochastic updates (5.34)–(5.36) replacing the high-complexity updates (5.30), (5.32), and (5.33), a scalable VI algorithm for tensor CPD can be obtained. More specifically, at each iteration, mini-batch data $\{\mathcal{Y}(n_1), \mathcal{Y}(n_2), \ldots, \mathcal{Y}(n_S)\}$ are sampled with index $n_i \sim \mathrm{Uniform}(1, 2, 3, \ldots, I_{P+1})$. Then, Eqs. (5.31), (5.34)–(5.36) are executed to update the model parameters. The process is repeated until the algorithm is deemed to have converged.

5.3.1 *Summary of Iterative Algorithm*

In the updating equations (5.31), (5.34)–(5.36), there are several expectation computations, and computing these expectations involve the properties of tensor algebra. To keep the presentation concise, the final iterative algorithm is summarized in **Algorithm 7**. Details on various expressions can be found in [18].

5.3.2 *Further Discussions*

To gain more insights from the above scalable tensor CPD algorithm, discussions of rank and factor matrices estimation, feature matrix recovery, computational complexity, memory requirement, convergence property, and selection of a threshold for column pruning are presented in the following.

5.3.2.1 Rank and Factor Matrices Estimation

For rank estimation, although we do not have an explicit variable for the unknown rank, we do have variables γ_l^{-1} for modeling the "power" of the lth column of the factor matrices. More specifically, after convergence, some $\mathbb{E}_{Q^*(\gamma_l)}[\gamma_l] = a_l^t/b_l^t$ will become very large, e.g., 10^6, indicating that the power of the corresponding columns in the factor matrices becomes zero. Therefore, these columns can be safely pruned out, and the remaining number of columns is the estimated tensor rank R. Meanwhile, since $Q^*(\Xi^{(p)})$ is a Gaussian distribution, factor matrices $\{\Xi^{(p)}\}_{p=1}^P$ are estimated by the mean $\{\boldsymbol{M}^{(p,t)}\}_{p=1}^P$ of $\{Q^*(\Xi^{(p)})\}_{p=1}^P$ at convergence.

5.3.2.2 Estimation of Feature Matrix

The factor matrices $\{\Xi^{(p)}\}_{p=1}^P$ are obtained after **Algorithm 7** terminates, while the feature matrix $\Xi^{(P+1)}$ is not explicitly recovered. In fact, in many applications such as the frequency estimation and video background modeling [13], since the goal is to find the subspace of signals but not the features, directly applying **Algorithm 7** is sufficient. Even in some applications where the feature matrix is desired, it can be easily learned in a scalable way. In particular, after executing **Algorithm 7**, the data $\mathcal{Y}(n)$ for $n = 1, 2, 3, \ldots, I_{P+1}$ can be fed to (5.39) sequentially and $\boldsymbol{\mu}(n)$ is the estimate of the nth row of the feature matrix $\Xi^{(P+1)}$.

Algorithm 7 Scalable VI CPD (SVI-S CPD)

Initialization: Choose $L > R$, $1 \le S \le I_{P+1}$, and initialize $\{\boldsymbol{M}^{(p,0)} \in \mathbb{R}^{I_p \times L}, \boldsymbol{\Sigma}^{(p,0)} \in \mathbb{R}^{L\times L}\}_{p=1}^{P}$. Choose a sequence $\{\rho_t\}_t$ that satisfying $\sum_t \rho_t = \infty$ and $\sum_t \rho_t^2 < \infty$. Let $\{\boldsymbol{F}^{(p,0)} = \left[\boldsymbol{\Sigma}^{(p,0)}\right]^{-1} \boldsymbol{M}^{(p,0)}\}_{p=1}^{P}$, $a_l^0 = 10^{-6}$, $b_l^0 = 10^{-6}$ for $l = 1, 2, 3, \ldots, L$; $c^0 = d^0 = 10^{-6}$.
Iterations: For the tth iteration ($t \ge 0$)
Sample $\{n_i\}_{i=1}^S$ uniformly $n_i \sim \text{Uniform}(1, 2, 3, \ldots, I_{P+1})$.

Update the parameter of $Q^*(\boldsymbol{\xi}(n_i))^{t+1}$:

$$\boldsymbol{\alpha}_{\boldsymbol{\xi}(n_i)}^{t+1} = \left[\left[\boldsymbol{\Phi}^t\right]^{-1} \boldsymbol{\mu}(n_i)^T; \tfrac{1}{2}\text{vec}\left(\left[\boldsymbol{\Phi}^t\right]^{-1}\right)\right] \tag{5.37}$$

with

$$\left[\boldsymbol{\Phi}^t\right]^{-1} = \boldsymbol{diag}\left\{\frac{a_1^t}{b_1^t}, \ldots, \frac{a_R^t}{b_R^t}\right\} + \frac{c^t}{d^t}\left(\bigodot_{p=1}^{P}\left[\left[\boldsymbol{M}^{(p,t)}\right]^T \boldsymbol{M}^{(p,t)} + I_p \boldsymbol{\Sigma}^{(p,t)}\right]\right) \tag{5.38}$$

$$\boldsymbol{\mu}(n_i) = \frac{c^t}{d^t}\text{vec}\left(\mathcal{Y}(n_i)\right)^T \left(\mathop{\diamond}_{p=1}^{P} \boldsymbol{M}^{(p,t)}\right)\boldsymbol{\Phi}^t \tag{5.39}$$

Update the parameter of each $Q^*(\boldsymbol{\Xi}^{(p)})^{t+1}$:

$$\boldsymbol{\alpha}_p^{t+1} = \left[\text{vec}\left(\boldsymbol{F}^{(p,t+1)}\right); \tfrac{1}{2}\text{vec}\left(\left[\boldsymbol{\Sigma}^{(p,t+1)}\right]^{-1}\right)\right] \tag{5.40}$$

with

$$\begin{aligned}\left[\boldsymbol{\Sigma}^{(p,t+1)}\right]^{-1} &= (1-\rho_t)\left[\boldsymbol{\Sigma}^{(p,t)}\right]^{-1} + \frac{\rho_t}{S}\sum_{i=1}^{S}\left(\boldsymbol{diag}\left\{\frac{a_1^t}{b_1^t}, \ldots, \frac{a_R^t}{b_R^t}\right\}\right. \\ &\left. + \frac{I_{P+1}c^t}{d^t}\left(\bigodot_{m=1, m\neq p}^{P}\left[\boldsymbol{M}^{(m,t)T}\boldsymbol{M}^{(m,t)} + I_m \boldsymbol{\Sigma}^{(m,t)}\right] \odot \left[\boldsymbol{\mu}(n_i)^T \boldsymbol{\mu}(n_i) + \boldsymbol{\Phi}^t\right]\right)\right)\end{aligned} \tag{5.41}$$

$$\boldsymbol{F}^{(p,t+1)} = (1-\rho_t)\boldsymbol{F}^{(p,t)} + \frac{\rho_t I_{P+1}}{S}\sum_{i=1}^{S}\left(\frac{c^t}{d^t}\left[\mathcal{Y}(n_i)\right]^{(p)}\left(\boldsymbol{\mu}(n_i) \diamond \left(\mathop{\diamond}_{m=1, m\neq p}^{P}\boldsymbol{M}^{(m,t)}\right)\right)\right) \tag{5.42}$$

$$\boldsymbol{M}^{(p,t+1)} = \boldsymbol{F}^{(p,t+1)T}\boldsymbol{\Sigma}^{(p,t+1)}. \tag{5.43}$$

Update the parameter of $Q^*(\boldsymbol{\gamma})^{t+1}$:

$$\boldsymbol{\lambda}_{\boldsymbol{\gamma}}^{t+1} = \left[-\boldsymbol{b}^{t+1}; (a_l^{t+1} - 1)\boldsymbol{1}\right] \tag{5.44}$$

where $a_l^{t+1} = 10^{-6} + \frac{1}{2}\left(\sum_{p=1}^P I_p + I_{P+1}\right)$ and $\boldsymbol{b}^{t+1} = [b_1^{t+1}, b_2^{t+1}, \ldots, b_L^{t+1}]^T$ with

$$b_l^{t+1} = (1-\rho_t)b_l^t + \frac{\rho_t}{S}\sum_{i=1}^{S}\left(10^{-6} + \frac{1}{2}\left(\sum_{p=1}^{P}\left[\boldsymbol{M}_{:,l}^{(p,t+1)T}\boldsymbol{M}_{:,l}^{(p,t+1)} + I_p \boldsymbol{\Sigma}_{l,l}^{(p,t+1)}\right] + I_{P+1}\left[|\boldsymbol{\mu}(n_i)_l|^2 + \boldsymbol{\Phi}_{l,l}^t\right]\right)\right). \tag{5.45}$$

Update the parameter of $Q^*(\beta)^{t+1}$:

$$\boldsymbol{\alpha}_{\beta}^{t+1} = \left[-d^{t+1}; c^{t+1} - 1 \right] \tag{5.46}$$

where

$$c^{t+1} = (1-\rho_t)(c^t - 1) + \frac{\rho_t}{S} \sum_{i=1}^{S} \left(10^{-6} - 1 + I_{P+1} \frac{\prod_{p=1}^{P} I_p}{2} \right) + 1 \tag{5.47}$$

$$d^{t+1} = (1-\rho_t) d^t + \frac{\rho_t}{S} \sum_{i=1}^{S} \left(10^{-6} + \frac{I_{P+1}}{2} f_i \right) \tag{5.48}$$

$$\begin{aligned} f_i &= \| \mathcal{Y}(n_i) \|_F^2 - 2\mathrm{vec}(\mathcal{Y}(n_i))^T \Big(\mathop{\diamond}_{p=1}^{P} \boldsymbol{M}^{(p,t+1)} \Big) \boldsymbol{\mu}(n_i)^T \\ &+ \mathrm{Tr}\Big(\Big[\boldsymbol{\mu}(n_i)^T \boldsymbol{\mu}(n_i) + \boldsymbol{\Phi}^t \Big] \Big[\mathop{\odot}_{p=1}^{P} \Big(\boldsymbol{M}^{(p,t+1)} \boldsymbol{M}^{(p,t+1)T} + I_p \boldsymbol{\Sigma}^{(p,t+1)} \Big) \Big] \Big). \end{aligned} \tag{5.49}$$

Until Convergence

5.3.2.3 Computational Complexity and Memory Requirements

At each iteration, the complexity is dominated by the update of each factor matrix, costing $O((P+1)L^2 \prod_{p=1}^{P} I_p S + \sum_{p=1}^{P} I_p S L^3)$ operations. From this expression, it is clear that **Algorithm 7** only incurs a linear computational complexity with respect to the input tensor size $\prod_{p=1}^{P} I_p S$. Compared to the corresponding computational cost of the batch-mode algorithm $O((P+1)L^2 \prod_{p=1}^{P+1} I_p + L^3(\sum_{p=1}^{P+1} I_p))$, which is linear with respect to the input tensor size $\prod_{p=1}^{P+1} I_p$, the computational cost of the scalable algorithm at each iteration is much smaller than that of the batch-mode algorithm as I_{P+1} is much larger than S. Furthermore, the maximum memory required for **Algorithm 7** at each iteration is $\prod_{p=1}^{P} I_p S$, which again depends only on the mini-batch size S, but not the whole data size I_{P+1}. Thus, **Algorithm 7** scales well to large dataset if S is much smaller than I_{P+1}.

5.3.2.4 Convergence Property

Although the functional minimization of the KL divergence is non-convex over the mean-field family $Q(\boldsymbol{\Theta}) = \prod_k Q(\Theta_k)$, it is convex with respect to a single variational density $Q(\Theta_k)$ when other $\{Q(\Theta_j) | j \neq k\}$ are fixed. Therefore, **Algorithm 7**, which iteratively updates the optimal solution for each Θ_k, is essentially a coordinate descent algorithm in the functional space of variational distributions with each update optimizing a convex subproblem. Furthermore, in the probabilistic tensor decomposition model in Sect. 5.1.1, the corresponding natural gradient for problem (5.7) can be computed as (5.26)–(5.29) for each $Q(\boldsymbol{\Theta}_k)$. If the step size is appropriately chosen, the gradient descent step using an unbiased estimate of the natural gradient (5.34)–(5.36) will decrease the objective function of the subproblem (5.7) on average. Notice that the optimal solution of each subproblem is unique, as there is only one solution that makes the derivative zero. Furthermore, since each subproblem is continuously differentiable, convex, and has a bounded and compact feasible set, the proposed algorithm is guaranteed to converge to at least a stationary point on average [17].

5.3.2.5 Selection of Threshold for Pruning Columns

The specific threshold for pruning columns in general depends on the application. A widely used choice is 10^6. However, this fixed threshold might not give the best performance for all applications, especially when the SNR is low. Another choice is to set the threshold at 100 times the minimal value of γ_l. This strategy has been shown empirically to work well in practice.

5.4 Numerical Examples

In this section, numerical results from both synthetic data and real-world datasets are presented to assess the performance of **Algorithm 7**, with the batch-mode algorithm in (**Algorithm 5** in Chap. 3, labeled as Batch VI) serving as a benchmark. For **Algorithm 7**, the dataset $\mathcal{Y}$ is sliced along the last dimension. The initial factor matrix $\boldsymbol{M}^{(p,0)}$ is set as the first iteration output of the batch-mode algorithm with $\{\mathcal{Y}(n_1), \mathcal{Y}(n_2), \ldots, \mathcal{Y}(n_S)\}$ being the tensor input, the initial $\boldsymbol{\Sigma}^{(p,0)} = \boldsymbol{I}_L$ with $L = \min\{I_1, I_2, \ldots, I_P\}$. The step size sequence is chosen as $\rho_t = (t+1)^{-0.9}$ [2]. Each result in this section is obtained by averaging 100 Monte Carlo runs. All the experiments were conducted in MATLAB R2015b with an Intel Core i5-4570 CPU at 3.2 GHz and 8 GB RAM.

5.4.1 Convergence Performance on Synthetic Data

A 4D data tensor $[\![\boldsymbol{A}^{(1)}, \boldsymbol{A}^{(2)}, \boldsymbol{A}^{(3)}, \boldsymbol{A}^{(4)}]\!] \in \mathbb{R}^{20\times 20\times 20\times 1000}$ with rank $R = 5$ is considered. The factor matrices $\{\boldsymbol{A}^{(p)}\}_{p=1}^{3}$ are drawn from $\mathcal{MN}(\boldsymbol{A}^{(p)} | \, \boldsymbol{0}_{20\times 5}, \boldsymbol{I}_{20\times 20}, \boldsymbol{I}_{5\times 5})$, and matrix $\boldsymbol{A}^{(4)} \sim \mathcal{MN}(\boldsymbol{A}^{(4)} | \, \boldsymbol{0}_{1000\times 5}, \boldsymbol{I}_{1000\times 1000}, \boldsymbol{I}_{5\times 5})$. The signal-to-noise ratio (SNR) is defined as $10\log_{10}(\| [\![\boldsymbol{A}^{(1)}, \boldsymbol{A}^{(2)}, \boldsymbol{A}^{(3)}, \boldsymbol{A}^{(4)}]\!] \|_F^2 \, / \, \| \dot{\mathcal{W}} \|_F^2)$. To see how the choice of mini-batch size affects the accuracy and speed, we consider three scenarios: $S = 1$, $S = 5$, and $S = 10$, and the corresponding algorithms are labeled as SVI-1, SVI-5, and SVI-10, respectively. The upper bound on the rank for **Algorithm 7** and batch-mode VI algorithm is chosen as $L = 20$.

A commonly used criterion for assessing how close the estimated $\boldsymbol{M}^{(p)}$ is to the ground truth factor matrix $\Xi^{(p)}$ is the averaged largest principal angle (LPA) $\frac{1}{3}\sum_{p=1}^{3} \text{LPA}(\boldsymbol{M}^{(p)}, \Xi^{(p)})$. LPA$(\boldsymbol{A}, \boldsymbol{B})$ is a measure of the "distance" between two subspaces spanned by the columns of matrices $\boldsymbol{A}$ and $\boldsymbol{B}$, defined as $\cos^{-1}\{\sigma_{\min}\{\text{orth}\{\boldsymbol{A}\}^H \text{orth}\{\boldsymbol{B}\}\}\}$, where the operator $\sigma_{\min}\{\boldsymbol{Q}\}$ denotes the smallest singular value of the matrix $\boldsymbol{Q}$, and orth$\{\boldsymbol{Q}\}$ is an orthonormal basis for the subspace spanned by the columns of $\boldsymbol{Q}$. The LPA is also called the angle of subspaces, and can be computed in MATLAB by the function "subspace". Notice that while LPA computation allows $\boldsymbol{M}^{(p)}$ and $\Xi^{(p)}$ to have a different number of columns (i.e., estimated rank differs from true rank), this criterion makes the most sense when $\boldsymbol{M}^{(p)}$ and $\Xi^{(p)}$ have the same number of columns. This can be seen from an example that if $\boldsymbol{M}^{(p)}$ is randomly generated with many more columns than that of $\Xi^{(p)}$, the true subspace generated by columns of $\Xi^{(p)}$ would be embedded in the subspace generated by columns of $\boldsymbol{M}^{(p)}$, thus giving a zero LPA. However, since $\Xi^{(p)}$ and $\boldsymbol{M}^{(p)}$ have very different sizes, $\boldsymbol{M}^{(p)}$ would not be a good estimate of $\Xi^{(p)}$. In order to avoid the misleadingly small LPA under incorrect rank estimate, if the tensor rank is overestimated in $\boldsymbol{M}^{(p)}$, only the principal columns of $\boldsymbol{M}^{(p)}$ with the R largest powers are used for computing the LPA. Similarly, if the tensor rank is underestimated in $\boldsymbol{M}^{(p)}$ with $R' < R$, only the principal columns of $\Xi^{(p)}$ with the R' largest powers are used.

The LPA assesses the performance of subspace recovery. However, due to the uniqueness property introduced in Chap. 5, tensor CPD in fact returns the columns of each factor matrix up to an unknown permutation and scaling ambiguity, which is stronger than simply subspace recovery. To directly assess the accuracies of factor matrices recovery (rather than subspace recovery), the best sum congruence, which involves computing the normalized correlation between the true columns in $\Xi^{(p)}$ and the estimated columns in $\boldsymbol{M}^{(p)}$, is a more suitable criterion. Mathematically, it is defined as

$$\max_{\pi(\cdot)} \sum_{r=1}^{R} \frac{\Xi_{:,r}^{(1)H} \boldsymbol{M}_{:,\pi(r)}^{(1)}}{\| \Xi_{:,r}^{(1)} \| \| \boldsymbol{M}_{:,\pi(r)}^{(1)} \|} \frac{\Xi_{:,r}^{(2)H} \boldsymbol{M}_{:,\pi(r)}^{(2)}}{\| \Xi_{:,r}^{(2)} \| \| \boldsymbol{M}_{:,\pi(r)}^{(2)} \|} \frac{\Xi_{:,r}^{(3)H} \boldsymbol{M}_{:,\pi(r)}^{(3)}}{\| \Xi_{:,r}^{(3)} \| \| \boldsymbol{M}_{:,\pi(r)}^{(3)} \|}$$

where the maximum is taken with respect to all possible permutations $\pi(\cdot)$ of the factor matrices' columns, which can be found using greedy search algorithm [16]. Similar to LPA, best sum congruence was previously applied when the rank estimate is accurate. In order to incorporate inaccurate rank estimate in $\boldsymbol{M}^{(p)}$, if the rank is overestimated in $\boldsymbol{M}^{(p)}$, the R columns of $\boldsymbol{M}^{(p)}$ that match the columns of $\Xi^{(p)}$ most after permutation $\pi(\cdot)$ are used. On the other hand, if the rank is underestimated in $\boldsymbol{M}^{(p)}$ with $R' < R$, only R' columns of $\Xi^{(p)}$ that match the columns of $\boldsymbol{M}^{(p)}$ most are used in the computation of best sum congruence.

Figure 5.2 shows the convergence performance of **Algorithm 7** in terms of LPA. It can be seen that for both SNR $=$ 10 dB and SNR $=$ 20 dB, the performance of **Algorithm 7** converges to that of the batch-mode algorithm. In particular, LPA decreases quickly at the beginning of the iterations, and gets pretty close to the LPA of batch-mode processing within a few hundred iterations. After that, the performance

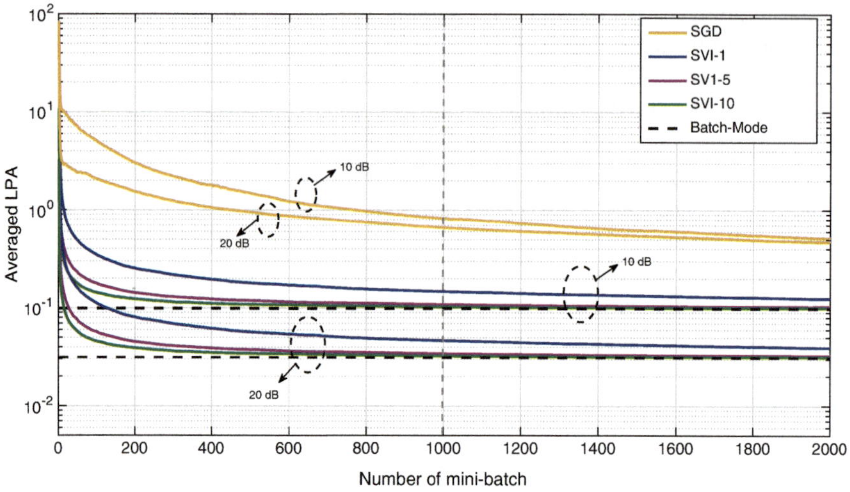

Fig. 5.2 Convergence performance of **Algorithm 7** with SNR = 10 dB and 20 dB

improvement becomes smaller and smaller. This is due to the learning rate ρ_t going to a very small value as the iteration number becomes large. Furthermore, after processing the first 1000 mini-batch data, the LPA gaps between the SVI-5/SVI-10 algorithms and the batch-mode algorithm are nearly unnoticeable. Even for the SVI-1 algorithm, the LPA gap is just 0.0492° and 0.0157° at SNR = 10 dB and SNR = 20 dB, respectively, showing that SVI-1 provides a very good approximation to the batch-mode result.

In addition to **Algorithm 7**, Fig. 5.2 also shows the performance of the algorithm in [1] (labeled as SGD)[3] with an upper bound on the rank ($L = 20$), and other parameter settings following those described in [1]. From Fig. 5.2, it is clear that **Algorithm 7** exhibits much better performance. This is because the algorithm in [1] is not capable of recovering the correct rank from the provided upper bound, thus the superfluous basis vectors in the factor matrices degrade the subspace recovery accuracy. This shows the importance of accurate rank information in tensor CPD.

Next, the best sum congruences obtained from different CPD algorithms are presented in Table 5.1. From the table, it can be seen that when SNR = 10 dB, **Algorithm 7** under different mini-batch sizes performs nearly the same as the batch-mode algorithm in terms of the best sum congruence after processing just 100 mini-batches of data, and they perform significantly better than the SGD algorithm. When the number of mini-batches is increased to 1000 or beyond, all the schemes would have the best sum congruence equal or very close to the maximum value (which is 5 for this particular simulation setting). Similar conclusions can be drawn for SNR =

[3] Notice that the algorithm in [1] was originally developed for 3D tensors. We extended it to work in 4D tensor data.

Table 5.1 The best sum congruence obtained by different algorithms

Algorithm	SNR = 10 dB				SNR = 20 dB			
	100 mini-batches	500 mini-batches	1000 mini-batches	2000 mini-batches	100 mini-batches	500 mini-batches	1000 mini-batches	2000 mini-batches
SVI-1	4.9999	5.0	5.0	5.0	5.0	5.0	5.0	5.0
SVI-5	5.0	5.0	5.0	5.0	5.0	5.0	5.0	5.0
SVI-10	5.0	5.0	5.0	5.0	5.0	5.0	5.0	5.0
Batch-mode	5.0	5.0	5.0	5.0	5.0	5.0	5.0	5.0
SGD	2.9853	4.9980	4.9994	4.9997	4.9972	4.9994	4.9996	4.9997

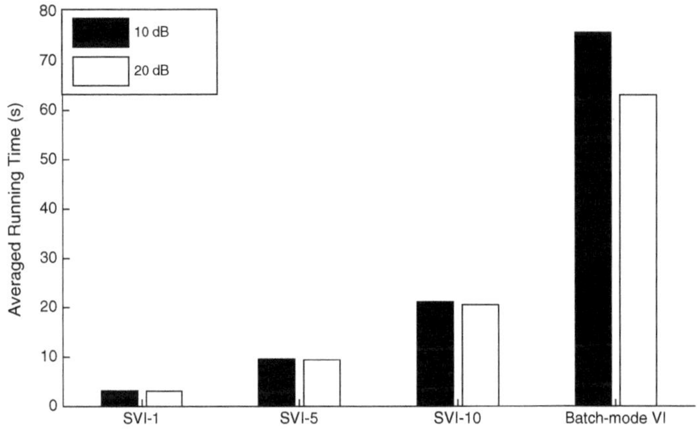

Fig. 5.3 Averaged running time of **Algorithm 7** and batch-mode VI

20 dB, except various schemes reach the maximum best sum congruence at an even smaller number of mini-batches.

On the other hand, the averaged running time of **Algorithm 7** (after processing 1000 mini-batch data) and the batch-mode VI algorithm over 100 Monte Carlo runs are presented in Fig. 5.3. It is seen that the SVI-1, SVI-5, and SVI-10 algorithms are around 20, 8, and 4 times faster than the batch-mode algorithm,[4] respectively. This shows the excellent scalability of the proposed algorithm for massive data.

5.4.2 Tensor Rank Estimation on Synthetic Data

Since the LPA and best sum congruence only give a partial picture of the performance of **Algorithm 7** under unknown rank, the rank determination ability is assessed

[4] The termination criterion of the batch-mode algorithm follows the recommendation from [13].

explicitly in this subsection. In particular, the tensor rank learned by **Algorithm 7** under different SNRs is shown in Fig. 5.4, with each vertical bar showing the mean and the error bars showing the standard deviation of tensor rank estimates. The blue horizontal dotted line indicates the true tensor rank. From Fig. 5.4, it is seen that **Algorithm 7** with different mini-batch sizes recovers the true tensor rank with 100% accuracy for a wide range of SNRs, in particular when SNR $\geq$ 10 dB. Even though the performance under SNR $\leq$ 5 dB is not 100% accurate, it can be observed that **Algorithm 7** still gives estimates close to the true tensor rank with a high probability. For a comprehensive performance assessment, Table 5.2 presents the percentages of correct rank estimation, overestimation, and underestimation. From Table 5.2, it can be seen that while the percentages of overestimation and underestimation are not equal in general, they are relatively small even for SNR at 0 dB and 5 dB. Notice that when the SNR is larger than 5 dB, all of the underestimation and overestimation percentages are zeros. Thus there is no need to include these results in the table.

One may wonder whether the performance of **Algorithm 7** is better than the classical algorithm starting with a small subset of data and then followed by the deterministic sequential algorithm in [1]. In particular, the classical method for finding the tensor rank is to run multiple ALS-based CPD algorithms with different assumed ranks, and then choose the model with the smallest rank that fits the data well. After the rank is estimated with a small subset of data, the deterministic sequential algorithm in [1] is used to update the subspace when more data come in (but with a fixed estimated rank). We refer to such a classical scheme as ALS-SGD. On the other hand, with the automatic rank estimation ability in the recent batch-mode probabilistic CPD [13], one could also consider a hybrid scheme in which the multiple ALSs are replaced by the recent batch-mode probabilistic CPD. We refer to this hybrid scheme as Probabilistic CPD + SGD.

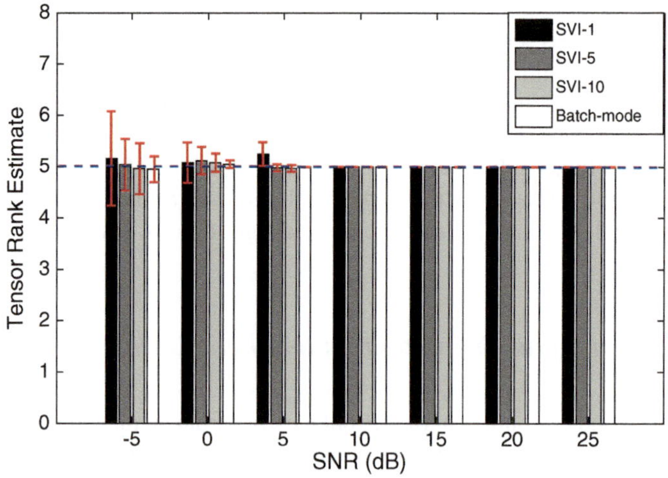

Fig. 5.4 Rank estimates of **Algorithm 7** and batch-mode VI

Table 5.2 Percentages of correct rank estimation, underestimation, and overestimation by different algorithms

Algorithm	SNR = −5 dB			SNR = 0 dB			SNR = 5 dB		
	Underestimation percentage (%)	Overestimation percentage (%)	Correct estimation percentage (%)	Underestimation percentage (%)	Overestimation percentage (%)	Correct estimation percentage (%)	Underestimation percentage (%)	Overestimation percentage (%)	Correct estimation percentage (%)
SVI-1	19	39	42	6	8	86	3	4	93
SVI-5	21	27	52	4	6	90	3	1	96
SVI-10	28	12	60	4	2	94	2	0	98
Batch-mode	26	10	64	4	2	94	0	0	100

Table 5.3 Performance of different rank estimation schemes

Algorithm	SNR = 0 dB				SNR = 20 dB			
	LPA (degree)	Best sum congru-ence	Running time (s)	Rank accuracy (%)	LPA (degree)	Best sum congru-ence	Running time (s)	Rank accuracy (%)
ALS + SGD	0.5094	4.2438	78.3712	79	0.0482	5.0	26.4884	100
Probabilistic CPD + SGD	0.5094	4.2655	6.1294	79	0.0482	5.0	4.1097	100
SVI-1	0.4484	4.9071	5.2435	86	0.0489	5.0	3.6034	100

To compare the performance of the three schemes, (1) ALS + SGD, (2) Probabilistic CPD + SGD, and (3) SVI-1,[5] we conduct experiments on synthetic tensor data with size $20 \times 20 \times 20 \times 1000$. For the classical scheme ALS + SGD and the Probabilistic CPD + SGD, a subset of data of size $20 \times 20 \times 20 \times 20$ was used for the estimation of the rank. The results are shown in Table 5.3. It can be seen that when the SNR is high (SNR = 20 dB), all three schemes achieve 100% rank estimation accuracy, the same best sum congruence, and nearly the same LPA. This is because when the SNR is high, a small amount of data is sufficient to obtain an accurate rank estimate. On the other hand, when the SNR is low (SNR = 0 dB), both the ALS + SGD and Probabilistic CPD + SGD schemes achieve lower rank estimation accuracy along with inferior LPA and best sum congruence compared to those of the SVI-1 algorithm. This is due to the fact that when the SNR is low, the small subset of data at the beginning is not sufficient for accurate rank estimation, which in turn hurts the subsequent subspace learning based on the fixed estimated rank. In contrast, the SVI-1 algorithm makes use of all the data in both rank and subspace estimations, thus giving better performance. In terms of computation time, since ALS + SGD needs to run multiple ALS algorithms, it requires a much higher computation time than the SVI-1 algorithm. For the hybrid scheme, since it is capable of estimating the rank automatically, its complexity is much smaller than that of the classical scheme ALS + SGD, but still slightly higher than that of SVI-1.

From the above results, it is obvious that the probabilistic approach (in both Probabilistic CPD + SGD and SVI-1) is instrumental in reducing the complexity of rank estimation compared to the classical scheme ALS + SGD. Furthermore, the SVI-1 has the added advantages of accurate rank estimation, small LPA, and large best sum congruence under low SNR, since it makes use of all data in rank estimation.

The simulation results presented so far are for well-conditioned tensors, i.e., the columns in each of the factor matrices are independently generated. In order to fully assess the rank learning ability of **Algorithm 7**, we consider another 4D data tensor $[\![\bar{\boldsymbol{A}}^{(1)}, \boldsymbol{A}^{(2)}, \boldsymbol{A}^{(3)}, \boldsymbol{A}^{(4)}]\!] \in \mathbb{R}^{20\times20\times20\times1000}$ with rank $R = 5$. The factor matrix $\bar{\boldsymbol{A}}^{(1)} = 0.1 \times \mathbf{1}_{20\times5} + 2^{-s} \times \boldsymbol{A}^{(1)}$. The factor matrices $\{\boldsymbol{A}^{(p)}\}_{p=1}^{3}$ are drawn from $\mathcal{MN}(\boldsymbol{A}^{(p)}|$

[5] Since the algorithm in [1] was developed for a mini-batch data with size 1 only, we focus on the comparison with SVI-1.

Table 5.4 Rank estimate accuracy when columns in all factor matrices are correlated

s	0	1	2	3
Batch-mode (%)	100	100	86	0
SVI-1 (%)	100	100	0	0
SVI-5 (%)	100	100	0	0
SVI-10 (%)	100	100	16	0

$\mathbf{0}_{20\times 5}, \boldsymbol{I}_{20\times 20}, \boldsymbol{I}_{5\times 5})$, and the matrix $\boldsymbol{A}^{(4)} \sim \mathcal{MN}(\boldsymbol{A}^{(4)} | \mathbf{0}_{1000\times 5}, \boldsymbol{I}_{1000\times 1000}, \boldsymbol{I}_{5\times 5})$. According to the definition of the tensor condition number [11, 12], when s increases, the correlation among the columns in the factor matrix $\bar{\boldsymbol{A}}^{(1)}$ increases, and the tensor condition number becomes larger. In particular, when s goes to infinity, the condition number of the tensor goes to infinity too. Experiments were conducted to evaluate the rank estimation accuracy of **Algorithm 7** and the batch-mode algorithm under SNR $= 20$ dB and $s \in \{0, 1, 10, 10^2, 10^3, 10^4\}$. It was found that both **Algorithm 7** and the batch-mode algorithm achieve 100% accuracy when the correlation between columns only happens in one factor matrix, even the tensor condition number is very large.

Next, we consider an extreme case when the columns in all factor matrices are highly correlated. In particular, a 4D data tensor $[\![\bar{\boldsymbol{A}}^{(1)}, \bar{\boldsymbol{A}}^{(2)}, \bar{\boldsymbol{A}}^{(3)}, \bar{\boldsymbol{A}}^{(4)}]\!] \in \mathbb{R}^{20\times 20\times 20\times 1000}$ with rank $R = 5$ is constructed, with $\bar{\boldsymbol{A}}^{(p)} = 0.1 \times \mathbf{1}_{20\times 5} + 2^{-s} \times \boldsymbol{A}^{(p)}$ for $p = 1, 2, 3, 4$. The factor matrices $\{\boldsymbol{A}^{(p)}\}_{p=1}^{3}$ are drawn from $\mathcal{MN}(\boldsymbol{A}^{(p)} | \mathbf{0}_{20\times 5}, \boldsymbol{I}_{20\times 20}, \boldsymbol{I}_{5\times 5})$, and the matrix $\boldsymbol{A}^{(4)} \sim \mathcal{MN}(\boldsymbol{A}^{(4)} | \mathbf{0}_{1000\times 5}, \boldsymbol{I}_{1000\times 1000}, \boldsymbol{I}_{5\times 5})$. The performance of the tensor rank estimation is shown in Table 5.4. It can be seen that as s increases (i.e., the columns are more correlated in all factor matrices), the performance of **Algorithm 7**, and the batch-mode algorithm degrades accordingly.

5.4.3 Video Background Modeling

One of the most important tasks in video surveillance is to separate the foreground objects from the background. Since the background remains the same along the video frames, it can be modeled as the low-rank subspace of the video data. In particular, after organizing the video data into a 3D tensor (pixels in a frame $\times$ RGB $\times$ frame) [13] as shown in Fig. 5.5, the background can be extracted using the tensor CPD.

We conduct experiments on the popular shopping mall video data.[6] Each frame is with size 256×320, and there are 100 frames. After organizing the data into a 3D tensor with dimension $81920 \times 3 \times 100$ [13], the batch-mode VI algorithm and the SVI-1 are run to learn the subspace (background). In this application, the SVI-1 is terminated after processing 100 mini-batches (i.e., the number of video frames).

[6] http://pages.cs.wisc.edu/jiaxu/projects/gosus/supplement/.

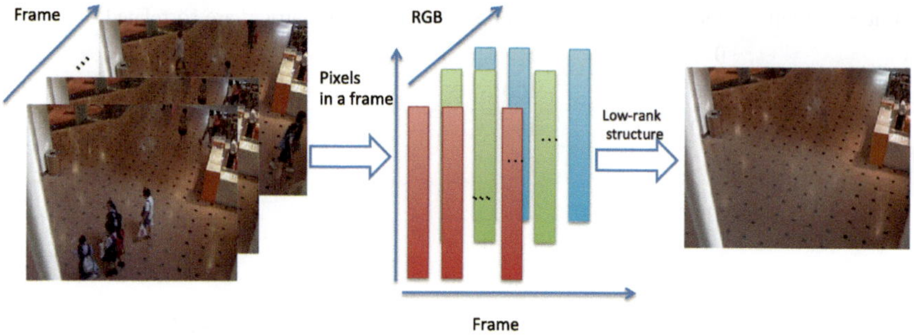

Fig. 5.5 Background modeling with a 3D tensor

Fig. 5.6 **a** Background estimation using the Batch-mode VI and **b** Background estimation using the SVI-1 algorithm

The learning results of the two algorithms are shown in Fig. 5.6. The value of L used in this scenario is 3 and the estimated rank is 2, both for the SVI-1 algorithm and the batch-mode algorithm. It can be seen that the two algorithms achieve nearly the same performance in background estimation (with mean square error per pixel being 0.3613). To quantify the difference in the estimation results, Fig. 5.7 shows the LPA between the subspaces estimated by the batch-mode algorithm and the SVI-1 algorithm. It can be seen that the LPA decreases as more mini-batch data is processed. At the beginning, the LPA is 10.02°. But after processing 100 mini-batch data, the LPA decreases to 0.0776°. However, the SVI-1 algorithm only costs 1.6317 s, while the batch-mode VI algorithm consumes 24.2254 s in running time. Obviously, the SVI-1 algorithm achieves nearly the same performance as the batch-mode VI algorithm while saving 93.2% computation time. Notice that the video data might not be generated from a Gaussian distribution, and the CPD factors' columns modeling the background are not really independent, thus deviating from the prior assumption (5.5) of the probabilistic model. However, after the Bayesian inference, the CPD factors' columns become correlated with covariance matrices shown in (5.41). In some sense, the inaccurate prior assumption could be remedied by learning from data, which is validated by the good result shown in Fig. 5.6.

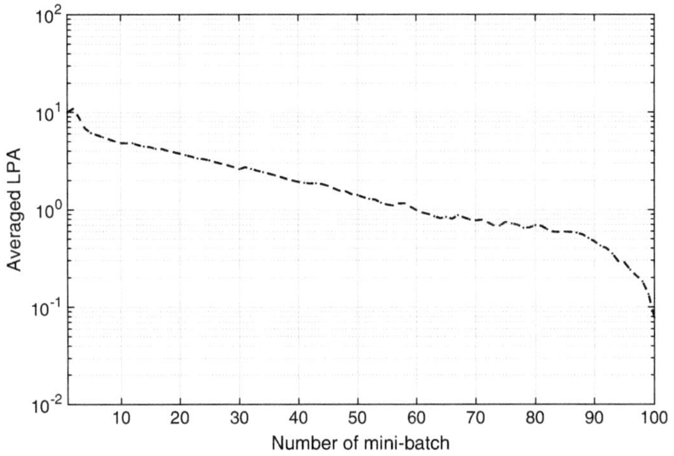

Fig. 5.7 LPA evolution for video data processing

5.4.4 Image Feature Extraction

In this subsection, we conduct experiments on 1000 images from the CIFAR-10 dataset,[7] representing 2 different objects (500 images for the dog and 500 images for the automobile). In particular, each image is of size 32×32, and the training data can be naturally represented by a fourth-order tensor $\mathbb{R}^{32\times 32\times 3\times 1000}$. The batch-mode algorithm and the SVI-1 algorithm (terminated after 1000 mini-batches) are run to learn the basis matrices and the feature vectors of these 1000 training images. With Fig. 5.8 showing the LPA between the subspace estimated by the batch-mode algorithm and the SVI-1 algorithm, it can be seen that after processing 1000 mini-batch data, the LPA reduces from 90° to only 0.0503°. Furthermore, using the obtained feature vectors of the training images, a 1-nearest neighbors (1-NN) classifier is trained, and the trained classifier is executed to recognize the objects for another 100 testing images projected onto the basis matrices. The accuracy of the 1-NN classifier using features from the SVI-1 algorithm and the batch-mode algorithm are both 78%, showing that the 0.0503° LPA does not affect the classification performance. However, the SVI-1 algorithm only costs 4.0533 s while the batch-mode algorithm costs 36.7521 s.

[7] https://www.cs.toronto.edu/~kriz/cifar.html.

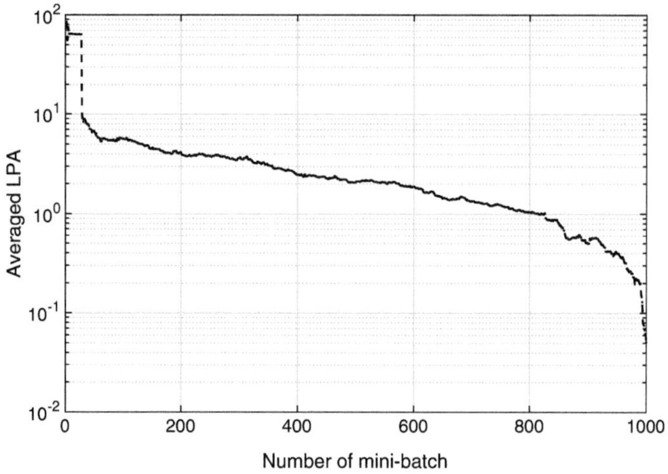

Fig. 5.8 LPA evolution for image feature extraction

References

1. M. Mardani, G. Mateos, G.B. Giannakis, Subspace learning and imputation for streaming big data matrices and tensors. IEEE Trans. Signal Process. **63**(10), 2663–2677 (2015)
2. J.C. Spall, *Introduction to Stochastic Search and Optimization: Estimation, Simulation, and Control* (Wiley, New York, 2003)
3. M. Hoffman, D. Blei, J. Paisley, C. Wang, Stochastic variational inference. J. Mach. Learn. Res. **14**, 1303–1347 (2013)
4. K.P. Murphy, *Machine Learning: A Probabilistic Perspective* (MIT Press, Cambridge, 2012)
5. M.J. Wainwright, M.I. Jordan, Graphical models, exponential families, and variational inference. Found. Trends Mach. Learn. **1**(102), 1–305 (2008)
6. B. Leon, Large-scale machine learning with stochastic gradient descent, in *Proceedings of COMPSTAT 2010* (Physica-Verlag HD, 2010), pp. 177–186
7. M. Li, D. Andersen, A. Smola, K. Yu, Communication efficient distributed machine learning with the parameter server, in *Proceedings of Neural Information Processing Systems (NIPS)* (2014)
8. M. Li, T. Zhang, Y. Chen, A. Smola, Efficient mini-batch training for stochastic optimization, in *Proceedings of ACM SIGKDD Conference on Knowledge Discovery and Data Mining (KDD)* (2014)
9. C. Udriste, *Convex Functions and Optimization Methods on Riemannian Manifolds* (Springer Science and Business Media Press, Berlin, 1994)
10. S. Amari, S.C. Douglas, Why natural gradient?, in *Proceedings of the IEEE International Conference on Acoustics, Speech and Signal Processing (ICASSP)*, no. 2 (1998), pp. 1213–1216
11. P. Breiding, N. Vannieuwenhoven, The condition number of join decompositions. SIAM J. Matrix Anal. Appl. **39**(1), 287–309 (2018)
12. N. Vannieuwenhoven, Condition numbers for the tensor rank decomposition. Linear Algebra Appl. **535**, 35–86 (2017)
13. Q. Zhao, G. Zhou, L. Zhang, A. Cichocki, S.I. Amari, Bayesian robust tensor factorization for incomplete multiway data. IEEE Trans. Neural Netw. Learn. Syst. **27**(4), 736–748 (2016)

14. J. Paisley, D. Blei, M. Jordan, Bayesian nonnegative matrix factorization with stochastic variational inference, in *Handbook of Mixed Membership Models and Their Applications* (Chapman and Hall/CRC, London, 2014)
15. L. Cheng, Y.-C. Wu, H.V. Poor, Probabilistic tensor canonical polyadic decomposition with orthogonal factors. IEEE Trans. Signal Process. **65**(3), 663–676 (2017)
16. N.D. Sidiropoulos, G.B. Giannakis, R. Bro, Blind PARAFAC receivers for DS-CDMA systems. IEEE Trans. Signal Process. **48**(3), 810–823 (2000)
17. P. Tseng, Convergence of a block coordinate descent method for nondifferentiable minimization. J. Optim. Theory Appl. **109**(3), 475–494
18. L. Cheng, Y.-C. Wu, H.V. Poor, Scaling probabilistic tensor canonical polyadic decomposition to massive data. IEEE Trans. Signal Process. **66**(21), 5534–5548 (2018)

Chapter 6
Bayesian Tensor CPD with Nonnegative Factors

Abstract In previous chapters, the probabilistic modeling and inference algorithm for the CPD with no constraint are discussed. In practical data analysis, one usually has additional prior structural information for the factor matrices, e.g., nonnegativeness and orthogonality. Encoding this structural information into the probabilistic tensor modeling while still achieving tractable inference remains a critical challenge. In this chapter, we introduce the development of Bayesian tensor CPD with nonnegative factors, with an integrated feature of automatic tensor rank learning. We will also connect the algorithm to the inexact block coordinate descent (BCD) to obtain a fast algorithm.

6.1 Tensor CPD with Nonnegative Factors

In the previous three chapters, the factor matrices of CPD are without constraints. If the nonnegative constraints are added, the resultant model allows only additions among the R latent rank-1 components. This leads to a parts-based representation of the data, in the sense that each rank-1 component is a part of the data, thus further enhancing the model interpretability. Below is a motivating application.

6.1.1 Motivating Example—Social Group Clustering

Social group clustering could be benefited from tensor data analysis, by which multiple views of social networks are provided [1, 2]; see Fig. 6.1. For example, consider a 3D email dataset $\mathcal{Y} \in \mathbb{R}^{I \times J \times K}$ with each element $\mathcal{Y}(i, j, k)$ denoting the number of emails sent from person i to person j at the kth day. Each frontal slice $\mathcal{Y}(:, :, k)$ represents the connection intensity among different pairs of peoples in the kth day, while each slice $\mathcal{Y}(:, j, :)$ shows the temporal evolution of the number of received mails for the person j from each of the other person in the dataset. Consequently, decomposing the tensor $\mathcal{Y}$ into latent CPD factors $\{\boldsymbol{A} \in \mathbb{R}^{I \times R}, \boldsymbol{B} \in \mathbb{R}^{J \times R}, \boldsymbol{C} \in \mathbb{R}^{K \times R}\}$ reveals different clustering groups from different views (i.e., different tensor dimensions).

L. Cheng et al., *Bayesian Tensor Decomposition for Signal Processing and Machine Learning*, https://doi.org/10.1007/978-3-031-22438-6_6

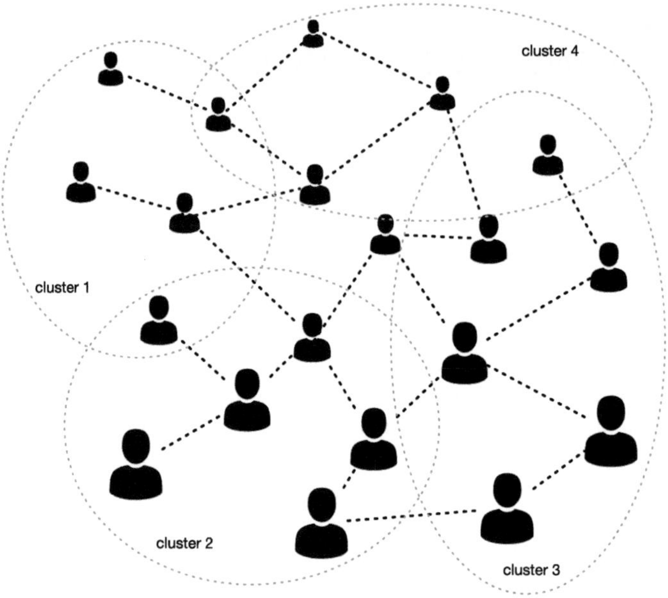

Fig. 6.1 An application example of CPD: social group clustering

In particular, using the unfolding property of tensor CPD, we have

$$\mathcal{Y}^{(1)} = (\boldsymbol{C} \diamond \boldsymbol{B})\boldsymbol{A}^T, \tag{6.1}$$

$$\mathcal{Y}^{(3)} = (\boldsymbol{B} \diamond \boldsymbol{A})\boldsymbol{C}^T, \tag{6.2}$$

where $\mathcal{Y}^{(k)}$ is a matrix obtained by unfolding the tensor $\mathcal{Y}$ along its kth dimension, and symbol $\diamond$ denotes the Khatri–Rao product (i.e., column-wise Kronecker product, see **Definition 1.3**). From (6.1), each column vector $\mathcal{Y}^{(1)}(:, i) \in \mathbb{R}^{JK\times 1}$ can be written as $\mathcal{Y}^{(1)}(:, i) = \sum_{r=1}^{R} (\boldsymbol{C} \diamond \boldsymbol{B})_{:,r}\boldsymbol{A}_{i,r}$, which is a linear combination of column vectors in matrix $(\boldsymbol{C} \diamond \boldsymbol{B}) \in \mathbb{R}^{JK\times R}$ with coefficients $\{\boldsymbol{A}_{i,r}\}_{r=1}^{R}$, and it represents the email sending pattern of person i. Thus, each column vector in matrix $\boldsymbol{C} \diamond \boldsymbol{B}$ can be interpreted as one of the R underlying email sending patterns, and $\boldsymbol{A}_{i,r}$ is the linear combining coefficient to generate the person i's email pattern. Similarly, from (6.2), each column in $\boldsymbol{B} \diamond \boldsymbol{A}$ can be interpreted as a temporal pattern and $\boldsymbol{C}_{k,r}$ is the coefficient of the rth temporal pattern for generating the kth day's pattern.

Obviously, due to the complicated multi-view clustering structure, tensor CPD is superior to the matrix-based models such as k-means or Gaussian mixture model. To find the latent factor matrices from the social network data $\mathcal{Y}$, the following problem is usually solved:

$$\begin{aligned} &\min_{\boldsymbol{A},\boldsymbol{B},\boldsymbol{C}} \parallel \mathcal{Y} - [\![\boldsymbol{A}, \boldsymbol{B}, \boldsymbol{C}]\!] \parallel_F^2 \\ &\text{s.t. } \boldsymbol{A} \geq \boldsymbol{0}_{I\times R}, \boldsymbol{B} \geq \boldsymbol{0}_{J\times R}, \boldsymbol{C} \geq \boldsymbol{0}_{K\times R}, \end{aligned} \tag{6.3}$$

where the nonnegative constraints are added to allow only additions among the R latent rank-1 components. This leads to a parts-based representation of the data, which enhances the model interpretability.

6.1.2 General Problem and Challenges

The general problem, which decomposes a N-dimensional tensor $\mathcal{Y} \in \mathbb{R}^{J_1\times J_2\times\cdots\times J_N}$ into a set of nonnegative factor matrices $\{\boldsymbol{\Xi}^{(n)}\}_{n=1}^N$, is formulated as:

$$\begin{aligned} &\min_{\{\boldsymbol{\Xi}^{(n)}\}_{n=1}^N} \|\mathcal{Y} - \underbrace{\sum_{r=1}^{R} \boldsymbol{\Xi}_{:,r}^{(1)} \circ \boldsymbol{\Xi}_{:,r}^{(2)} \circ \cdots \circ \boldsymbol{\Xi}_{:,r}^{(N)}}_{\triangleq [\![\boldsymbol{\Xi}^{(1)},\boldsymbol{\Xi}^{(2)},\ldots,\boldsymbol{\Xi}^{(N)}]\!]} \|_F^2 \\ &\text{s.t. } \boldsymbol{\Xi}^{(n)} \geq \boldsymbol{0}_{J_n\times R}, \quad n = 1, 2, \ldots, N, \end{aligned} \tag{6.4}$$

where symbol $\circ$ denotes the vector outer product.

In problem (6.4), there are two significant challenges. Firstly, nonnegative factor matrices $\{\boldsymbol{\Xi}^{(n)}\}_{n=1}^N$ are complicatedly coupled, resulting in a difficult non-convex optimization problem. To tackle this challenge, alternating optimization is one of the most commonly used techniques. In each iteration, after fixing all but one factor matrices, problem (6.4) will become a standard nonlinear least-squares (NLS) subproblem, for which there are various off-the-shelf algorithms for solving it [3], including interior point methods and augmented Lagrangian methods. To handle big tensor data, first-order methods, such as Nesterov-type projected gradient descent, have been proposed to replace the interior point methods to solve each subproblem [4, 5].

Although pioneering works [4, 5] allow the learning of nonnegative factors from big multi-dimensional data, they still face the second critical challenge of problem (6.4): how to automatically learn the tensor rank R from the data? With the physical meaning of tensor rank being the number of components/groups inside the data, this value is usually unknown in practice and its estimation has been shown to be NP-hard. If the tensor CPD has at least one factor matrix without nonnegative constraint [6–8], this problem can be solved by applying Gaussian-gamma prior (GG) prior or the Generalized Hyperbolic (GH) prior to the factor matrix without constraint, which is a slight variation of the standard Bayesian CPD in Chap. 3 (note that a standard CPD is one having no constraint on all factor matrices).

However, when all the factor matrices have nonnegative constraints, GG or GH prior is not applicable since they are in the form of Gaussian scale mixture (GSM, see Sect. 2.3) and the support of the Gaussian probability density function (pdf) is not restricted to the nonnegative region. This calls for a different prior distribution modeling. An immediate idea might be to replace the Gaussian distribution with the truncated Gaussian distribution with a nonnegative support. However, a closer inspection is needed since there is no existing work discussing whether a gamma distribution or generalized inverse Gaussian (GIG) is conjugate to a truncated Gaussian distribution with a nonnegative support. This chapter focuses on demonstrating the conjugacy of gamma distribution to the truncated Gaussian distribution.

6.2 Probabilistic Modeling for CPD with Nonnegative Factors

6.2.1 Properties of Nonnegative Gaussian-Gamma Prior

Recall from Chap. 2 that for a model parameter vector $\boldsymbol{w} \in \mathbb{R}^{M\times 1}$ consisting of S non-overlapped blocks with each block denoted by $\boldsymbol{w}_s \in \mathbb{R}^{M_s\times 1}$. The Gaussian-gamma prior can be expressed as

$$p(\boldsymbol{w}|\{\alpha_s\}_{s=1}^S) = \prod_{s=1}^S p(\boldsymbol{w}_s|\alpha_s) = \prod_{s=1}^S \mathcal{N}(\boldsymbol{w}_s|\boldsymbol{0}_{M_s\times 1}, \alpha_s^{-1}\boldsymbol{I}_{M_s}), \tag{6.5}$$

$$p(\{\alpha_s\}_{s=1}^S) = \prod_{s=1}^S \text{gamma}(\alpha_s|a_s, b_s), \tag{6.6}$$

where α_s is the precision parameter (i.e., the inverse of variance, also called weight decay rate) that controls the relevance of model block $\boldsymbol{w}_s$ in data interpretation, and $\{a_s, b_s\}_{s=1}^S$ are pre-determined hyper-parameters. There are two important properties of Gaussian-gamma prior that lead to its success and prevalence in a variety of applications. Firstly, after integrating the gamma hyper-prior, the marginal distribution of model parameter $p(\boldsymbol{w})$ is a student's t distribution, which is strongly peaked at zero and with heavy tails, thus promoting sparsity. Secondly, the gamma hyper-prior (6.6) is conjugate to the Gaussian prior (6.5). This conjugacy permits the closed-form solution of the variational inference, which has recently come up as a major tool in inferring complicated probabilistic models with inexpensive computations.

Unfortunately, the support of the Gaussian pdf in (6.5) is unbounded and thus cannot model nonnegativeness. On the other hand, the truncated Gaussian prior with a nonnegative support for each model block $\boldsymbol{w}_s$ can be written as:

$$p^+(\boldsymbol{w}|\{\alpha_s\}_{s=1}^S) = \prod_{s=1}^S p^+(\boldsymbol{w}_s|\alpha_s)$$
$$= \prod_{s=1}^S \frac{\mathcal{N}(\boldsymbol{w}_s|\boldsymbol{0}_{M_s\times 1}, \alpha_s^{-1}\boldsymbol{I}_{M_s})}{\int_{\boldsymbol{0}_{M_s\times 1}}^{\infty} \mathcal{N}(\boldsymbol{w}_s|\boldsymbol{0}_{M_s\times 1}, \alpha_s^{-1}\boldsymbol{I}_{M_s})} \mathrm{U}\left(\boldsymbol{w}_s \geq \boldsymbol{0}_{M_s\times 1}\right), \tag{6.7}$$

where the function $\mathrm{U}\left(\boldsymbol{w}_s \geq \boldsymbol{0}_{M_s\times 1}\right)$ is a unit-step function with value one when $\boldsymbol{w}_s \geq \boldsymbol{0}_{M_s\times 1}$, and with a value of zero otherwise. Together with the gamma distributions (6.6) for modeling the precision parameters $\{\alpha_s\}_{s=1}^S$, we have the nonnegative Gaussian-gamma prior. Even though it is clear that nonnegative Gaussian-gamma prior can model the nonnegativeness of model parameters due to the unit-step function $\mathrm{U}\left(\boldsymbol{w}_s \geq \boldsymbol{0}_{M_s\times 1}\right)$, whether it enjoys the advantages of the vanilla Gaussian-gamma prior needs further inspection. In the following, two properties of the nonnegative Gaussian-gamma prior are presented, and their proofs can be found in [9].

Property 6.1. The gamma distribution in (6.6) is conjugate to the nonnegative Gaussian distribution in (6.7).

Property 6.2. After integrating out the precision parameters $\{\alpha_s\}_{s=1}^S$, the marginal pdf of model parameter $\boldsymbol{w}$ is

$$p^+(\boldsymbol{w}) = \int p^+(\boldsymbol{w}|\{\alpha_s\}_{s=1}^S)p(\{\alpha_s\}_{s=1}^S)d\{\alpha_s\}_{s=1}^S$$
$$= \prod_{s=1}^S 2^{M_s}\left(\frac{1}{\pi}\right)^{\frac{M_s}{2}} \frac{\Gamma(a_s + M_s/2)}{(2b_s)^{-a_s}\Gamma(a_s)}(2b_s + \boldsymbol{w}_s^T\boldsymbol{w}_s)^{-a_s - M_s/2}$$
$$\times \mathrm{U}\left(\boldsymbol{w}_s \geq \boldsymbol{0}_{M_s\times 1}\right). \tag{6.8}$$

It is a product of multivariate truncated student's t distributions, each of which is with a nonnegative support.

The shape of the marginal distribution $p^+(\boldsymbol{w})$ is determined by the hyper-parameters $\{a_s, b_s\}_{s=1}^S$. Usually, their values are chosen to be a very small value (e.g., $\epsilon = 10^{-6}$), since as $a_s \to 0$ and $b_s \to 0$, a Jeffrey's non-informative prior $p(\alpha_s) \propto \alpha_s^{-1}$ can be obtained. After letting the hyper-parameters $\{a_s, b_s\}_{s=1}^S$ in (6.8) go to zero, it is easy to obtain the following property.

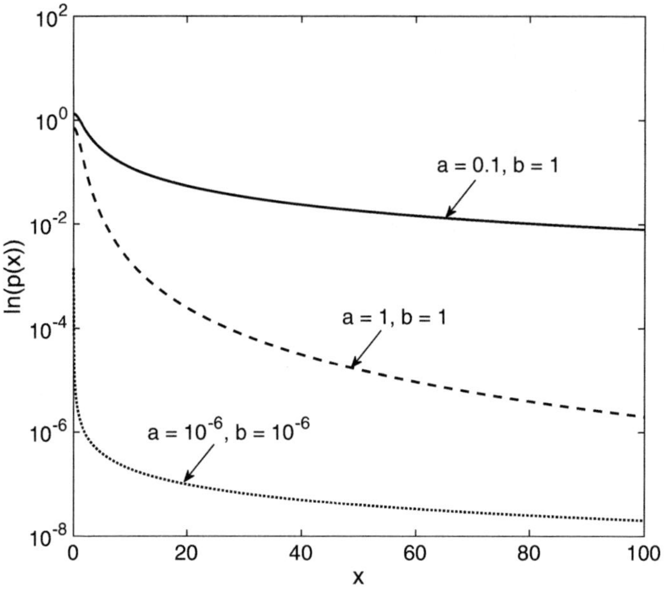

Fig. 6.2 Univariate marginal probability density functions with different parameters (© [2020] IEEE. Reprinted, with permission, from [L. Cheng, X. Tong, S. Wang, Y.-C. Wu, and H. V. Poor, Learning Nonnegative Factors From Tensor Data: Probabilistic Modeling and Inference Algorithm, IEEE Transactions on Signal Processing, Jan 2020]. It also applies to Figs. 6.3, 6.4, 6.5, 6.6, 6.7, 6.8, 6.9, 6.10 and Tables 6.1, 6.2, 6.3, 6.4, 6.5)

Property 6.3. If $a_s \to 0$ and $b_s \to 0$, the marginal pdf (6.8) becomes

$$p^{+}(\boldsymbol{w}) \propto \prod_{s=1}^{S} 2^{M_s} \left(\frac{1}{||\boldsymbol{w}_s||_2} \right)^{M_s} \mathrm{U}\left(\boldsymbol{w}_s \geq \boldsymbol{0}_{M_s \times 1}\right), \tag{6.9}$$

which is highly peaked at zeros.

As an illustration for **Properties 6.2 and 6.3**, univariate marginal pdfs with different hyper-parameters $\{a_s, b_s\}_{s=1}^{S}$ are plotted in Fig. 6.2, from which it is clear that the nonnegative Gaussian-gamma prior is sparsity-promoting. Further with the conjugacy property revealed in **Property 6.1**, the nonnegative Gaussian-gamma prior is a good candidate for probabilistic modeling with nonnegative model parameters.

6.2.2 Probabilistic Modeling of CPD with Nonnegative Factors

In the CPD problem with nonnegative factors in (6.4), the lth column group $\{\Xi^{(n)}_{:,l}\}_{n=1}^{N}$, which consists of the lth column of all the factor matrices, can be treated as a model block since their outer product contributes a rank-1 tensor. Therefore, with the principle of nonnegative Gaussian-gamma prior in the previous subsection, the lth column group $\{\Xi^{(n)}_{:,l}\}_{n=1}^{N}$ can be modeled using (6.7), but with $\boldsymbol{w}_s$ replaced by $\{\Xi^{(n)}_{:,l}\}_{n=1}^{N}$ and α_s replaced by γ_l. Considering the independence among different column groups in $\{\Xi^{(n)}\}_{n=1}^{N}$, we have

$$p(\{\Xi^{(n)}\}_{n=1}^{N}|\{\gamma_l\}_{l=1}^{L}) = \prod_{n=1}^{N}\prod_{l=1}^{L}\frac{\mathcal{N}\left(\Xi^{(n)}_{:,l}|\boldsymbol{0}_{J_n\times 1}, \gamma_l^{-1}\boldsymbol{I}_L\right)}{\int_{\boldsymbol{0}_{J_n\times 1}}^{\infty}\mathcal{N}\left(\Xi^{(n)}_{:,l}|\boldsymbol{0}_{J_n\times 1}, \gamma_l^{-1}\boldsymbol{I}_L\right)d\Xi^{(n)}_{:,l}}\mathrm{U}\left(\Xi^{(n)}_{:,l} \geq \boldsymbol{0}_{J_n\times 1}\right), \tag{6.10}$$

$$p(\{\gamma_l\}_{l=1}^{L}|\boldsymbol{\lambda}_{\gamma}) = \prod_{l=1}^{L}\mathrm{gamma}(\gamma_l|c_l^0, d_l^0), \tag{6.11}$$

where the precision γ_l is modeled as a gamma distribution. From discussions below **Property 6.2**, c_l^0 and d_l^0 can be chosen to be very small values (e.g., $c_l^0 = d_l^0 = \epsilon = 10^{-6}$) to approach a non-informative prior of precision parameter γ_l.

On the other hand, the least-squares objective function in the nonnegative tensor CPD problem (6.4) motivates the use of a Gaussian likelihood [6–8]:

$$p\left(\mathcal{Y} \mid \Xi^{(1)}, \Xi^{(2)}, \ldots, \Xi^{(N)}, \beta\right) \propto \exp\left(-\frac{\beta}{2} \parallel \mathcal{Y} - [\![\Xi^{(1)}, \Xi^{(2)}, \ldots, \Xi^{(N)}]\!] \parallel_F^2\right), \tag{6.12}$$

in which the parameter β represents the inverse of noise power. Since there is no information about it, a gamma distribution $p(\beta|\boldsymbol{\alpha}_{\beta}) = \mathrm{gamma}(\beta|\epsilon, \epsilon)$ with very small ϵ is employed, making $p(\beta|\boldsymbol{\alpha}_{\beta})$ approaches Jeffrey's non-informative prior.

The Gaussian likelihood function in (6.12) is with an unbounded support over the real space, and thus it is suitable for applications such as fluorescence data analysis and blind speech separation, in which the observed data $\mathcal{Y}$ could be both positive and negative. On the other hand, if the data $\mathcal{Y}$ are all nonnegative and continuous (e.g., the email dataset [1, 2] after pre-processing), a truncated Gaussian likelihood can be used to model the data:

$$p\left(\mathcal{Y} \mid \Xi^{(1)}, \Xi^{(2)}, \ldots, \Xi^{(N)}, \beta\right) \propto \exp\left(-\frac{\beta}{2} \parallel \mathcal{Y} - [\![\Xi^{(1)}, \Xi^{(2)}, \ldots, \Xi^{(N)}]\!] \parallel_F^2\right)\mathrm{U}(\mathcal{Y} \geq 0). \tag{6.13}$$

Finally, the complete probabilistic model is a three-layer Bayes network and is illustrated in Fig. 6.3.

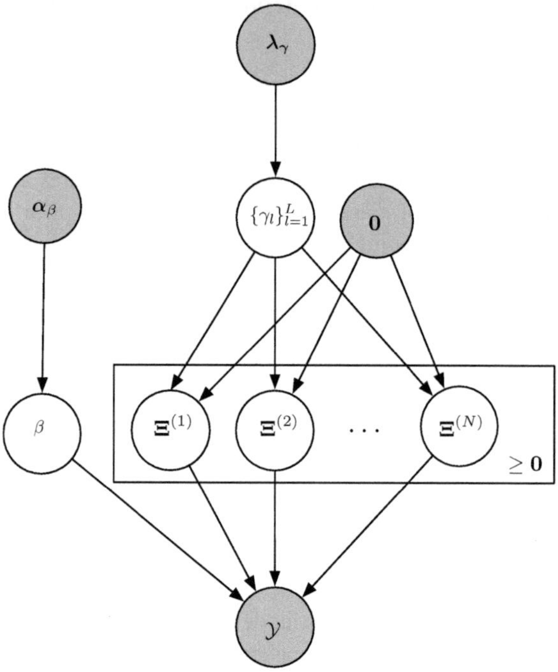

Fig. 6.3 Probabilistic model for tensor CPD with nonnegative factors

6.3 Inference Algorithm for Tensor CPD with Nonnegative Factors

The unknown parameter set $\boldsymbol{\Theta}$ includes the factor matrices $\{\Xi^{(n)}\}_{n=1}^N$, the noise power β^{-1}, and the precision parameter $\{\gamma_l\}_{l=1}^L$. Recall from Chaps. 2 and 3 that under mean-field assumption $Q(\boldsymbol{\Theta}) = \prod_{k=1}^K Q(\boldsymbol{\Theta}_k)$, the optimal variational pdf is obtained by solving

$$\min_{Q(\boldsymbol{\Theta}_k)} \int Q(\boldsymbol{\Theta}_k)(-\mathbb{E}_{\prod_{j\neq k} Q(\boldsymbol{\Theta}_j)}\left[\ln p(\boldsymbol{\Theta}, \mathcal{Y})\right] + \ln Q(\boldsymbol{\Theta}_k))d\boldsymbol{\Theta}_k, \tag{6.14}$$

and its solution is

$$Q^*\left(\boldsymbol{\Theta}_k\right) = \frac{\exp\left(\mathbb{E}_{\prod_{j\neq k} Q(\boldsymbol{\Theta}_j)}\left[\ln p\left(\boldsymbol{\Theta}, \mathcal{Y}\right)\right]\right)}{\int \exp\left(\mathbb{E}_{\prod_{j\neq k} Q(\boldsymbol{\Theta}_j)}\left[\ln p\left(\boldsymbol{\Theta}, \mathcal{Y}\right)\right]\right) d\boldsymbol{\Theta}_k}. \tag{6.15}$$

Nevertheless, even under the mean-field family assumption, the unknown parameter $\Xi^{(k)}$ is still difficult to be inferred since its moments cannot be easily computed. In particular, in the proposed probabilistic model, if no functional assumption is made for variational pdf $Q(\Xi^{(k)})$, after using (6.15), a multivariate truncated Gaussian distribution would be obtained, of which the moments are known to be very difficult to be computed due to the multiple integrations involved [10]. In this case, the variational pdf $Q(\Xi^{(k)})$ could be further restricted to be a Dirac delta function $Q(\Xi^{(k)}) = \delta(\Xi^{(k)} - \hat{\Xi}^{(k)})$, where $\hat{\Xi}^{(k)}$ is the point estimate of the parameter $\Xi^{(k)}$. After substituting this functional form into problem (6.14), the optimal point estimate $\hat{\Xi}^{(k)*}$ is obtained by

$$\hat{\Xi}^{(k)*} = \arg\max \mathbb{E}_{\prod_{\Theta_j \neq \Xi^{(k)}} Q(\Theta_j)} \left[\ln p\left(\Theta, \mathcal{Y}\right)\right]. \tag{6.16}$$

This is indeed the framework of variational expectation maximization (EM), in which the factor matrices $\{\Xi^{(k)}\}_{k=1}^{N}$ are treated as global parameters and other variables are treated as latent variables (see also the discussion around (2.7)).

In (6.15) and (6.16), the log of the joint pdf $\ln p\left(\Theta, \mathcal{Y}\right)$ needs to be evaluated. If the Gaussian likelihood function (6.12) is adopted, it is expressed as

$$\begin{aligned}
\ln p^{\dagger}\left(\Theta, \mathcal{Y}\right) = & \sum_{n=1}^{N} \ln\left(\mathrm{U}(\Xi^{(n)} \geq \mathbf{0}_{J_n \times L})\right) + \frac{\prod_{n=1}^{N} J_n}{2} \ln \beta \\
& - \frac{\beta}{2} \parallel \mathcal{Y} - [\![\Xi^{(1)}, \Xi^{(2)}, \ldots, \Xi^{(N)}]\!] \|_F^2 + \sum_{n=1}^{N} \frac{J_n}{2} \sum_{l=1}^{L} \ln \gamma_l \\
& - \sum_{n=1}^{N} \frac{1}{2} \mathrm{Tr}\left(\Xi^{(n)} \mathbf{\Gamma} \Xi^{(n)T}\right) + \sum_{l=1}^{L} \left[(10^{-6} - 1) \ln \gamma_l - 10^{-6} \gamma_l\right] \\
& + (10^{-6} - 1) \ln \beta - 10^{-6} \beta + \text{const},
\end{aligned} \tag{6.17}$$

where $\mathbf{\Gamma} = \mathrm{diag}\{\gamma_1, \gamma_2, \ldots, \gamma_L\}$. On the other hand, if the truncated Gaussian likelihood (6.13) is used, the log of the joint pdf $\ln p^{\sharp}\left(\Theta, \mathcal{Y}\right)$ takes the following form:

$$\ln p^{\sharp}\left(\Theta, \mathcal{Y}\right) = \ln p^{\dagger}\left(\Theta, \mathcal{Y}\right) - \ln \Phi\left(\{\Xi^{(n)}\}_{n=1}^{N}, \beta\right) + \text{const}. \tag{6.18}$$

where

$$\Phi\left(\{\Xi^{(n)}\}_{n=1}^{N}, \beta\right) = \int_{\mathbf{0}}^{\infty} \left(\frac{\beta}{2\pi}\right)^{\prod_{n=1}^{N} \frac{J_n}{2}} \exp\left(\frac{-\beta}{2} \parallel \mathcal{Y} - [\![\Xi^{(1)}, \Xi^{(2)}, \ldots, \Xi^{(N)}]\!] \|_F^2\right) d\mathcal{Y}. \tag{6.19}$$

In (6.18), the term $\ln \Phi\left(\{\Xi^{(n)}\}_{n=1}^{N}, \beta\right)$, which arises from the truncated Gaussian likelihood in (6.13), is very difficult to evaluate and differentiate, due to the multiple integrations involved. Fortunately, for most applications in Bayesian nonnegative matrix/tensor decomposition, the confidence of the low-rank matrix/tensor model is relatively high, in the sense that the noise power $1/\beta$ is small compared to the average element in signal tensor $[\![\Xi^{(1)}, \Xi^{(2)}, \ldots, \Xi^{(N)}]\!]$. Under this assumption, it is easy to see $\ln \Phi\left(\{\Xi^{(n)}\}_{n=1}^{N}, \beta\right) \approx \ln 1 = 0$, since Gaussian pdf

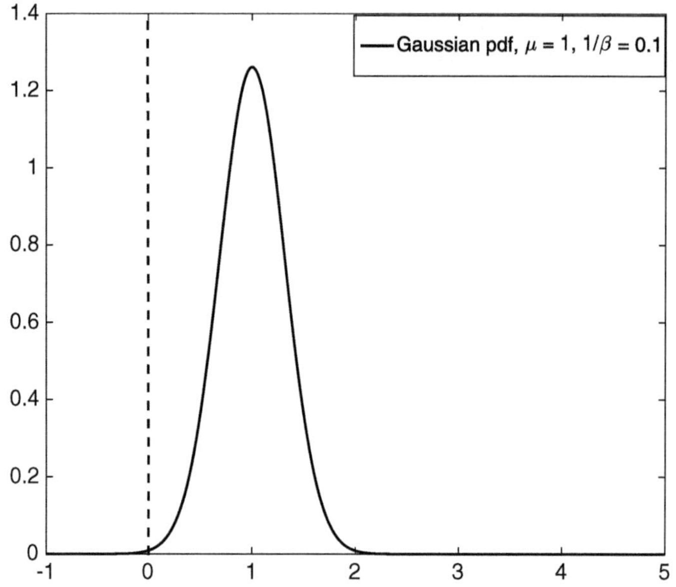

Fig. 6.4 Illustration of a univariate Gaussian probability density function with its mean much larger than the variance

$p(\mathcal{Y}) = \left(\frac{\beta}{2\pi}\right)^{\frac{\prod_n J_n}{2}} \exp\left(-\beta \parallel \mathcal{Y} - [\![\Xi^{(1)}, \Xi^{(2)}, \ldots, \Xi^{(N)}]\!] \parallel_F^2\right) d\mathcal{Y}$ decays very rapidly and thus most densities are over the region $\mathcal{Y} \geq 0$. As an illustration, a univariate Gaussian pdf with its mean much larger than the variance is plotted in Fig. 6.4, in which the probability density in the negative region is negligible. This suggests that the log of the joint pdf $\ln p^{\sharp}(\boldsymbol{\Theta}, \mathcal{Y})$ in (6.18) can be well approximated by $\ln p^{\dagger}(\boldsymbol{\Theta}, \mathcal{Y})$ in (6.17), and therefore, algorithm derivations are unified for the two likelihoods. This also explains why the previous Bayesian nonnegative matrix/tensor decompositions all employ the Gaussian likelihood function.

6.3.1 Derivation for Variational Pdfs

As discussed at the beginning of this section, the mean-field approximation is employed to enable closed-form expression for each variational pdf. For the precision parameter γ_l, by substituting (6.17) into (6.15) and only taking the terms relevant to γ_l, the variational pdf $Q(\gamma_l)$ can be found to take the same functional form as that of the gamma distribution, i.e., $Q(\gamma_l) = \text{gamma}(\gamma_l | c_l, d_l)$ with

$$c_l = \sum_{n=1}^{N} \frac{J_n}{2} + \epsilon, \tag{6.20}$$

$$d_l = \sum_{n=1}^{N} \frac{1}{2} \mathbb{E}_{Q(\boldsymbol{\Xi}^{(n)})} \left[\boldsymbol{\Xi}_{:,l}^{(n)T} \boldsymbol{\Xi}_{:,l}^{(n)} \right] + \epsilon. \tag{6.21}$$

Since the variational pdf $Q(\gamma_l)$ is determined by parameters c_l and d_l, its update is equivalent to the update of the two parameters in (6.20) and (6.21).

Similarly, using (6.15) and (6.17), the variational pdf $Q(\beta)$ can be found to be a gamma distribution $Q(\beta) = \text{gamma}(\beta|e, f)$, where

$$e = \frac{\prod_{n=1}^{N} J_n}{2} + \epsilon, \tag{6.22}$$

$$f = \frac{1}{2} \mathbb{E}_{\prod_{\Theta_j \neq \beta} Q(\Theta_j)} \left[\| \mathcal{Y} - [\![\boldsymbol{\Xi}^{(1)}, \boldsymbol{\Xi}^{(2)}, \ldots, \boldsymbol{\Xi}^{(N)}]\!] \|_F^2 \right] + \epsilon. \tag{6.23}$$

On the other hand, by substituting (6.17) into (6.16), the point estimate of $\hat{\boldsymbol{\Xi}}^{(k)}$ can be obtained via solving the following problem:

$$\max \mathbb{E}_{\prod_{\Theta_j \neq \boldsymbol{\Xi}^{(k)}} Q(\Theta_j)} \Big[\ln \left(\mathrm{U}(\boldsymbol{\Xi}^{(k)} \geq \mathbf{0}_{J_k \times L}) \right)$$
$$- \frac{\beta}{2} \| \mathcal{Y} - [\![\boldsymbol{\Xi}^{(1)}, \boldsymbol{\Xi}^{(2)}, \ldots, \boldsymbol{\Xi}^{(N)}]\!] \|_F^2 - \frac{1}{2} \mathrm{Tr}\left(\boldsymbol{\Xi}^{(k)} \boldsymbol{\Gamma} \boldsymbol{\Xi}^{(k)T} \right) \Big]. \tag{6.24}$$

After distributing the expectations, expanding the Frobenius norm, and utilizing the fact that $\ln(0) = -\infty$, problem (6.24) can be equivalently shown to be

$$\min f(\boldsymbol{\Xi}^{(k)})$$
$$\text{s.t. } \boldsymbol{\Xi}^{(k)} \geq \mathbf{0}_{J_k \times L}, \tag{6.25}$$

where

$$f(\boldsymbol{\Xi}^{(k)}) = \frac{1}{2} \mathrm{Tr}\Big(\boldsymbol{\Xi}^{(k)} \mathbb{E}_{\prod_{\Theta_j \neq \boldsymbol{\Xi}^{(k)}} Q(\Theta_j)} \left[\beta \mathfrak{B}^{(k)} \mathfrak{B}^{(k)T} + \boldsymbol{\Gamma} \right] \boldsymbol{\Xi}^{(k)T}$$
$$- 2\beta \boldsymbol{\Xi}^{(k)} \mathbb{E}_{\prod_{\Theta_j \neq \boldsymbol{\Xi}^{(k)}} Q(\Theta_j)} \left[\mathfrak{B}^{(k)} \right] \mathcal{Y}^{(k)T} \Big). \tag{6.26}$$

In (6.26), the term $\mathfrak{B}^{(k)} = \left(\diamond_{n=1, n\neq k}^{N} \boldsymbol{\Xi}^{(n)} \right)^T$, with the multiple Khatri–Rao products $\diamond_{n=1, n\neq k}^{N} \boldsymbol{A}^{(n)} = \boldsymbol{A}^{(N)} \diamond \boldsymbol{A}^{(N-1)} \diamond \cdots \diamond \boldsymbol{A}^{(k+1)} \diamond \boldsymbol{A}^{(k-1)} \diamond \cdots \diamond \boldsymbol{A}^{(1)}$. It is easy to see that problem (6.25) is a quadratic programming (QP) problem with nonnegative constraints. Since each diagonal element γ_l in the diagonal matrix $\boldsymbol{\Gamma}$ is larger than zero, the Hessian matrix of the function $f(\boldsymbol{\Xi}^{(k)})$, with the expression being

$$\boldsymbol{H}^{(k)} = \mathbb{E}_{\prod_{\Theta_j \neq \Xi^{(k)}} Q(\Theta_j)} \left[\beta \mathfrak{B}^{(k)} \mathfrak{B}^{(k)T} + \boldsymbol{\Gamma} \right] \tag{6.27}$$

is positive definite. This implies that problem (6.25) is a convex problem, and its solutions have been investigated for decades. In particular, first-order methods have recently received much attention due to their scalability in big data applications. Within the class of first-order methods, a simple gradient projection method is introduced as follows.

In each iteration of the gradient projection method, the update equation is of the form:

$$\boldsymbol{\Xi}^{(k,t+1)} = \left[\boldsymbol{\Xi}^{(k,t)} - \alpha_t \nabla f(\boldsymbol{\Xi}^{(k,t)}) \right]_+, \tag{6.28}$$

where the gradient $\nabla f(\boldsymbol{\Xi}^{(k,t)})$ is computed as

$$\begin{aligned} \nabla f(\boldsymbol{\Xi}^{(k,t)}) = {} & \boldsymbol{\Xi}^{(k,t)} \mathbb{E}_{\prod_{\Theta_j \neq \Xi^{(k)}} Q(\Theta_j)} \left[\beta \mathfrak{B}^{(k)} \mathfrak{B}^{(k)T} + \boldsymbol{\Gamma} \right] \\ & - \boldsymbol{\mathcal{Y}}^{(k)} \mathbb{E}_{\prod_{\Theta_j \neq \Xi^{(k)}} Q(\Theta_j)} \left[\beta \mathfrak{B}^{(k)T} \right]. \end{aligned} \tag{6.29}$$

In (6.29), the symbol $[\cdot]_+$ denotes projecting each element of $\boldsymbol{\Xi}^{(k)}$ to $[0, \infty)$ (i.e., $[x]_+ = x$ if $x \geq 0$ and $[x]_+ = 0$ otherwise) and $\alpha_t \geq 0$ is the step size. During the inference, due to the sparsity-promoting property of the nonnegative Gaussian-gamma prior, some of the precision parameters will go to very large numbers while some of them will tend to be zero. This will result in a very large condition number of the Hessian matrix $\boldsymbol{H}^{(k)}$. In this case, applying the diminishing rule[1] to the step size α_t still enjoys a good convergence performance [3] and thus is set as the default step size rule in the algorithm.

6.3.2 Summary of the Inference Algorithm

From Eqs. (6.20)–(6.29), it can be seen that we need to compute various expectations. In particular, for expectations $\mathbb{E}_{Q(\Xi^{(n)})}[\boldsymbol{\Xi}^{(n)}]$, $\mathbb{E}_{Q(\gamma_l)}[\gamma_l]$, and $\mathbb{E}_{Q(\beta)}[\beta]$, their computations are very straightforward, i.e., $\mathbb{E}_{Q(\Xi^{(n)})}[\boldsymbol{\Xi}^{(n)}] = \hat{\boldsymbol{\Xi}}^{(n)}$, $\mathbb{E}_{Q(\gamma_l)}[\gamma_l] = c_l/d_l$, and $\mathbb{E}_{Q(\beta)}[\beta] = e/f$. However, when updating $\hat{\boldsymbol{\Xi}}^{(n)}$ using (6.25) and (6.26), there is one complicated expectation $\mathbb{E}_{\prod_{\Theta_j \neq \Xi^{(k)}} Q(\Theta_j)} \left[\mathfrak{B}^{(k)} \mathfrak{B}^{(k)T} \right]$. Fortunately, it can be shown that $\mathbb{E}_{\prod_{\Theta_j \neq \Xi^{(k)}} Q(\Theta_j)} \left[\mathfrak{B}^{(k)} \mathfrak{B}^{(k)T} \right] = \bigodot_{n=1, n \neq k}^{N} \hat{\boldsymbol{\Xi}}^{(n)T} \hat{\boldsymbol{\Xi}}^{(n)}$, where the multiple Hadamard products $\bigodot_{n=1, n \neq k}^{N} \boldsymbol{A}^{(n)} = \boldsymbol{A}^{(N)} \odot \boldsymbol{A}^{(N-1)} \odot \cdots \odot \boldsymbol{A}^{(k+1)} \odot \boldsymbol{A}^{(k-1)} \odot \cdots \odot \boldsymbol{A}^{(1)}$. Since the computation of one variational pdf needs the statistics of other

[1] In the diminishing rule [3], the step size α_t needs to satisfy $\alpha_t \to 0$ and $\sum_{t=0}^{\infty} \alpha_t = \infty$.

Algorithm 8 Probabilistic Tensor CPD with Nonnegative Factors (PNCPD)

Initializations: Choose $L > R$ and initial values $\{\hat{\Xi}^{(n,0)}\}_{n=1}^{N}, \epsilon$.
Iterations: For the sth iteration ($s \geq 0$),
Update the parameter of $Q(\Xi^{(k)})^{(s+1)}$:

Set initial value $\Xi^{(k,0)} = \hat{\Xi}^{(k,s)}$.
Iterations: For the tth iteration ($t \geq 0$), compute

$$\Xi^{(k,t+1)} = \Big[\Xi^{(k,t)} - \alpha_t \nabla f(\Xi^{(k,t)})\Big]_+ ,$$

where

$$\nabla f(\Xi^{(k,t)}) = \Xi^{(k,t)}\Big[\frac{e^s}{f^s} \bigodot_{n=1,n\neq k}^{N} \hat{\Xi}^{(n,s)T}\hat{\Xi}^{(n,s)} + \mathrm{diag}\{\frac{c_1^s}{d_1^s}, \ldots, \frac{c_L^s}{d_L^s}\}\Big] - \frac{e^s}{f^s}\mathcal{Y}^{(k)}\Big(\mathop{\diamond}_{n=1,n\neq k}^{N} \hat{\Xi}^{(n,s)}\Big),$$

and α_t is chosen by the diminishing rule [42, p. 227].
Until Convergence
Set $\hat{\Xi}^{(k,s+1)} = \Xi^{(k,t+1)}$.
Update the parameter of $Q(\gamma_l)^{s+1}$:

$$c_l^{s+1} = \sum_{n=1}^{N} \frac{J_n}{2} + \epsilon,$$

$$d_l^{s+1} = \sum_{n=1}^{N} \frac{1}{2} \hat{\Xi}_{:,l}^{(n,s+1)T} \hat{\Xi}_{:,l}^{(n,s+1)} + \epsilon.$$

Update the parameter of $Q(\beta)^{s+1}$:

$$e^{s+1} = \epsilon + \frac{\prod_{n=1}^{N} J_n}{2},$$

$$f^{s+1} = \epsilon + \frac{1}{2} \parallel \mathcal{Y} - [\![\hat{\Xi}^{(1,s+1)}, \hat{\Xi}^{(2,s+1)}, \ldots, \hat{\Xi}^{(N,s+1)}]\!] \parallel_F^2 .$$

Until Convergence

variational pdfs, alternating update is needed, resulting in the iterative algorithm summarized in **Algorithm 8**.

6.3.3 Discussions and Insights

To gain further insight from the inference algorithm, discussions of its convergence property, automatic rank determination, relationship to the NALS algorithm, computational complexity, and scalability improvement are presented in the following.

6.3.3.1 Convergence Property

Algorithm 8 is derived under the framework of mean-field variational inference, where a variational pdf $Q(\boldsymbol{\Theta}) = \prod_k Q(\boldsymbol{\Theta}_k)$ is sought that minimizes the KL divergence KL $(Q(\boldsymbol{\Theta})||p(\boldsymbol{\Theta}|\mathcal{Y}))$. Even though this problem is known to be non-convex due to the non-convexity of the mean-field family set, it is convex with respect to a single variational pdf $Q(\boldsymbol{\Theta}_k)$ after fixing other variational pdfs $\{Q(\boldsymbol{\Theta}_j), j \neq k\}$ [3]. Therefore, the iterative algorithm, in which a single variational pdf is optimized in each iteration with other variational pdfs fixed, is essentially a coordinate descent algorithm in the functional space of variational pdfs. Since the subproblem in each iteration is not only convex but also has a unique solution, the limit point generated by the coordinate descent steps over the functional space of variational pdfs is guaranteed to be at least a stationary point of the KL divergence [3].

6.3.3.2 Automatic Rank Determination

During the inference, the expectations of some precision parameters $\{\gamma_l\}$, i.e., $\{\frac{c_l^s}{d_l^s}\}$ will go to a very large value. It indicates that the corresponding columns in the factor matrices are close to zero vectors, thus playing no role in data interpretation. As a result, after convergence, those columns can be pruned out and the number of remaining columns in each factor matrix gives the estimate of tensor rank.

In practice, to reduce the computational complexity, the pruning would be executed during the iteration. In particular, in each iteration, after the precision estimates $\{\frac{c_l^s}{d_l^s}\}$ exceed a certain threshold (e.g., 10^6), the associated columns are safely pruned out. After every pruning, it is equivalent to starting minimization of the KL divergence of a new (but smaller) probabilistic model, with the current variational distributions acting as the initialization of the new minimization. Therefore, the pruning steps will not affect the convergence and are widely used in recent related works [6–8].

Usually, the hyper-parameters $\{c_l^0, d_l^0\}$ of the prior gamma distribution are set to be a very small number $\epsilon = 10^{-6}$ to approach a non-informative prior. Otherwise, their values might affect the behavior of tensor rank estimate. For example, if we prefer a high value of the tensor rank, the initial value d_l^0 can be set to be very large while the initial value c_l^0 can be set to be small, so that the update of $\frac{c_l}{d_l}$ can be steered toward a small value in order to promote a high tensor rank. However, how to set the hyper-parameters to accurately control the degree of low rank is challenging and deserves future investigation.

6.3.3.3 Relationship to NALS Algorithm

The mean-field variational inference for tensor CPD problem could be interpreted as alternating optimizations over the Riemannian space (in which the Euclidean space is a special case). This insight has been revealed in previous works [8, 11], and can also be found in **Algorithm 8**. For example, for the precision parameters and the noise power parameter, the variational pdfs are with no constraint on the functional form, and thus, the corresponding alternating optimization is over the Riemannian space due to the exponentially conjugacy property of the probabilistic model [8, 11]. On the other hand, for unknown factor matrices, since the variational pdfs to be optimized are with a delta functional form, the corresponding alternating optimization is over the Euclidean space, thus is similar to the conventional NALS step. However, there is a significant difference. In **Algorithm 8**, there is a shrinkage term $\mathbf{\Gamma}$ appeared in the Hessian matrix in (6.27), and $\mathbf{\Gamma}$ will be updated together with other parameters in the algorithm. This intricate interaction is due to the employed Bayesian framework and cannot be revealed by NALS framework. Consequently, **Algorithm 8** is a generalization of the NALS algorithm, with the additional novel feature in automatic rank determination achieved via optimization in Riemannian space.

6.3.3.4 Computational Complexity

For each iteration, the computational complexity is dominated by computing the gradient of each factor matrix in (6.28), costing $O(\prod_{n=1}^{N} J_n L)$. From this expression, it is clear that the computational complexity in each iteration is linear with respect to the tensor dimension product $\prod_{n=1}^{N} J_n$. Consequently, the complexity of the algorithm is $O(q(\prod_{n=1}^{N} J_n L))$ where q is the iteration number at convergence.

6.4 Algorithm Accelerations

From **Algorithm 8**, it is clear that the bottleneck of the algorithm computation is the update of factor matrices $\{\Xi^{(n)}\}_{n=1}^{N}$ via solving problem (6.25). In particular, for each subproblem for solving $\Xi^{(k)}$, projected gradient may take a large number of iterations to converge. Fortunately, if the problem is well conditioned, in the sense that the condition number of the Hessian matrix $\boldsymbol{H}^{(k)}$ in (6.27) is smaller than a threshold (e.g., 100), acceleration schemes, including variants of the Nesterov scheme [4], can be utilized to significantly reduce the required number of iterations for solving problem (6.25), thus speeding up **Algorithm 8**.

On the other hand, since $\Xi^{(k)}$ for $k = 1, 2, \ldots, n$ are updated alternatively under the framework of block coordinate descent (BCD), besides reducing the iteration number for solving each $\Xi^{(k)}$, we can also borrow the idea of inexact BCD, where each subproblem for $\Xi^{(k)}$ is not exactly solved, thus avoiding the large number of iterations required for convergence. In general, as long as the solution of each $\Xi^{(k)}$ leads to a reduction of the overall optimization objective function, inexact BCD reduces the computation times significantly while maintaining the solution quality. Below, we will give the details on how to connect **Algorithm 8** to inexact BCD.

More specifically, **Algorithm 8** can be viewed as solving the following deterministic optimization problem:

$$\min_{\{\boldsymbol{\Xi}^{(n)} \geq \mathbf{0}\}_{n=1}^{N}, \{\gamma_l\}_{l=1}^{L}, \beta} g(\boldsymbol{\Theta}), \tag{6.30}$$

where

$$g(\boldsymbol{\Theta}) = -\frac{\prod_{n=1}^{N} J_n}{2} \ln \beta + \frac{\beta}{2} \parallel \mathcal{Y} - [\![\boldsymbol{\Xi}^{(1)}, \boldsymbol{\Xi}^{(2)}, \ldots, \boldsymbol{\Xi}^{(N)}]\!]\|_F^2 - \sum_{n=1}^{N} \frac{J_n}{2} \sum_{l=1}^{L} \ln \gamma_l$$
$$+ \sum_{n=1}^{N} \frac{1}{2} \mathrm{Tr}\left(\boldsymbol{\Xi}^{(n)} \boldsymbol{\Gamma} \boldsymbol{\Xi}^{(n)T}\right) - \sum_{l=1}^{L} \left[c_l^0 \ln \gamma_l - d_l^0 \gamma_l\right] - e^0 \ln \beta + f^0 \beta, \tag{6.31}$$

with $\boldsymbol{\Gamma} = \mathrm{diag}\{\gamma_1, \ldots, \gamma_L\}$. To see the connection between solving (6.30) and **Algorithm 8**, the BCD method could be employed. That is, in each iteration, after fixing other unknown parameters $\{\boldsymbol{\Theta}_j\}_{j \neq k}$ at their last updated values, $\boldsymbol{\Theta}_k$ is updated as follows:

$$\boldsymbol{\Theta}_k^{t+1} = \arg\min_{\boldsymbol{\Theta}_k} g(\boldsymbol{\Theta}_1^{t+1}, \ldots, \boldsymbol{\Theta}_{k-1}^{t+1}, \boldsymbol{\Theta}_k, \boldsymbol{\Theta}_{k+1}^{t}, \ldots, \boldsymbol{\Theta}_K^{t}). \tag{6.32}$$

This connection is formally stated in **Proposition 6.1**.

Proposition 6.1. Assume each initial value of the unknown parameter $\boldsymbol{\Theta}_k^0$ in (6.32) equal to the expectation of $\boldsymbol{\Theta}_k$ with respect to the initial variational pdf $Q^0(\boldsymbol{\Theta}_k)$ in **Algorithm 1**. With the same update schedule for various parameter blocks, in each iteration, the result of the block minimization update (6.32) for parameter $\boldsymbol{\Theta}_k$ equals to the expectation of $\boldsymbol{\Theta}_k$ with respect to the variational pdf $Q(\boldsymbol{\Theta}_k)$ from **Algorithm 8**, i.e., $\boldsymbol{\Theta}_k^t = \mathbb{E}_{Q^t(\boldsymbol{\Theta}_k)}[\boldsymbol{\Theta}_k]$.

Therefore, **Algorithm 8** is indeed related to parameter block minimization of the optimization problem (6.30). This new interpretation opens up the possibility of using inexact BCD to accelerate **Algorithm 8** and its detailed derivations are presented as follows.

For updating noise precision parameter β in the iteration $t + 1 (t \geq 0)$, after fixing other parameters in problem (6.30), the subproblem is expressed as

$$\min_{\beta} h^{t+1}(\beta), \tag{6.33}$$

where

$$h^{t+1}(\beta) = -\left(\frac{\prod_{n=1}^{N} J_n}{2} + e^0\right) \ln \beta$$
$$+ \beta \left(\frac{1}{2} \parallel \mathcal{Y} - [\![\boldsymbol{\Xi}^{(1),t+1}, \boldsymbol{\Xi}^{(2),t+1}, \ldots, \boldsymbol{\Xi}^{(N),t+1}]\!] \parallel_F^2 + f^0\right). \tag{6.34}$$

However, the objective function $h^{t+1}(\beta)$ is not strongly convex since the second-order derivative $\nabla^2_\beta h^{t+1}(\beta) = \left(\frac{\prod_{n=1}^N J_n}{2} + e^0\right)\frac{1}{\beta^2}$ can be arbitrarily close to zero as $\beta \to \infty$. Consequently, the block minimization scheme cannot be used. To guarantee convergence, as suggested in [3, 12–15], a proximal term $\frac{\mu_\beta}{2}(\beta - \beta^t)^2$ is added to the objective function in (6.33), giving the following optimization problem:

$$\min_\beta h^{t+1}(\beta) + \frac{\mu_\beta}{2}(\beta - \beta^t)^2, \tag{6.35}$$

with parameter $0 < \mu_\beta < \infty$. After setting the derivative of the objective function of (6.35) to zero, it can be shown that the optimal solution takes a closed form:

$$\beta^{t+1} = \frac{-\left(f^{t+1} - \mu_\beta \beta^t\right) + \sqrt{\left(f^{t+1} - \mu_\beta \beta^t\right)^2 + 4\mu_\beta e^{t+1}}}{2\mu_\beta}, \tag{6.36}$$

in which

$$e^{t+1} = \frac{\prod_{n=1}^N J_n}{2} + e^0, \tag{6.37}$$

$$f^{t+1} = \frac{1}{2} \parallel \mathcal{Y} - [\![\Xi^{(1),t+1}, \Xi^{(2),t+1}, \ldots, \Xi^{(N),t+1}]\!] \parallel_F^2 + f^0. \tag{6.38}$$

Similarly, for updating parameter γ_l, the subproblem is

$$\min_{\gamma_l} h^{t+1}(\gamma_l), \tag{6.39}$$

where

$$h^{t+1}(\gamma_l) = -\left(\sum_{n=1}^N \frac{J_n}{2} + c_l^0\right) \ln \gamma_l + \gamma_l \left(d_l^0 + \sum_{n=1}^N \frac{1}{2}\left[\Xi^{(n),t+1}_{:,l}\right]^T \Xi^{(n),t+1}_{:,l}\right) \tag{6.40}$$

is not strongly convex. Therefore, a proximal term $\frac{\mu_{\gamma_l}}{2}(\gamma_l - \gamma_l^t)^2$ with $0 < \mu_{\gamma_l} < \infty$ is added to the objective function in (6.39) as follows [3, 12–15]:

$$\min_{\gamma_l} h^{t+1}(\gamma_l) + \frac{\mu_{\gamma_l}}{2}(\gamma_l - \gamma_l^t)^2, \tag{6.41}$$

of which the optimal solution can be shown to be

$$\gamma_l^{t+1} = \frac{-\left(d_l^{t+1} - \mu_{\gamma_l}\gamma_l^t\right) + \sqrt{\left(d_l^{t+1} - \mu_{\gamma_l}\gamma_l^t\right)^2 + 4\mu_{\gamma_l} c_l^{t+1}}}{2\mu_{\gamma_l}}, \tag{6.42}$$

where parameters c_l^{t+1} and d_l^{t+1} are

$$c_l^{t+1} = \sum_{n=1}^{N} \frac{J_n}{2} + c_l^0, \tag{6.43}$$

$$d_l^{t+1} = \sum_{n=1}^{N} \frac{1}{2} \left[\Xi_{:,l}^{(n),t+1} \right]^T \Xi_{:,l}^{(n),t+1} + d_l^0. \tag{6.44}$$

For updating each nonnegative factor in the iteration $t+1 (t \geq 0)$, after fixing other parameters, the subproblem can be formulated as:

$$\min_{\Xi^{(n)} \geq \mathbf{0}} h^{t+1}(\Xi^{(n)}), \tag{6.45}$$

where

$$h^{t+1}(\Xi^{(n)}) = \frac{\beta^t}{2} \parallel \mathcal{Y} - [\![\Xi^{(1),t+1}, \ldots, \Xi^{(n)}, \ldots, \Xi^{(N),t}]\!] \|_F^2 + \frac{1}{2} \mathrm{Tr}\Big(\Xi^{(n)} \boldsymbol{\Gamma}^t \Xi^{(n)T} \Big). \tag{6.46}$$

After expanding the Frobenius norm and some algebraic operations, problem (6.45) can be equivalently formulated as:

$$\min_{\Xi^{(n)} \geq \mathbf{0}} c^{t+1}(\Xi^{(n)}), \tag{6.47}$$

where

$$c^{t+1}(\Xi^{(n)}) = \frac{1}{2} \mathrm{Tr}\Big(\Xi^{(n)} \Big[\beta^t \left(\mathbf{B}^{(n),t}\right)^T \mathbf{B}^{(n),t} + \boldsymbol{\Gamma}^t \Big] \Xi^{(n)T} - 2\beta^t \Xi^{(n)} \left(\mathbf{B}^{(n),t}\right)^T \mathcal{Y}^{(n)} \Big). \tag{6.48}$$

In (6.48), matrix $\mathbf{B}^{(n),t} = \Xi^{(N),t} \diamond \cdots \diamond \Xi^{(n+1),t} \diamond \Xi^{(n-1),t+1} \diamond \cdots \diamond \Xi^{(1),t+1}$, where $\diamond$ denotes the Khatri–Rao product (see **Definition 1.3**), and $\mathcal{Y}^{(k)}$ is a matrix obtained by unfolding the tensor $\mathcal{Y}$ along its kth dimension (see **Definition 1.1**).

In the inexact BCD framework, it is not necessary to obtain the optimal solution of (6.47). Instead, we can construct a prox-linear update step with careful extrapolations and monotonicity-guaranteed modifications. In particular, in the iteration $t+1$, the extrapolation parameter w_n^t is computed by [12, 14–16]:

$$w_n^t = \min \left(\hat{w}_n^t, p_w \sqrt{\frac{L_n^{t-1}}{L_n^t}} \right), \tag{6.49}$$

where $p_w < 1$ is preselected, $\hat{w}_n^t = \frac{s_t - 1}{s_{t+1}}$ with $s_0 = 1$, $s_{t+1} = \frac{1}{2}(1 + \sqrt{1 + 4s_t^2})$, and L_n^t is assigned to be the spectral norm of the Hessian matrix of $c^{t+1}(\Xi^{(n)})$, i.e.,

$$L_n^t = ||\beta^t \left(\mathbf{B}^{(n),t}\right)^T \mathbf{B}^{(n),t} + \mathbf{\Gamma}^t||_2. \tag{6.50}$$

Using (6.49), the extrapolated factor matrix $\hat{\boldsymbol{M}}^{(n),t}$ is with the expression:

$$\hat{\boldsymbol{M}}^{(n),t} = \Xi^{(n),t} + w_n^t(\Xi^{(n),t} - \Xi^{(n),t-1}). \tag{6.51}$$

Based on (6.51), the prox-linear update can be expressed as [12]:

$$\Xi^{(n),t+1} = \arg\min_{\Xi^{(n)} \geq \mathbf{0}} \left\langle \nabla c^{t+1}(\hat{\boldsymbol{M}}^{(n),t}), \Xi^{(n)} - \hat{\boldsymbol{M}}^{(n),t} \right\rangle + \frac{L_n^t}{2}||\Xi^{(n)} - \hat{\boldsymbol{M}}^{(n),t}||_F^2, \tag{6.52}$$

where the gradient is

$$\nabla c^{t+1}(\Xi^{(n)}) = \Xi^{(n)} \left[\beta^t \left(\mathbf{B}^{(n),t}\right)^T \mathbf{B}^{(n),t} + \mathbf{\Gamma}^t\right] - \beta^t \mathcal{Y}^{(n)T} \mathbf{B}^{(n),t}, \tag{6.53}$$

and $\langle \cdot, \cdot \rangle$ denotes the inner product of the arguments. It can be shown that the solution of (6.52) takes the following closed form:

$$\Xi^{(n),t+1} = \left[\hat{\boldsymbol{M}}^{(n),t} - \frac{1}{L_n^t} \nabla c^{t+1}(\hat{\boldsymbol{M}}^{(n),t})\right]_+, \tag{6.54}$$

from which it can be seen that the prox-linear update only needs one-step computation and thus can be computed very fast.

Besides the extrapolation, monotonicity-guaranteed modification is needed to ensure the convergence [12]. More specifically, after updating all the parameters in $\mathbf{\Theta}$, whether the objective function of problem (6.30) is decreased (i.e., $g(\mathbf{\Theta}^{t+1}) \leq g(\mathbf{\Theta}^t)$) is tested. If not, prox-linear update (6.54) for each factor matrix $\Xi^{(n),t+1}$ will be re-executed without the extrapolation, i.e.,

$$\Xi^{(n),t+1} = \left[\Xi^{(n),t} - \frac{1}{L_n^t} \nabla c^{t+1}(\Xi^{(n),t})\right]_+. \tag{6.55}$$

Using (6.36), (6.42), (6.54), and (6.55), the inexact BCD-based algorithm for probabilistic tensor CPD with nonnegative factors is summarized in **Algorithm 9**.

Algorithm 9 Inexact BCD-Based Probabilistic Tensor CPD with Nonnegative Factors

Initializations: Choose $L > R$, $p_w < 1$, $\{\mu_{\gamma_l}\}_{l=1}^L$, μ_β and initial values $\{\boldsymbol{\Xi}^{(n),0}, \boldsymbol{\Xi}^{(n),-1}\}_{n=1}^N$, $\{c_l^0, d_l^0\}_{l=1}^L$, e^0, f^0.
Iterations: For the iteration $t+1$ $(t \geq 0)$,
For $n = 1$ **to** N
Update factor matrix $\boldsymbol{\Xi}^{(n),t+1}$:
Compute extrapolation parameter w_n^t using (6.49).
Compute extrapolated factor matrix:

$$\hat{\boldsymbol{M}}^{(n),t} = \boldsymbol{\Xi}^{(n),t} + w_n^t(\boldsymbol{\Xi}^{(n),t} - \boldsymbol{\Xi}^{(n),t-1}).$$

Update factor matrix:

$$\boldsymbol{\Xi}^{(n),t+1} = \left[\hat{\boldsymbol{M}}^{(n),t} - \frac{1}{L_n^t}\nabla c^{t+1}(\hat{\boldsymbol{M}}^{(n),t})\right]_+,$$

where $\nabla c^{t+1}(\hat{\boldsymbol{M}}^{(n),t})$ is computed using (6.53), and L_n^t is computed using (6.50).
End
Update parameter γ_l^{t+1}:

$$\gamma_l^{t+1} = \frac{-\left(d_l^{t+1} - \mu_{\gamma_l}\gamma_l^t\right) + \sqrt{\left(d_l^{t+1} - \mu_{\gamma_l}\gamma_l^t\right)^2 + 4\mu_{\gamma_l}c_l^{t+1}}}{2\mu_{\gamma_l}},$$

where c_l^{t+1} and d_l^{t+1} are computed via (6.43) and (6.44), respectively.
Update parameter β^{t+1}:

$$\beta^{t+1} = \frac{-\left(f^{t+1} - \mu_\beta\beta^t\right) + \sqrt{\left(f^{t+1} - \mu_\beta\beta^t\right)^2 + 4\mu_\beta e^{t+1}}}{2\mu_\beta},$$

where e^{t+1} and f^{t+1} are computed via (6.37) and (6.38), respectively.
Monotonicity check:
Let $\boldsymbol{\Theta}^{t+1} = \{\{\boldsymbol{\Xi}^{(n),t+1}\}_{n=1}^N, \{\gamma_l^{t+1}\}_{l=1}^L, \beta^{t+1}\}$.
If $g(\boldsymbol{\Theta}^{t+1}) > g(\boldsymbol{\Theta}^t)$:

$$\boldsymbol{\Xi}^{(n),t+1} = \left[\boldsymbol{\Xi}^{(n),t} - \frac{1}{L_n^t}\nabla c^{t+1}(\boldsymbol{\Xi}^{(n),t})\right]_+.$$

Until Convergence

6.5 Numerical Results

In this section, numerical results using synthetic data are firstly presented to assess the performance of **Algorithm 8** in terms of convergence property, factor matrix recovery, tensor rank estimation, and running time. Next, **Algorithm 8** is utilized to analyze two real-world datasets (the amino acids fluorescence data and the ENRON email corpus), for the demonstration on blind source separation and social group clustering. For all the simulated algorithms, the initial factor matrix $\hat{\Xi}^{(k,0)}$ is set as the singular value decomposition (SVD) approximation $U_{:,1:L}\left(S_{1:L,1:L}\right)^{\frac{1}{2}}$ where $[U, S, V] = \text{SVD}[\mathcal{Y}^{(k)}]$ and $L = \min\{J_1, J_2, \ldots, J_N\}$. The parameter ϵ is set to be 10^{-6}. The algorithms are deemed to be converged when $\| [\![\hat{\Xi}^{(1,s+1)}, \hat{\Xi}^{(2,s+1)}, \ldots, \hat{\Xi}^{(N,s+1)}]\!] - [\![\hat{\Xi}^{(1,s)}, \hat{\Xi}^{(2,s)}, \ldots, \hat{\Xi}^{(N,s)}]\!] \|_F^2 < 10^{-6}$. All experiments were conducted in Matlab R2015b with an Intel Core i7 CPU at 2.2 GHz.

6.5.1 Validation on Synthetic Data

A 3D tensor $\mathcal{X} = [\![M^{(1)}, M^{(2)}, M^{(3)}]\!] \in \mathbb{R}^{100\times 100\times 100}$ with rank $R = 10$ is considered as the noise-free data tensor. Each element in factor matrix $M^{(n)}$ is independently drawn from a uniform distribution over [0, 1] and thus is nonnegative. On the other hand, two observation data tensors are considered: (1) the data $\mathcal{X}$ is corrupted by a noise tensor $\mathcal{W} \in \mathbb{R}^{100\times 100\times 100}$, i.e., $\mathcal{Y} = \mathcal{X} + \mathcal{W}$, with each element of noise tensor $\mathcal{W}$ being independently drawn from a zero-mean Gaussian distribution with variance σ_w^2, and this corresponds to the Gaussian likelihood model (6.12); (2) the data $\mathcal{Y}^+$ is obtained by setting the negative elements of $\mathcal{Y}$ to zero, i.e., $\mathcal{Y}^+_{i_1,i_2,i_3} = \mathcal{Y}_{i_1,i_2,i_3}\text{U}(\mathcal{Y}_{i_1,i_2,i_3} \geq 0)$, and the truncated Gaussian likelihood model (6.13) is employed to fit these data. The SNR is defined as $10\log_{10}\left(\| \mathcal{X} \|_F^2 / \mathbb{E}_{p(\mathcal{W})}\left[\| \mathcal{W} \|_F^2\right]\right) = 10\log_{10}\left(\| \mathcal{X} \|_F^2 / (100^3\sigma_w^2)\right)$. For **Algorithm 8**, the step size sequence is chosen as $\alpha_t = 10^{-3}/(t+1)$ [3], and the gradient projection update is terminated when $||f(\Xi^{(k,t)}) - f(\Xi^{(k,t-1)})||_F \leq 10^{-3}$. Each result in this subsection is obtained by averaging 100 Monte Carlo runs.

Figure 6.5 presents the convergence performances of **Algorithm 8** under different SNRs and different test data, where the mean square error (MSE) $\| [\![\hat{\Xi}^{(1,s)}, \hat{\Xi}^{(2,s)}, \hat{\Xi}^{(3,s)}]\!] - \mathcal{X} \|_F^2$ is chosen as the assessment criterion. From Fig. 6.5a, it can be seen that for test data $\mathcal{Y}$, the MSEs of **Algorithm 8**, which assumes no knowledge of tensor rank, decrease significantly in the first few iterations and converge to the MSE of the NALS algorithm [17] (with exact tensor rank) under SNR = 10 dB and SNR = 20 dB. Similar convergence performances can be observed for the test data $\mathcal{Y}^+$ in Fig. 6.5b. This is of no surprise because approximating (6.18) by (6.17) does not make any changes in the algorithm framework of variational inference, and thus, the excellent convergence performance of **Algorithm 8** is expected.

The MSE in Fig. 6.5 measures the performance of low-rank tensor recovery as a whole but does not indicate the accuracy at factor matrices level. On the other hand, due to the uniqueness property of tensor CPD [18], each factor matrix can be recovered up to an unknown permutation and scaling ambiguity. To assess the accuracies of factor matrices recovery, the best congruence ratio (BCR), which involves

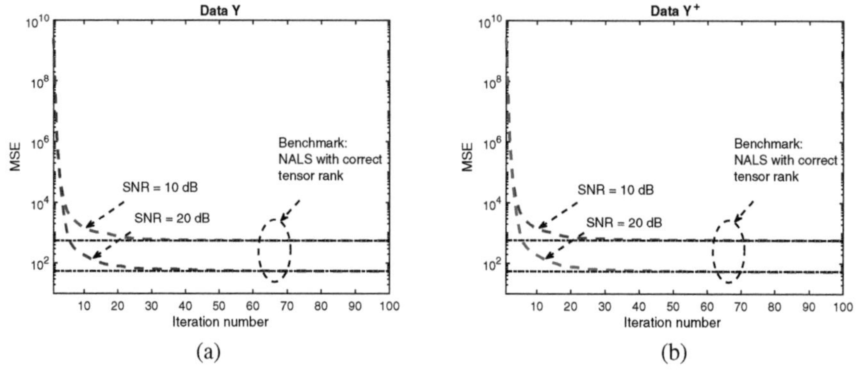

Fig. 6.5 Convergence of the proposed algorithm for different test data

computing the MSE between the true factor matrix $\boldsymbol{M}^{(k)}$ and the estimated factor matrix $\hat{\boldsymbol{\Xi}}^{(k)}$, is widely used as the assessment criterion. It is defined as

$$\sum_{k=1}^{3} \min_{\boldsymbol{\Delta}^{(k)}, \boldsymbol{P}^{(k)}} \frac{||\boldsymbol{M}^{(k)} - \hat{\boldsymbol{\Xi}}^{(k)} \boldsymbol{P}^{(k)} \boldsymbol{\Delta}^{(k)}||_F}{||\boldsymbol{M}^{(k)}||_F},$$

where the diagonal matrix $\boldsymbol{\Delta}^{(k)}$ and the permutation matrix $\boldsymbol{P}^{(k)}$ are found via the greedy least-squares column matching algorithm [19]. From Fig. 6.6, it is seen that both **Algorithm 8** (labeled as PNCPD) and the NALS algorithm (with exact tensor rank) achieve much better factor matrix recovery than the ALS algorithm (with exact tensor rank) [18]. This shows the importance of incorporating the nonnegative constraint into the algorithm design. Furthermore, the factor matrix recovery performances of **Algorithm 8** under test data $\mathcal{Y}$ and $\mathcal{Y}^+$ are indistinguishable under SNR $= 20$ dB. This shows that when SNR is high, Eq. (6.17) gives a quite good approximation to Eq. (6.18), thus leading to remarkably accurate factor matrices recovery. Although the BCR of **Algorithm 8** is higher for the data $\mathcal{Y}^+$ than that for the data $\mathcal{Y}$, it is with nearly the same performance as that of the NALS algorithm (with exact tensor rank).

On the other hand, the tensor rank estimates of **Algorithm 8** under different SNRs are presented in Fig. 6.7, with each vertical bar showing the mean and the error bars showing the standard derivation of tensor rank estimates. The blue horizontal dashed line indicates the true tensor rank. From Fig. 6.7, it is seen that **Algorithm 8** recovers the true tensor rank with 100% accuracy for a wide range of SNRs, in particular when SNR is 10 dB or higher. When SNR is 0 and 5 dB, even though the performance is not 100% accurate, the estimated tensor rank is still close to the true tensor rank with a high probability for test data $\mathcal{Y}$. However, under these two low SNRs, the rank estimation performances of the proposed algorithm for the data $\mathcal{Y}^+$ degrade significantly. This is because Eq. (6.18) cannot be well approximated by Eq. (6.17)

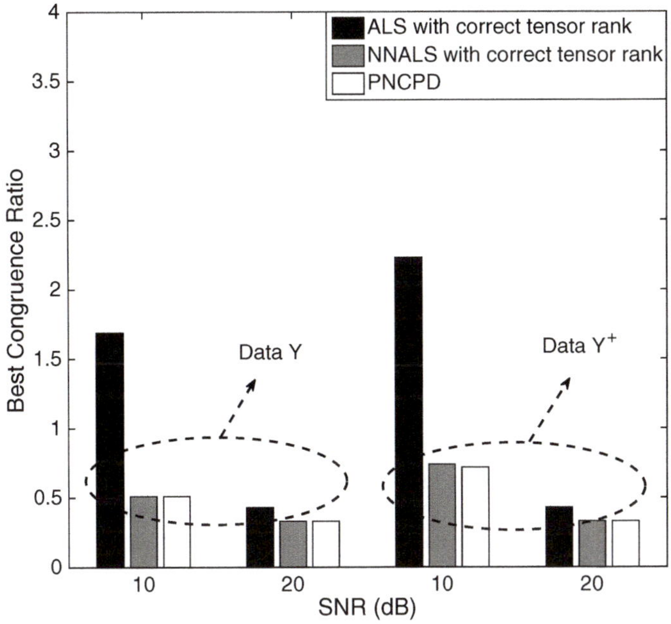

Fig. 6.6 Best congruence ratios of the proposed algorithm for different test data

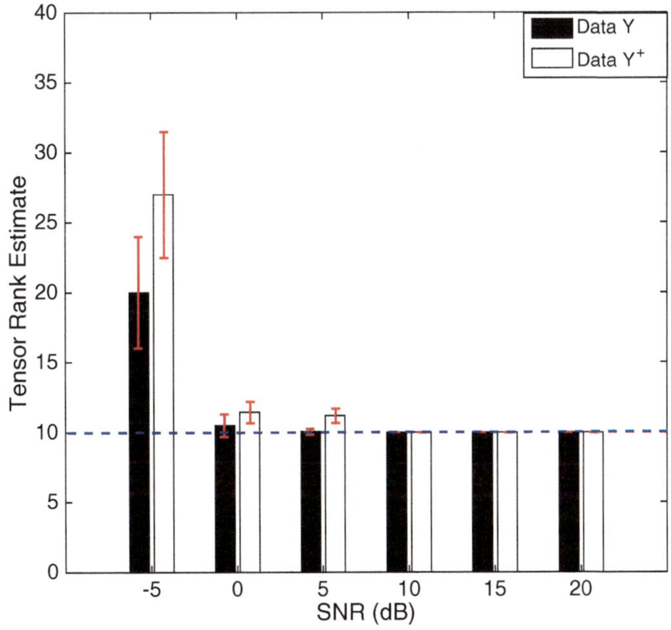

Fig. 6.7 Tensor rank estimates of the proposed algorithm for different test data

Table 6.1 Performance of tensor rank estimation versus different true tensor ranks for tensor data $\mathcal{Y}$ when SNR = 20 dB

True tensor rank	Mean of tensor rank estimates	Standard derivation of tensor rank estimates	Percentage of correct tensor rank estimates (%)
10	10	0	100
30	29.6	1.27	90
50	28.15	18.29	25

under very low SNRs. Furthermore, **Algorithm 8** fails to give correct rank estimates when SNR is lower than −5 dB for both two test data sets, since the noise with very large power masks the low-rank structure of the data.

To assess the tensor rank estimation accuracy when the tensor is with a larger true rank, we apply **Algorithm 8** to the tensor data $\mathcal{Y}$ with the true rank $R = \{10, 30, 50\}$ and SNR = 20 dB. The rank estimation performance is presented in Table 6.1. From Table 6.1, it can be seen that **Algorithm 8** recovers the rank accurately when the true rank is 10 and 30. However, when $R = 50$, **Algorithm 8** fails to accurately recover the tensor rank.

The simulation results presented so far are for well-conditioned tensors, i.e., the columns in each of the factor matrices are independently generated. In order to fully assess the rank learning ability of **Algorithm 8**, we consider another noise-free 3D tensor $\bar{\mathcal{X}} = [\![\bar{\boldsymbol{M}}^{(1)}, \boldsymbol{M}^{(2)}, \boldsymbol{M}^{(3)}]\!] \in \mathbb{R}^{100\times100\times100}$ with rank $R = 10$. The factor matrix is set as $\bar{\boldsymbol{M}}^{(1)} = 0.1\mathbf{1}_{100\times10} + 2^{-s}\boldsymbol{M}^{(1)}$, and each element in factor matrices $\{\boldsymbol{M}^{(n)}\}_{n=1}^{3}$ is independently drawn from a uniform distribution over [0, 1]. According to the definition of the tensor condition number [20, 21], when s increases, the correlation among the columns in the factor matrix $\bar{\boldsymbol{M}}^{(1)}$ increases, and the tensor condition number becomes larger. In particular, when s goes to infinity, the condition number of the tensor goes to infinity too. We apply the proposed algorithm to $\bar{\mathcal{X}}$ corrupted with noise: $\bar{\mathcal{Y}} = \bar{\mathcal{X}} + \bar{\mathcal{W}}$, where each element of noise tensor $\bar{\mathcal{W}}$ is independently drawn from a zero-mean Gaussian distribution with variance $\bar{\sigma}_w^2$. Table 6.2 shows the rank estimation accuracy of the proposed algorithm when SNR = 20 dB. It can be seen that the proposed algorithm can correctly estimate the tensor rank when $s < 5$. But as the tensor conditional number increases (i.e., the columns are more correlated in the factor matrix $\bar{\boldsymbol{M}}^{(1)}$), the tensor rank estimation performance decreases significantly.

Next, we consider an extreme case in which the columns in all factor matrices are highly correlated: $\tilde{\mathcal{X}} = [\![\bar{\boldsymbol{M}}^{(1)}, \bar{\boldsymbol{M}}^{(2)}, \bar{\boldsymbol{M}}^{(3)}]\!] \in \mathbb{R}^{100\times100\times100}$ with rank $R = 10$, where each factor matrix $\bar{\boldsymbol{M}}^{(n)} = 0.1\mathbf{1}_{100\times10} + 2^{-s}\boldsymbol{M}^{(n)}$, and each element in factor matrices $\{\boldsymbol{M}^{(n)}\}_{n=1}^{3}$ is independently drawn from a uniform distribution over [0, 1]. With the same observation data model as $\bar{\mathcal{Y}}$ as before and when SNR = 20 dB, the percentages of the correct tensor rank estimate are shown in Table 6.2. It can be seen

Table 6.2 Performance of tensor rank estimation when the columns are correlated and SNR = 20 dB

s	0	1	3	5	100
Columns in one factor matrix are correlated (%)	100	100	100	25	5
Columns in all factor matrices are correlated (%)	100	40	0	0	0

that it is difficult for **Algorithm 8** to accurately estimate the tensor rank even when $s = 1$.

Finally, acceleration schemes could be incorporated to speed up **Algorithm 8**. In particular, in the first few iterations, since some precision parameters are learned to be very large while some of them are with very small values, the average condition number of the Hessian matrix of problem (6.25), i.e., $\frac{1}{3}\sum_{k=1}^{3} \text{condition\_number}\left(\boldsymbol{H}^{(k)}\right)$, is very large. After several iterations, **Algorithm 8** has gradually recovered the tensor rank, and then the remaining precision parameters are with comparable values, leading to a well-conditioned Hessian matrix $\boldsymbol{H}^{(k)}$ of problem (6.25). This phenomenon can be observed in Fig. 6.8. When the Hessian matrix is well-conditioned (e.g., when the condition number is smaller than 100), the Nesterov scheme can be utilized for the problem (6.25) to speed up the convergence [4]. Consequently, even with the similar MSE and BCR performances, the accelerated algorithm (denoted as PNCPD-A) is much faster than **Algorithm 8**[2] as presented in Table 6.3. Moreover, from Table 6.3, the inexact BCD acceleration PNCPD-I (**Algorithm 9**) also achieves similar MSE and BCR, but with computation time further reduced.

6.5.2 Fluorescence Data Analysis

In this subsection, the application of Fluorescence Data Analysis from Chap. 4 is revisited. Since factor matrices A, B, and C in this application represent excitation values, emission values, and concentrations, respectively, strictly speaking, they should be nonnegative. **Algorithm 8** (PNCPD algorithm) is utilized to analyze the amino acids fluorescence data[3] [22] as in Sect. 4.2.1. The fluorescence excitation-emission measured (EEM) data collected is with size $5 \times 201 \times 61$ and should be representable with a CPD model with rank 3, since there are three different types of amino acid and each individual amino acid gives a rank-one CPD component. The samples were corrupted by Gaussian noise with SNR = 0 dB.

The PNCPD algorithm (**Algorithm 8**) is run to decompose the EEM tensor data with initial rank $L = 201$, for which, the step size sequence is chosen as $\alpha_t = 10^{-2}/(t + 1)$ [3], and the gradient projection update is terminated when the

[2] The presented time for the accelerated scheme includes the time for computing the condition numbers.

[3] http://www.models.life.ku.dk.

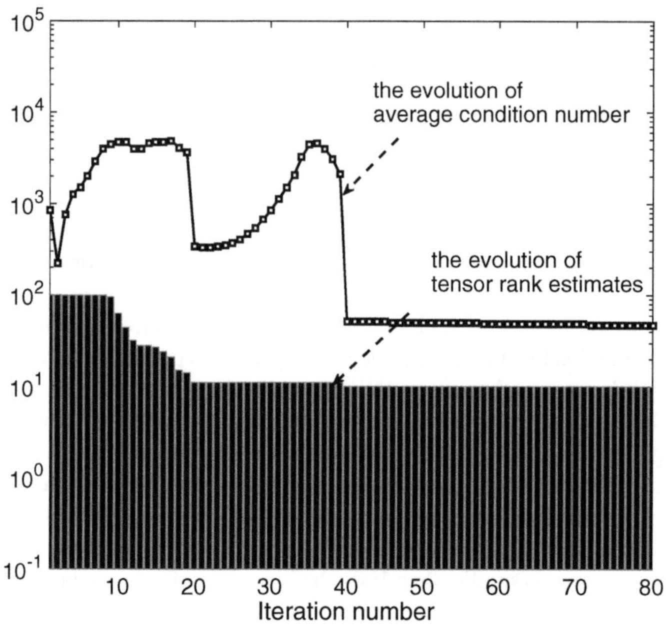

Fig. 6.8 The average condition number of the Hessian matrix of problem (6.25) and the tensor rank estimate versus number of iterations for test data $\mathcal{Y}$ when SNR = 20 dB

Table 6.3 Comparisions between different algorithm accelerations. The accelerated version using the Nesterov scheme is labeled as PNCPD-A. The inexact BCD version is labeled as PNCPD-I

	SNR = 10 dB					
	$\mathcal{Y}$			$\mathcal{Y}^+$		
	PNCPD	PNCPD-A	PNCPD-I	PNCPD	PNCPD-A	PNCPD-I
MSE	545.5608	545.7800	545.8255	576.7381	576.7603	576.7990
BCR	0.4798	0.4886	0.4892	0.7616	0.7673	0.7680
Running time	134.9246	74.8712	**74.0267**	119.2018	94.1182	**84.7767**
	SNR = 20 dB					
	$\mathcal{Y}$			$\mathcal{Y}^+$		
	PNCPD	PNCPD-A	PNCPD-I	PNCPD	PNCPD-A	PNCPD-I
MSE	54.8459	54.8485	54.8448	55.7471	55.7484	56.3062
BCR	0.2398	0.2413	0.2412	0.2438	0.2450	0.2453
Running time	80.0258	61.9218	**44.3579**	90.7224	82.7590	**47.7782**

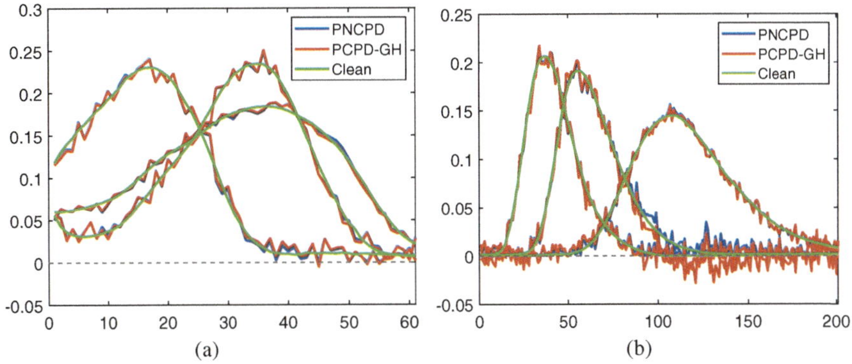

Fig. 6.9 The estimates of **a** excitation spectra and **b** emission spectra recovered by PNCPD and PCPD-GH algorithms, with the clean data spectra serving as the benchmark

gradient norm is smaller than 10^{-3}. At convergence, the PNCPD algorithm identified the correct tensor rank $R = 3$ and the corresponding Fit value is 91.5582, which is higher than that of PCPD-GH (**Algorithm 4**) without nonnegative constraint (90.8433 in Table 4.3). Furthermore, the emission spectra and the excited spectra of three amino acids recovered by PNCPD and PCPD-GH algorithms are plotted in Fig. 6.9, which are obtained from the decomposed factor matrices [22], with the clean data spectra[4] serving as the benchmark. From Fig. 6.9, it can be seen that some excitation/emission values recovered by PCPD-GH are negative under the noise corruption, while that of PNCPD are strictly nonnegative, which explains why PNCPD yields a higher Fit value and more interpretable results.

6.5.3 *ENRON E-mail Data Mining*

In this subsection, **Algorithm 8** is demonstrated in the social group clustering (see Sect. 6.1.1). In particular, the ENRON Email corpus[5] [1] was analyzed. This dataset $\mathcal{Y}$ is with size $184 \times 184 \times 44$ (i.e., $I = 184$, $J = 184$, $K = 44$ in the notation of Sect. 6.1.1) and contains e-mail communication records between 184 people within 44 months, in which each entry denotes the number of e-mails exchanged between two particular people within a particular month. Before fitting the data to the proposed algorithms, the same pre-processing as in [1, 2] is applied to the non-zero data to compress the dynamic range. Then, **Algorithm 8** is utilized to fit the data, with the initial rank set as $L = 44$, the step size sequence being $\alpha_t = 1/(t+1)$ [3], and the

[4] The clean data spectra are obtained by decomposing the clean data using the NALS algorithm with correct tensor rank $R = 3$.

[5] The original source of the data is from [1], and we greatly appreciate Prof. Vagelis Papalexakis for sharing the data with us.

Table 6.4 Social groups with people in the top 10 scores in each group for the ENRON e-mail data using the PNCPD algorithm

Legal	Pipeline
Tana Jones (tana.jones) Employee Financial Trading Group ENA Legal	Michelle Lokay (michelle.lokay) Admin. Asst. Transwestern Pipeline Company (ETS)
Sara Shackleton (sara.shackleton) Employee ENA Legal	Kimberly Watson (kimberly.watson) Employee Transwestern Pipeline Company (ETS)
Mark Taylor (mark.taylor) Manager Financial Trading Group ENA Legal	Lynn Blair (lynn.blair) Employee Northern Natural Gas Pipeline (ETS)
Stephanie Panus (stephanie.panus) Senior Legal Specialist ENA Legal	Shelley Corman (shelley.corman) VP Regulatory Affairs
Marie Heard (marie.heard) Senior Legal Specialist ENA Legal	Drew Fossum (drew.fossum) VP Transwestern Pipeline Company (ETS)
Mark Haedicke (mark.haedicke) Managing Director ENA Legal	Lindy Donoho (lindy.donoho) Employee Transwestern Pipeline Company (ETS)
Susan Bailey (susan.bailey) Legal Assistant ENA Legal	Kevin Hyatt (kevin.hyatt) Director Asset Development TW Pipeline Business (ETS)
Louise Kitchen (louise.kitchen) President Enron Online	Darrell Schoolcraft (darrell.schoolcraft) Employee Gas Control (ETS)
Kay Mann (kay.mann) Lawyer	Rod Hayslett (rod.hayslett) VP Also CFO and Treasurer
Debra Perlingiere (debra.perlingiere) Legal Specialist ENA Legal	Susan Scott (susan.scott) Employee Transwestern Pipeline Company (ETS)
Trading/executive	**Government affairs**
Michael Grigsby (mike.grigsby) Director West Desk Gas Trading	Jeff Dasovich (jeff.dasovich) Employee Government Relationship Executive
Kevin Presto (m..presto) VP East Power Trading	James Steffes (james.steffes) VP Government Affairs
Mike McConnell (mike.mcconnell) Executive VP Global Markets	Steven Kean (steven.kean) VP Chief of Staff
John Arnold (john.arnold) VP Financial Enron Online	Richard Shapiro (richard.shapiro) VP Regulatory Affairs
Louise Kitchen (louise.kitchen) President Enron Online	David Delainey (david.delainey) CEO ENA and Enron Energy Services
David Delainey (david.delainey) CEO ENA and Enron Energy Services	Richard Sanders (richard.sanders) VP Enron Wholesale Services
John Lavorato (john.lavorato) CEO Enron America	Shelley Corman (shelley.corman) VP Regulatory Affairs
Sally Beck (sally.beck) COO	Margaret Carson (margaret.carson) Employee Corporate and Environmental Policy
Joannie Williamson (joannie.williamson) Executive Assistant	Mark Haedicke (mark.haedicke) Managing Director ENA Legal
Liz Taylor (liz.taylor) Executive Assistant to Greg Whalley	Vince Kaminski (vince.kaminski) Manager Risk Management Head

Table 6.5 Social groups with people in the top 10 scores in each group for the ENRON e-mail data using the PCPD-GH algorithm

Legal	Pipeline
Tana Jones (tana.jones) Employee Financial Trading Group ENA Legal Sara Shackleton (sara.shackleton) Employee ENA Legal Mark Taylor (mark.taylor) Manager Financial Trading Group ENA Legal Stephanie Panus (stephanie.panus) Senior Legal Specialist ENA Legal Marie Heard (marie.heard) Senior Legal Specialist ENA Legal Mark Haedicke (mark.haedicke) Managing Director ENA Legal Susan Bailey (susan.bailey) Legal Assistant ENA Legal Louise Kitchen (louise.kitchen) President Enron Online Kay Mann (kay.mann) Lawyer Debra Perlingiere (debra.perlingiere) Legal Specialist ENA Legal	Drew Fossum (drew.fossum) VP Transwestern Pipeline Company (ETS) Susan Scott (susan.scott) Employee Transwestern Pipeline Company (ETS) Shelley Corman (shelley.corman) VP Regulatory Affairs Michelle Lokay (michelle.lokay) Admin. Asst. Transwestern Pipeline Company (ETS) Lindy Donoho (lindy.donoho) Employee Transwestern Pipeline Company (ETS) Jeff Dasovich (jeff.dasovich) Employee Government Relationship Executive Kevin Hyatt (kevin.hyatt) Director Asset Development TW Pipeline Business (ETS) Kimberly Watson (kimberly.watson) Employee Transwestern Pipeline Company (ETS) Lynn Blair (lynn.blair) Employee Northern Natural Gas Pipeline (ETS) Rod Hayslett (rod.hayslett) VP Also CFO and Treasurer
Executives	Goverment affairs
John Lavorato (john.lavorato) CEO Enron America Louise Kitchen (louise.kitchen) President Enron Online Sally Beck (sally.beck) COO Liz Taylor (liz.taylor) Executive Assistant to Greg Whalley David Delainey (david.delainey) CEO ENA and Enron Energy Services Vince Kaminski (vince.kaminski) Manager Risk Management Head John Arnold (john.arnold) VP Financial Enron Online Joannie Williamson (joannie.williamson) Executive Assistant Mike McConnell (mike.mcconnell) Executive VP* Global Markets Jeffrey Shankman (jeffrey.shankman) President Enron Global Markets	Jeff Dasovich (jeff.dasovich) Employee Government Relationship Executive James Steffes (james.steffes) VP Government Affairs Steven Kean (steven.kean) VP Chief of Staff Richard Shapiro (richard.shapiro) VP Regulatory Affairs Richard Sanders (richard.sanders) VP Enron Wholesale Services David Delainey (david.delainey) CEO ENA and Enron Energy Services Shelley Corman (shelley.corman) VP Regulatory Affairs Margaret Carson (margaret.carson) Employee Corporate and Environmental Policy Mark Haedicke (mark.haedicke) Managing Director ENA Legal **Philip Allen (phillip.allen) VP West Desk Gas Trading**
Trading	
Michael Grigsby (mike.grigsby) Director West Desk Gas Trading Matthew Lenhart (matthew.lenhart) Analyst West Desk Gas Trading Monique Sanchez (monique.sanchez) Associate West Desk Gas Trader (EWS) Jane Tholt (m..tholt) VP West Desk Gas Trading Philip Allen (phillip.allen) VP West Desk Gas Trading Kam Keiser (kam.keiser) Employee Gas Mark Whitt (mark.whitt) Director Marketing Eric Bass (eric.bass) Trader Texas Desk Gas Trading **Jeff Dasovich (jeff.dasovich) Employee Government Relationship Executive** Chris Dorland (chris.dorland) Manager	

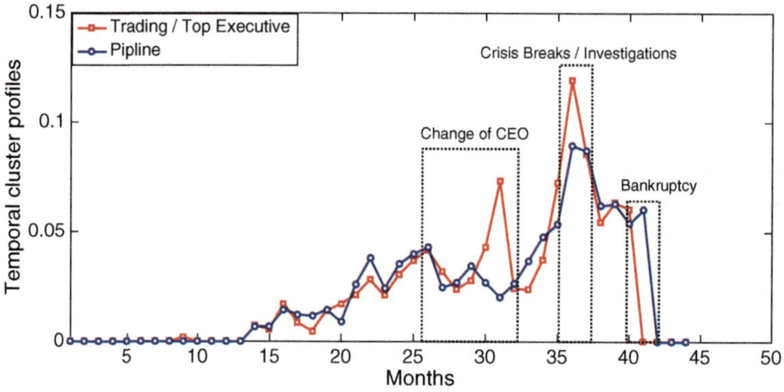

Fig. 6.10 Temporal cluster profiles (from the third factor matrix) for the ENRON e-mail dataset

gradient projection update terminated when the gradient norm is smaller than 10^{-3}. The estimated tensor rank has the physical meaning of the number of underlying social groups, and each element in the first factor matrix $\mathbf{A}$ can be interpreted as the score that a particular person belongs to a particular email sending group.

After executing PNCPD (**Algorithm 8**), the tensor rank estimate is found to be 4, indicating that there are four underlying social groups. This is consistent with the results from [1, 2], which are obtained via trail-and-error experiments. From the factor matrix $\mathbf{A}$, the people with top 10 scores in each social group are shown in Table 6.4. From the information of each person in Table 6.4, the clustering results can be clearly interpreted. For instance, the people in the first group are either in the legal department or lawyers, thus being clustered together. The clustering results of PNCPD are also consistent with the results from [1, 2], which are obtained via nonlinear programming methods assuming the knowledge of tensor rank. In contrast, the PCPD-GH algorithm (**Algorithm 4**) without nonnegative constraint identifies five social groups, where the detailed groupings are presented in Table 6.5. While the number of groups can also be considered reasonable [2], after a closer inspection, two people, namely Philip Allen and Jeff Dasovich (marked bold in Table 6.5) were put in the wrong groups. This demonstrates that the prior nonnegative structure is beneficial for social group clustering.

Finally, interesting patterns can be observed from the temporal cluster profiles, which are obtained from the third factor matrix $\mathbf{C}$ [1, 2], as illustrated in Fig. 6.10. It is clear that when the company has important events, such as the change of CEO, crisis breaks, and bankruptcy, distinct peaks appear.

References

1. B.W. Bader, R.A. Harshman, T.G. Kolda, Temporal analysis of social networks using three-way dedicom. Sandia National Laboratories (SNL), Albuquerque, NM, and Livermore, CA ..., Technical Report (2006)
2. E.E. Papalexakis, N.D. Sidiropoulos, R. Bro, From k-means to higher-way co-clustering: multilinear decomposition with sparse latent factors. IEEE Trans. Signal Process. **61**(2), 493–506 (2012)
3. D.P. Bertsekas, Nonlinear programming. J. Oper. Res. Soc. **48**(3), 334–334 (1997)
4. A.P. Liavas, G. Kostoulas, G. Lourakis, K. Huang, N.D. Sidiropoulos, Nesterov-based alternating optimization for nonnegative tensor factorization: algorithm and parallel implementation. IEEE Trans. Signal Process. **66**(4), 944–953 (2017)
5. K. Huang, N.D. Sidiropoulos, A.P. Liavas, A flexible and efficient algorithmic framework for constrained matrix and tensor factorization. IEEE Trans. Signal Process. **64**(19), 5052–5065 (2016)
6. Q. Zhao, L. Zhang, A. Cichocki, Bayesian cp factorization of incomplete tensors with automatic rank determination. IEEE Trans. Pattern Anal. Mach. Intell. **37**(9), 1751–1763 (2015)
7. L. Cheng, Y.-C. Wu, H.V. Poor, Probabilistic tensor canonical polyadic decomposition with orthogonal factors. IEEE Trans. Signal Process. **65**(3), 663–676 (2016)
8. L. Cheng, Y.-C. Wu, H.V. Poor, Scaling probabilistic tensor canonical polyadic decomposition to massive data. IEEE Trans. Signal Process. **66**(21), 5534–5548 (2018)
9. L. Cheng, X. Tong, S. Wang, Y.-C. Wu, H.V. Poor, Learning nonnegative factors from tensor data: probabilistic modeling and inference algorithm. IEEE Trans. Signal Process. **68**, 1792–1806 (2020)
10. J.C. Arismendi, Multivariate truncated moments. J. Multivar. Anal. **117**, 41–75 (2013)
11. M.D. Hoffman, D.M. Blei, C. Wang, J. Paisley, Stochastic variational inference. J. Mach. Learn. Res. (2013)
12. Y. Xu, W. Yin, A block coordinate descent method for regularized multiconvex optimization with applications to nonnegative tensor factorization and completion. SIAM J. Imag. Sci. **6**(3), 1758–1789 (2013)
13. L. Grippo, M. Sciandrone, On the convergence of the block nonlinear gauss-seidel method under convex constraints. Oper. Res. Lett. **26**(3), 127–136 (2000)
14. H.-J.M. Shi, S. Tu, Y. Xu, W. Yin, A primer on coordinate descent algorithms (2016). arXiv:1610.00040
15. M. Razaviyayn, M. Hong, Z.-Q. Luo, A unified convergence analysis of block successive minimization methods for nonsmooth optimization. SIAM J. Optim. **23**(2), 1126–1153 (2013)
16. P. Tseng, S. Yun, A coordinate gradient descent method for nonsmooth separable minimization. Math. Programm. **117**(1), 387–423 (2009)
17. A. Cichocki, R. Zdunek, A.H. Phan, S.-I. Amari, *Nonnegative Matrix and Tensor Factorizations: Applications to Exploratory Multi-way Data Analysis and Blind Source Separation* (Wiley, New York, 2009)
18. T.G. Kolda, B.W. Bader, Tensor decompositions and applications. SIAM Rev. **51**(3), 455–500 (2009)
19. N.D. Sidiropoulos, G.B. Giannakis, R. Bro, Blind parafac receivers for ds-cdma systems. IEEE Trans. Signal Process. **48**(3), 810–823 (2000)
20. P. Breiding, N. Vannieuwenhoven, The condition number of join decompositions. SIAM J. Matrix Anal. Appl. **39**(1), 287–309 (2018)
21. N. Vannieuwenhoven, Condition numbers for the tensor rank decomposition. Linear Algebra Appl. **535**, 35–86 (2017)
22. H.A. Kiers, A three-step algorithm for candecomp/parafac analysis of large data sets with multicollinearity. J. Chemom.: J. Chemom. Soc. **12**(3), 155–171 (1998)

References

Chapter 7
Complex-Valued CPD, Orthogonality Constraint, and Beyond Gaussian Noises

Abstract In previous chapters, Bayesian CPDs are developed for real-valued tensor data. They cannot deal with complex-valued tensor data, which, however, frequently occurs in applications including wireless communications and sensor array signal processing. In addition, we have not touched on the design of Bayesian CPD that incorporates the orthogonality structure and/or handles non-Gaussian noises. In this chapter, we present a unified Bayesian modeling and inference for complex-valued tensor CPD with/without orthogonal factors, under/not under Gaussian noises. Applications on blind receiver design and linear image coding are presented.

7.1 Problem Formulation

We consider a generalized problem in which the complex-valued tensor $\mathcal{Y} \in \mathbb{C}^{I_1 \times I_2 \times \cdots \times I_N}$ obeys the following model:

$$\mathcal{Y} = [\![\mathbf{A}^{(1)}, \mathbf{A}^{(2)}, \ldots, \mathbf{A}^{(N)}]\!] + \mathcal{W} + \mathcal{E}, \tag{7.1}$$

where $\mathcal{W}$ represents an additive noise tensor with each element being i.i.d. and $w_{i_1,i_2,\ldots,i_N} \sim \mathcal{CN}\left(w_{i_1,i_2,\ldots,i_N} | 0, \beta^{-1}\right)$; $\mathcal{E}$ denotes potential outliers in measurements with each element $e_{i_1,i_2,\ldots,i_N}$ taking an unknown complex value if an outlier emerges, and otherwise taking the value zero. Furthermore, it is assumed that $\{\mathbf{A}^{(n)}\}_{n=1}^{P}$ are known to be orthogonal where $P < N$, while the remaining factor matrices are unconstrained.

Due to the orthogonality structure of the first P factor matrices $\{\mathbf{A}^{(n)}\}_{n=1}^{P}$, they can be written as $\mathbf{A}^{(n)} = \mathbf{U}^{(n)} \mathbf{\Lambda}^{(n)}$ where $\mathbf{U}^{(n)}$ is an orthonormal matrix and $\mathbf{\Lambda}^{(n)}$ is a diagonal matrix describing the scale of each column. Putting $\mathbf{A}^{(n)} = \mathbf{U}^{(n)} \mathbf{\Lambda}^{(n)}$ for $1 \le n \le P$ into the definition of the tensor CPD, it is easy to show that

$$[\![\mathbf{A}^{(1)}, \mathbf{A}^{(2)}, \ldots, \mathbf{A}^{(N)}]\!] = [\![\mathbf{\Xi}^{(1)}, \mathbf{\Xi}^{(2)}, \ldots, \mathbf{\Xi}^{(N)}]\!] \tag{7.2}$$

with $\mathbf{\Xi}^{(n)} = \mathbf{U}^{(n)} \mathbf{\Pi}$ for $1 \le n \le P$, $\mathbf{\Xi}^{(n)} = \mathbf{A}^{(n)} \mathbf{\Pi}$ for $P+1 \le n \le N-1$, and $\mathbf{\Xi}^{(N)} = \mathbf{A}^{(N)} \mathbf{\Lambda}^{(1)} \mathbf{\Lambda}^{(2)} \cdots \mathbf{\Lambda}^{(P)} \mathbf{\Pi}$, where $\mathbf{\Pi} \in \mathbb{R}^{R \times R}$ is a permutation matrix. From

L. Cheng et al., *Bayesian Tensor Decomposition for Signal Processing and Machine Learning*, https://doi.org/10.1007/978-3-031-22438-6_7

(7.2), it can be seen that up to a scaling and permutation indeterminacy, the tensor CPD under orthogonal constraints is equivalent to that under orthonormal constraints. In general, the scaling and permutation ambiguity can be easily resolved using side information [1, 2]. Thus, without loss of generality, our goal is to estimate an N-tuplet of factor matrices $(\boldsymbol{\Xi}^{(1)}, \boldsymbol{\Xi}^{(2)}, \ldots, \boldsymbol{\Xi}^{(N)})$ with the first P (where $P < N$) of them being orthonormal, based on the observation $\mathcal{Y}$ and in the absence of the knowledge of noise power β^{-1}, outlier statistics, and the tensor rank R.

In particular, since we do not know the exact value of R, it is assumed that there are L columns in each factor matrix $\boldsymbol{\Xi}^{(n)}$, where L is the maximum possible value of the tensor rank R. Then, the problem to be solved can be stated as

$$\min_{\{\boldsymbol{\Xi}^{(n)}\}_{n=1}^{N}, \mathcal{E}} \beta \parallel \mathcal{Y} - [\![\boldsymbol{\Xi}^{(1)}, \boldsymbol{\Xi}^{(2)}, \ldots, \boldsymbol{\Xi}^{(N)}]\!] - \mathcal{E} \parallel_F^2 + \sum_{l=1}^{L} \gamma_l \Big(\sum_{n=1}^{N} \boldsymbol{\Xi}_{:,l}^{(n)H} \boldsymbol{\Xi}_{:,l}^{(n)} \Big)$$
$$\text{s.t.} \quad \boldsymbol{\Xi}^{(n)H} \boldsymbol{\Xi}^{(n)} = \mathbf{I}_L, \quad n = 1, 2, \ldots, P, \tag{7.3}$$

where the regularization term $\sum_{l=1}^{L} \gamma_l \left(\sum_{n=1}^{N} \boldsymbol{\Xi}_{:,l}^{(n)H} \boldsymbol{\Xi}_{:,l}^{(n)} \right)$ is added to control the complexity of the model and avoid overfitting of noise, since more columns (thus more degrees of freedom) in $\boldsymbol{\Xi}^{(n)}$ than the true model are introduced, and $\{\gamma_l\}_{l=1}^{L}$ are regularization parameters trading off the relative importance of the square error term and the regularization term.

Problem (7.3) is more general than the model in existing works [3] since an extra outlier term is present. Furthermore, the choice of regularization parameters plays an important role, since setting γ_l too large results in excessive residual squared error, while setting γ_l too small risks overfitting of noise. In general, determining the optimal regularization parameters (e.g., using cross-validation, or the L-curve) requires exhaustive search and thus is computationally demanding. To overcome these problems, a novel algorithm based on the framework of probabilistic inference is presented, which effectively mitigates the outliers $\mathcal{E}$ and automatically learns the regularization parameters and the tensor rank.

7.2 Probabilistic Modeling

Before solving problem (7.3), we first interpret different terms in (7.3) as probability density functions, based on which a probabilistic model that encodes our knowledge of the observation and the unknowns can be established.

Firstly, since the elements of the additive noise $\mathcal{W}$ is white, zero-mean, and circularly symmetric complex Gaussian, the squared error term in problem (7.3) can be interpreted as the negative log of the likelihood

$$p\left(\mathcal{Y} \mid \boldsymbol{\Xi}^{(1)}, \boldsymbol{\Xi}^{(2)}, \ldots, \boldsymbol{\Xi}^{(N)}, \mathcal{E}, \beta\right) \propto \exp\left(-\beta \parallel \mathcal{Y} - [\![\boldsymbol{\Xi}^{(1)}, \boldsymbol{\Xi}^{(2)}, \ldots, \boldsymbol{\Xi}^{(N)}]\!] - \mathcal{E} \parallel_F^2\right). \tag{7.4}$$

Secondly, the regularization term in problem (7.3) can be interpreted as arising from a circularly symmetric complex Gaussian prior distribution over the columns of the factor matrices, i.e., $\prod_{n=1}^{N}\prod_{l=1}^{L}\mathcal{CN}\Big(\Xi_{:,l}^{(n)}\,|\,\mathbf{0}_{I_n\times 1},\gamma_l^{-1}\mathbf{I}_L\Big)$. Note that this is analogue to (3.4) but is tailored to complex-valued factor matrices.

On the other hand, for the first P factor matrices $\{\Xi^{(n)}\}_{n=1}^{P}$, there are additional hard constraints in problem (7.3), which correspond to the Stiefel manifold [6] $\mathcal{V}_L(\mathbb{C}^{I_n}) = \{\mathbf{A}\in\mathbb{C}^{I_n\times L} : \mathbf{A}^H\mathbf{A} = \mathbf{I}_L\}$ for $1\le n\le P$. Since the orthonormal constraints result in $\Xi_{:,l}^{(n)H}\Xi_{:,l}^{(n)} = 1$, the hard constraints would dominate the Gaussian distribution of the columns in $\{\Xi^{(n)}\}_{n=1}^{P}$. Therefore, $\Xi^{(n)}$ can be interpreted as being uniformly distributed over the Stiefel manifold $\mathcal{V}_L(\mathbb{C}^{I_n})$ for $1\le n\le P$, and Gaussian distributed for $P+1\le n\le N$:

$$
\begin{aligned}
p(\Xi^{(1)},\Xi^{(2)},\ldots,\Xi^{(P)}) &\propto \prod_{n=1}^{P}\mathbb{I}_{\mathcal{V}_L(\mathbb{C}^{I_n})}(\Xi^{(n)}),\\
p(\Xi^{(P+1)},\Xi^{(P+2)},\ldots,\Xi^{(N)}) &= \prod_{n=P+1}^{N}\prod_{l=1}^{L}\mathcal{CN}\Big(\Xi_{:,l}^{(n)}|\mathbf{0}_{I_n\times 1},\gamma_l^{-1}\mathbf{I}_L\Big),
\end{aligned}
\tag{7.5}
$$

where $\mathbb{I}_{\mathcal{V}_L(\mathbb{C}^{I_n})}(\Xi^{(n)})$ is an indicator function with $\mathbb{I}_{\mathcal{V}_L(\mathbb{C}^{I_n})}(\Xi^{(n)}) = 1$ when $\Xi^{(n)}\in\mathcal{V}_L(\mathbb{C}^{I_n})$, and otherwise $\mathbb{I}_{\mathcal{V}_L(\mathbb{C}^{I_n})}(\Xi^{(n)}) = 0$. For the parameters β and $\{\gamma_l\}_{l=1}^{L}$, which correspond to the inverse noise power and the variances of columns in the factor matrices, since we have no information about their distributions, a non-informative Jeffrey's prior is imposed on them, i.e., $p(\beta)\propto\beta^{-1}$ and $p(\gamma_l)\propto\gamma_l^{-1}$ for $l = 1,\ldots,L$. This is equivalent to imposing gamma distribution on β and $\{\gamma_l\}_{l=1}^{L}$ but with their hyper-parameters approaching zero.

Finally, although the generative model for outliers $\mathcal{E}_{i_1,\ldots,i_N}$ is unknown, the rare occurrence of outliers motivates us to employ a complex-valued student's t distribution as its prior, i.e., $p(\mathcal{E}_{i_1,\ldots,i_N}) = \mathcal{T}(\mathcal{E}_{i_1,\ldots,i_N}|0, c_{i_1,\ldots,i_N}, d_{i_1,\ldots,i_N})$. Similar to the real-valued student's t distribution (see Table 2.1), complex-valued student's t distribution can also be equivalently represented as a Gaussian scale mixture [7]:

$$
\begin{aligned}
&\mathcal{T}\left(\mathcal{E}_{i_1,\ldots,i_N}\mid 0, c_{i_1,\ldots,i_N}, d_{i_1,\ldots,i_N}\right)\\
&= \int \mathcal{CN}\left(\mathcal{E}_{i_1,\ldots,i_N}\mid 0,\zeta_{i_1,\ldots,i_N}^{-1}\right)\text{gamma}\left(\zeta_{i_1,\ldots,i_N}\mid c_{i_1,\ldots,i_N}, d_{i_1,\ldots,i_N}\right)\,\mathrm{d}\zeta_{i_1,\ldots,i_N}.
\end{aligned}
\tag{7.6}
$$

This means that student's t distribution can be obtained by mixing an infinite number of zero-mean circularly symmetric complex Gaussian distributions where the mixing distribution on the precision $\zeta_{i_1,\ldots,i_N}$ is the gamma distribution with parameters $c_{i_1,\ldots,i_N}$ and $d_{i_1,\ldots,i_N}$. In addition, since the statistics of outliers such as means and correlations are generally unavailable in practice, we set the hyper-parameters $c_{i_1,\ldots,i_N}$ and $d_{i_1,\ldots,i_N}$

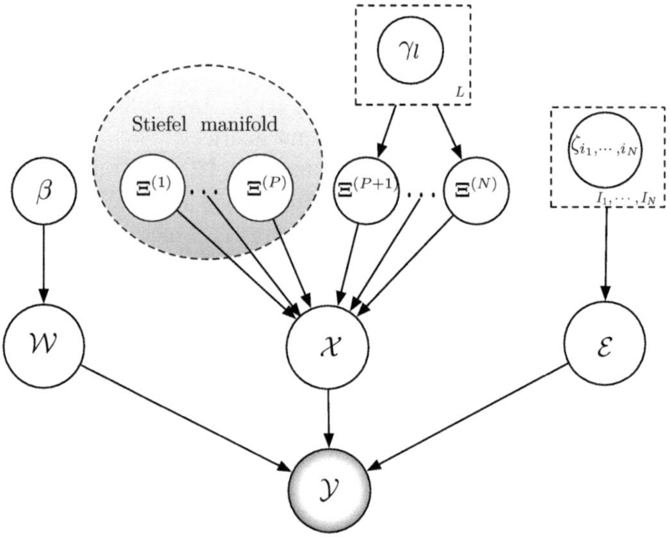

Fig. 7.1 Probabilistic model for complex-valued tensor CPD with orthogonal factors and outliers (© [2017] IEEE. Reprinted, with permission, from [L. Cheng, Y.-C. Wu, and H. V. Poor, Probabilistic Tensor Canonical Polyadic Decomposition With Orthogonal Factors, IEEE Transactions on Signal Processing, Feb 2017]. It applies all figures and tables in this chapter)

as 10^{-6} to produce a non-informative prior on $\mathcal{E}_{i_1,\ldots,i_N}$, and assume outliers are independent of each other:

$$p(\mathcal{E}) = \prod_{i_1=1}^{I_1} \cdots \prod_{i_N=1}^{I_N} \mathcal{T}\big(\mathcal{E}_{i_1,\ldots,i_N} \mid 0, c_{i_1,\ldots,i_N} = 10^{-6}, d_{i_1,\ldots,i_N} = 10^{-6}\big). \tag{7.7}$$

The complete probabilistic model is shown in Fig. 7.1.

7.3 Inference Algorithm Development

Let $\boldsymbol{\Theta}$ be a set containing the factor matrices $\{\boldsymbol{\Xi}^{(n)}\}_{n=1}^{N}$, and other variables $\mathcal{E}$, $\{\gamma_l\}_{l=1}^{L}$, $\{\zeta_{i_1,\ldots,i_N}\}_{i_1=1,\ldots,i_N=1}^{I_1,\ldots,I_N}$, β. From the probabilistic model established above, the marginal probability density functions of the unknown factor matrices $\{\boldsymbol{\Xi}\}_{n=1}^{N}$ are given by

$$p(\boldsymbol{\Xi}^{(n)}|\mathcal{Y}) = \int \frac{p(\mathcal{Y}, \boldsymbol{\Theta})}{p(\mathcal{Y})} d\boldsymbol{\Theta} \backslash \boldsymbol{\Xi}^{(n)}, \; n = 1, 2, \ldots, N, \tag{7.8}$$

where

$$
\begin{aligned}
p(\mathcal{Y}, \boldsymbol{\Theta}) \propto \prod_{n=1}^{P} \mathbb{I}_{V_L(\mathbb{C}^{I_n})}\left(\boldsymbol{\Xi}^{(n)}\right) \exp\Bigg\{ & \left(\prod_{n=1}^{N} I_n - 1\right) \ln \beta \\
& + \left(\sum_{n=P+1}^{N} I_n + 1\right) \sum_{l=1}^{L} \ln \gamma_l - \operatorname{Tr}\left(\boldsymbol{\Gamma} \sum_{n=P+1}^{N} \boldsymbol{\Xi}^{(n)H} \boldsymbol{\Xi}^{(n)}\right) \\
& + \sum_{i_1=1}^{I_1} \cdots \sum_{i_N=1}^{I_N} \left[(c_{i_1,\ldots,i_N} - 1) \ln \zeta_{i_1,\ldots,i_N} - d_{i_1,\ldots,i_N} \zeta_{i_1,\ldots,i_N}\right] \\
& + \sum_{i_1=1}^{I_1} \cdots \sum_{i_N=1}^{I_N} \left(\ln \zeta_{i_1,\ldots,i_N} - \zeta_{i_1,\ldots,i_N} \mathcal{E}^*_{i_1,\ldots,i_N} \mathcal{E}_{i_1,\ldots,i_N}\right) \\
& - \beta \parallel \mathcal{Y} - [\![\boldsymbol{\Xi}^{(1)}, \boldsymbol{\Xi}^{(2)}, \ldots, \boldsymbol{\Xi}^{(N)}]\!] - \mathcal{E} \parallel_F^2 \Bigg\}
\end{aligned}
\tag{7.9}
$$

with $\boldsymbol{\Gamma} = \operatorname{diag}\{\gamma_1, \ldots, \gamma_R\}$.

Since the factor matrices and other variables are nonlinearly coupled in (7.9), the multiple integrations in (7.8) are analytically intractable, which prohibits exact Bayesian inference. To handle this problem, we use variational inference as in previous chapters.

However, in addition to the mean-field approximation $Q(\boldsymbol{\Theta}) = \prod_k Q(\Theta_k)$ where $\Theta_k \in \boldsymbol{\Theta}$, to facilitate the manipulation of hard constraints on the first P factor matrices, their variational densities are further assumed to take a Dirac delta functional form $Q(\boldsymbol{\Xi}^{(k)}) = \delta(\boldsymbol{\Xi}^{(k)} - \hat{\boldsymbol{\Xi}}^{(k)})$ for $k = 1, 2, \ldots, P$, where $\hat{\boldsymbol{\Xi}}^{(k)}$ is a parameter to be derived.

Under these approximations, the probability density functions $Q(\Theta_k)$ of the variational distribution can be obtained via [5]

$$
Q(\boldsymbol{\Xi}^{(k)}) = \delta\Big(\boldsymbol{\Xi}^{(k)} - \underbrace{\arg\max_{\boldsymbol{\Xi}^{(k)}} \mathbb{E}_{\prod_{\Theta_j \neq \boldsymbol{\Xi}^{(k)}} Q(\Theta_j)}\big[\ln p\,(\mathcal{Y}, \boldsymbol{\Theta})\big]}_{\triangleq \hat{\boldsymbol{\Xi}}^{(k)}}\Big), k = 1, 2, \ldots, P,
\tag{7.10}
$$

and

$$
Q(\Theta_k) \propto \exp\left\{\mathbb{E}_{\prod_{j\neq k} Q(\Theta_j)}\big[\ln p\,(\mathcal{Y}, \boldsymbol{\Theta})\big]\right\}, \Theta_k \in \boldsymbol{\Theta} \setminus \{\boldsymbol{\Xi}^{(k)}\}_{k=1}^{P}.
\tag{7.11}
$$

Obviously, these variational distributions are coupled in the sense that the computation of the variational distribution of one parameter requires the knowledge of variational distributions of other parameters. Therefore, these variational distributions should be updated iteratively. In the following, an explicit expression for each $Q(\cdot)$ is derived.

7.3.1 Derivation for $Q(\Xi^{(k)}), 1 \le k \le P$

By substituting (7.9) into (7.11) and only keeping the terms relevant to $\Xi^{(k)}$ ($1 \le k \le P$), we directly have

$$\hat{\Xi}^{(k)} = \underset{\Xi^{(k)} \in V_L(\mathbb{C}^{I_k})}{\arg\max}\ \mathbb{E}_{\prod_{\Theta_j \neq \Xi^{(k)}} Q(\Theta_j)}\Big[-\beta \parallel \mathcal{Y} - \big\| [\![\Xi^{(1)}, \ldots, \Xi^{(N)}]\!] - \mathcal{E} \big\|_F^2 \Big]. \tag{7.12}$$

To expand the square of the Frobenius norm inside the expectation in (7.12), we use the result that $\|\mathcal{A}\|_F^2 = \|\mathcal{A}^{(k)}\|_F^2 = \operatorname{Tr}\left(\mathcal{A}^{(k)}\left(\mathcal{A}^{(k)}\right)^H\right)$, where $\mathcal{A}^{(k)}$ is the unfolding of an Nth-order tensor $\mathcal{A} \in \mathbb{C}^{I_1 \times \cdots \times I_N}$ along its kth mode (see **Definition 1.1**). After expanding the square of the Frobenius norm and taking expectations, the parameter $\hat{\Xi}^{(k)}$ for each variational density in $\{Q(\Xi^{(k)})\}_{k=1}^P$ can be obtained from the following problem:

$$\begin{aligned}
\hat{\Xi}^{(k)} = \underset{\Xi^{(k)} \in V_L(\mathbb{C}^{I_k})}{\arg\max}\ & \operatorname{Tr}\Big(\underbrace{\mathbb{E}_{Q(\beta)}[\beta]\Big[\mathcal{Y} - \mathbb{E}_{Q(\mathcal{E})}[\mathcal{E}]\Big]^{(k)}\Big(\mathop{\diamond}_{n=1, n\neq k}^{N} \mathbb{E}_{Q(\Xi^{(n)})}\Big[\Xi^{(n)}\Big]\Big)^{*}}_{\triangleq \mathbf{F}^{(k)}} \Xi^{(k)H} + \Xi^{(k)}\mathbf{F}^{(k)H}\Big) \\
& - \operatorname{Tr}\Big(\Xi^{(k)H}\Xi^{(k)} \underbrace{\Big[\mathbb{E}_{Q(\beta)}[\beta]\,\mathbb{E}_{\prod_{n=1,n\neq k}^{N} Q(\Xi^{(n)})}\Big[\Big(\mathop{\diamond}_{n=1,n\neq k}^{N} \Xi^{(n)}\Big)^{T}\Big(\mathop{\diamond}_{n=1,n\neq k}^{N} \Xi^{(n)}\Big)^{*}\Big] + \mathbb{E}_{\prod_{l=1}^{L} Q(\gamma_l)}[\mathbf{\Gamma}]\Big]}_{\triangleq \mathbf{G}^{(k)}}\Big).
\end{aligned} \tag{7.13}$$

Using the fact that the feasible set for parameter $\Xi^{(k)}$ is the Stiefel manifold $V_L(\mathbb{C}^{I_k})$, i.e., $\Xi^{(k)H}\Xi^{(k)} = \mathbf{I}_L$, the term $\mathbf{G}^{(k)}$ is irrelevant to the factor matrix of interest $\Xi^{(k)}$. Consequently, problem (7.13) is equivalent to

$$\hat{\Xi}^{(k)} = \underset{\Xi^{(k)} \in V_L(\mathbb{C}^{I_k})}{\arg\max}\ \operatorname{Tr}\Big(\mathbf{F}^{(k)}\Xi^{(k)H} + \Xi^{(k)}\mathbf{F}^{(k)H} \Big), \tag{7.14}$$

where $\mathbf{F}^{(k)}$ was defined in the first line of (7.13). Problem (7.14) is a non-convex optimization problem, as its feasible set $V_L(\mathbb{C}^{I_k})$ is non-convex [9]. While in general (7.14) can be solved by numerical iterative algorithms based on a geometric approach or the alternating direction method of multipliers [9], a closed-form optimal solution can be obtained by noticing that the objective function in (7.14) has the same functional form as the log of the von Mises–Fisher matrix distribution with parameter matrix $\mathbf{F}^{(k)}$, and the feasible set in (7.14) also coincides with the support of this von Mises–Fisher matrix distribution [8]. As a result, we have

$$\hat{\Xi}^{(k)} = \underset{\Xi^{(k)}}{\arg\max} \ln \mathcal{VMF}\left(\Xi^{(k)} \mid \mathbf{F}^{(k)}\right). \tag{7.15}$$

Then, the closed-form solution for problem (7.15) can be acquired using the property below, which has been proved in [6].

Property 6.1. *Suppose the matrix* $\mathbf{A} \in \mathbb{C}^{\kappa_1 \times \kappa_2}$ *follows a von Mises–Fisher matrix distribution with parameter matrix* $\mathbf{F} \in \mathbb{C}^{\kappa_1 \times \kappa_2}$*. If* $\mathbf{F} = \mathbf{U}\boldsymbol{\Phi}\mathbf{V}^H$ *is the SVD of the matrix* $\mathbf{F}$*, then the unique mode of* $\mathcal{VMF}\,(\mathbf{A} \mid \mathbf{F})$ *is* $\mathbf{UV}^H$*.*

From **Property 6.1**, it is easy to conclude that $\hat{\Xi}^{(k)} = \boldsymbol{\Upsilon}^{(k)}\boldsymbol{\Pi}^{(k)H}$, where $\boldsymbol{\Upsilon}^{(k)}$ and $\boldsymbol{\Pi}^{(k)}$ are the left-orthonormal matrix and right-orthonormal matrix from the SVD of $\mathbf{F}^{(k)}$, respectively.

7.3.2 Derivation for $Q(\Xi^{(k)})$, $P+1 \leq k \leq N$

Using (7.9) and (7.11), the variational density $Q\left(\Xi^{(k)}\right)$ $(P+1 \leq k \leq N)$ can be derived to be a circularly symmetric complex matrix normal distribution [8] as

$$Q\left(\Xi^{(k)}\right) = \mathcal{CMN}(\Xi^{(k)} \mid \mathbf{M}^{(k)}, \mathbf{I}_{I_k}, \boldsymbol{\Sigma}^{(k)}) \tag{7.16}$$

where

$$\boldsymbol{\Sigma}^{(k)} = \left(\mathbb{E}_{Q(\beta)}\,[\beta]\,\mathbb{E}_{\prod_{n=1,n\neq k}^{N} Q(\Xi^{(n)})}\Big[\big(\mathop{\diamond}_{n=1,n\neq k}^{N} \Xi^{(n)}\big)^T\big(\mathop{\diamond}_{n=1,n\neq k}^{N} \Xi^{(n)}\big)^*\Big] + \mathbb{E}_{\prod_{l=1}^{L} Q(\gamma_l)}\,[\boldsymbol{\Gamma}]\right)^{-1} \tag{7.17}$$

$$\mathbf{M}^{(k)} = \mathbb{E}_{Q(\beta)}\,[\beta]\Big[\mathcal{Y} - \mathbb{E}_{Q(\mathcal{E})}\,[\mathcal{E}]\Big]^{(k)}\Big(\mathop{\diamond}_{n=1,n\neq k}^{N} \mathbb{E}_{Q(\Xi^{(n)})}\Big[\Xi^{(n)}\Big]\Big)^*\boldsymbol{\Sigma}^{(k)}. \tag{7.18}$$

Due to the fact that $Q(\Xi^{(k)})$ is Gaussian, the parameter $\mathbf{M}^{(k)}$ is both the expectation and the mode of the variational density $Q\left(\Xi^{(k)}\right)$.

To calculate $\mathbf{M}^{(k)}$, some expectation computations are required as shown in (7.17) and (7.18). For those with the form $\mathbb{E}_{Q(\Theta_k)}\,[\Theta_k]$ where $\Theta_k \in \boldsymbol{\Theta}$, the value can be easily obtained if the corresponding $Q\,(\Theta_k)$ is available. The remaining challenge stems from the expectation $\mathbb{E}_{\prod_{n=1,n\neq k}^{N} Q(\Xi^{(n)})}\Big[\big(\mathop{\diamond}_{n=1,n\neq k}^{N} \Xi^{(n)}\big)^T\big(\mathop{\diamond}_{n=1,n\neq k}^{N} \Xi^{(n)}\big)^*\Big]$ in (7.17). But its calculation becomes straightforward after exploiting the orthonormal structure of $\{\hat{\Xi}^{(k)}\}_{k-1}^{P}$ and the property of multiple Khatri–Rao products, as presented in the following property.

Property 6.2. *Suppose the matrix* $\mathbf{A}^{(n)} \in \mathbb{C}^{\kappa_n \times \rho} \sim \delta(\mathbf{A}^{(n)} - \hat{\mathbf{A}}^{(n)})$ *for* $1 \leq n \leq P$*, where* $\hat{\mathbf{A}}^{(n)} \in V_\rho(\mathbb{C}^{\kappa_n})$ *and* $P < N$*, and the matrix* $\mathbf{A}^{(n)} \in \mathbb{C}^{\kappa_n \times \rho} \sim \mathcal{CMN}(\mathbf{A}^{(n)} \mid \mathbf{M}^{(n)}, \mathbf{I}_{\kappa_n}, \Sigma^{(n)})$ *for* $P + 1 \leq n \leq N$*. Then,*

$$\mathbb{E}_{\prod_{n=1,n\neq k}^{N} p(\mathbf{A}^{(n)})}\Big[\Big(\mathop{\diamond}_{n=1,n\neq k}^{N} \mathbf{A}^{(n)}\Big)^T \Big(\mathop{\diamond}_{n=1,n\neq k}^{N} \mathbf{A}^{(n)}\Big)^*\Big]$$
$$= \mathfrak{D}\Big[\mathop{\odot}_{n=P+1,n\neq k}^{N} \Big(\mathbf{M}^{(n)H}\mathbf{M}^{(n)} + \kappa_n \Sigma^{(n)}\Big)^*\Big] \tag{7.19}$$

where $\mathfrak{D}[\mathbf{A}]$ *is a diagonal matrix taking the diagonal element from* $\mathbf{A}$*, and the multiple Hadamard products* $\mathop{\odot}_{n=1,n\neq k}^{N} \mathbf{A}^{(n)} = \mathbf{A}^{(N)} \odot \cdots \odot \mathbf{A}^{(k+1)} \odot \mathbf{A}^{(k-1)} \odot \cdots \odot \mathbf{A}^{(1)}$.

7.3.3 Derivation for $Q(\mathcal{E})$

The variational density $Q(\mathcal{E})$ can be obtained by taking only the terms relevant to $\mathcal{E}$ after substituting (7.9) into (7.11) and is expressed as

$$Q(\mathcal{E}) \propto \prod_{i_1=1}^{I_1} \cdots \prod_{i_N=1}^{I_N} \exp\Big\{\mathbb{E}_{\prod_{\Theta_j \neq \mathcal{E}} Q(\Theta_j)}\Big[-\zeta_{i_1,\ldots,i_N}\Big|\mathcal{E}_{i_1,\ldots,i_N}\Big|^2$$
$$-\beta\Big|\mathcal{Y}_{i_1,\ldots,i_n} - \sum_{l=1}^{L}\prod_{n=1}^{N} \Xi_{i_n,l}^{(n)} - \mathcal{E}_{i_1,\ldots,i_N}\Big|^2\Big]\Big\}. \tag{7.20}$$

After taking expectations, the term inside the exponent of (7.20) is

$$-\mathcal{E}^*_{i_1,\ldots,i_N}\underbrace{\Big(\mathbb{E}_{Q(\beta)}[\beta] + \mathbb{E}_{Q(\zeta_{i_1,\ldots,i_N})}\big[\zeta_{i_1,\ldots,i_N}\big]\Big)}_{\triangleq p_{i_1,\ldots,i_N}}\mathcal{E}_{i_1,\ldots,i_N}$$
$$+ 2\Re\mathfrak{e}\Big[\mathcal{E}^*_{i_1,\ldots,i_N} p_{i_1,\ldots,i_N} \underbrace{\mathbb{E}_{Q(\beta)}[\beta]p^{-1}_{i_1,\ldots,i_N}\Big(\mathcal{Y}_{i_1,\ldots,i_N} - \sum_{l=1}^{L}\Big(\prod_{n=1}^{N}\mathbb{E}_{Q(\Xi^{(n)})}[\Xi^{(n)}_{i_n,l}]\Big)\Big)}_{\triangleq m_{i_1,\ldots,i_N}}\Big]. \tag{7.21}$$

Since (7.21) is a quadratic function with respect to $\mathcal{E}_{i_1,\ldots,i_N}$, it is easy to show that

$$Q(\mathcal{E}) = \prod_{i_1=1}^{I_1} \cdots \prod_{i_N=1}^{I_N} \mathcal{CN}\left(\mathcal{E}_{i_1,\ldots,i_N} \mid m_{i_1,\ldots,i_N}, p_{i_1,\ldots,i_N}^{-1}\right). \tag{7.22}$$

Notice that from (7.21), the computation of outlier mean $m_{i_1,\ldots,i_N}$ can be rewritten as $m_{i_1,\ldots,i_N} = \mathfrak{n}_1\mathfrak{n}_2$, where $\mathfrak{n}_1 = \frac{\left(\mathbb{E}_{Q(\zeta_{i_1,\ldots,i_N})}[\zeta_{i_1,\ldots,i_N}]\right)^{-1}}{\left(\mathbb{E}_{Q(\zeta_{i_1,\ldots,i_N})}[\zeta_{i_1,\ldots,i_N}]\right)^{-1} + \left(\mathbb{E}_{Q(\beta)}[\beta]\right)^{-1}}$ and $\mathfrak{n}_2 = \mathcal{Y}_{i_1,\ldots,i_N} - \sum_{l=1}^{L}\left(\prod_{n=1}^{N}\mathbb{E}_{Q(\Xi^{(n)})}[\Xi_{i_n,l}^{(n)}]\right)$. From the general data model in (7.1), it can be seen that $\mathfrak{n}_2$ consists of the estimated outliers plus noise. On the other hand, since $\left(\mathbb{E}_{Q(\zeta_{i_1,\ldots,i_N})}[\zeta_{i_1,\ldots,i_N}]\right)^{-1}$ and $\left(\mathbb{E}_{Q(\beta)}[\beta]\right)^{-1}$ can be interpreted as the estimated power of the outliers and the noise, respectively, $\mathfrak{n}_1$ represents the strength of the outliers in the estimated outliers plus noise. Therefore, if the estimated power of the outliers $\left(\mathbb{E}_{Q(\zeta_{i_1,\ldots,i_N})}[\zeta_{i_1,\ldots,i_N}]\right)^{-1}$ goes to zero, the outlier mean $m_{i_1,\ldots,i_N}$ becomes zero accordingly.

7.3.4 Derivations for $Q(\gamma_l)$, $Q(\zeta_{i_1,\ldots,i_N})$, and $Q(\beta)$

Using (7.9) and (7.11) again, the variational density $Q(\gamma_l)$ can be expressed as

$$Q(\gamma_l) \propto \exp\left\{\left(\underbrace{\sum_{n=P+1}^{N} I_n - 1}_{\triangleq \tilde{a}_l}\right)\ln\gamma_l - \gamma_l \underbrace{\left[\sum_{n=P+1}^{N}\mathbb{E}_{Q(\Xi^{(n)})}\left[\Xi_{:,l}^{(n)H}\Xi_{:,l}^{(n)}\right]\right]}_{\triangleq \tilde{b}_l}\right\}, \tag{7.23}$$

which has the same functional form as the probability density function of gamma distribution, i.e., $Q(\gamma_l) = \text{gamma}(\gamma_l \mid \tilde{a}_l, \tilde{b}_l)$. Since $\mathbb{E}_{Q(\gamma_l)}[\gamma_l] = \tilde{a}_l/\tilde{b}_l$ is required for updating the variational distributions of other variables in $\boldsymbol{\Theta}$, we need to compute $\tilde{a}_l$ and $\tilde{b}_l$. While the computation of $\tilde{a}_l$ is straightforward, the computation of $\tilde{b}_l$ can be facilitated by using the correlation property of the matrix normal distribution $\mathbb{E}_{Q(\Xi^{(n)})}[\Xi_{:,l}^{(n)H}\Xi_{:,l}^{(n)}] = \mathbf{M}_{:,l}^{(n)H}\mathbf{M}_{:,l}^{(n)} + I_n\boldsymbol{\Sigma}_{l,l}^{(n)}$ [8] for $P+1 \le n \le N$.

Similarly, using (7.9) and (7.11), the variational densities $Q\left(\zeta_{i_1,\ldots,i_N}\right)$ and $Q(\beta)$ can be found to be gamma distributions as

$$Q\left(\zeta_{i_1,\ldots,i_N}\right) = \text{gamma}\left(\zeta_{i_1,\ldots,i_N} \mid \tilde{c}_{i_1,\ldots,i_N}, \tilde{d}_{i_1,\ldots,i_N}\right), \tag{7.24}$$

$$Q(\beta) = \text{gamma}\left(\beta \mid \tilde{e}, \tilde{f}\right), \tag{7.25}$$

with parameters $\tilde{c}_{i_1,\ldots,i_N} = c_{i_1,\ldots,i_N} + 1$, $\tilde{d}_{i_1,\ldots,i_N} = d_{i_1,\ldots,i_N} + \left(m_{i_1,\ldots,i_N}\right)^* m_{i_1,\ldots,i_N} + p_{i_1,\ldots,i_N}^{-1}$, $\tilde{e} = \prod_{n=1}^{N} I_n$, and $\tilde{f} = \mathbb{E}_{\prod_{n=1}^{N} Q(\Xi^{(n)})Q(\mathcal{E})}\left[\ \|\mathcal{Y} - [\![\Xi^{(1)}, \ldots, \Xi^{(N)}]\!] - \mathcal{E}\|_F^2\ \right]$.

For $\tilde{c}_{i_1,\ldots,i_N}$, $\tilde{d}_{i_1,\ldots,i_N}$ and $\tilde{e}$, the computations are straightforward. Furthermore, $\tilde{f}$ is derived to be [14]

$$\tilde{f} = \| \mathcal{Y} - \mathcal{M} \|_F^2 + \sum_{i_1=1}^{I_1} \cdots \sum_{i_N=1}^{I_N} p_{i_1,\ldots,i_N}^{-1} + \mathrm{Tr}\Big(\mathfrak{D} \Big[\bigodot_{n=P+1}^{N} \Big(\mathbf{M}^{(n)H} \mathbf{M}^{(n)} + I_n \boldsymbol{\Sigma}^{(n)} \Big)^* \Big] \Big)$$
$$- 2\mathfrak{Re}\Big(\mathrm{Tr}\Big(\Big[\mathcal{Y} - \mathcal{M} \Big]^{(1)} \Big[\Big(\mathop{\diamond}_{n=P+1}^{N} \mathbf{M}^{(n)} \Big) \diamond \Big(\mathop{\diamond}_{n=2}^{P} \hat{\boldsymbol{\Xi}}^{(n)} \Big) \Big]^* \hat{\boldsymbol{\Xi}}^{(1)H} \Big) \Big), \tag{7.26}$$

where $\mathcal{M}$ is a tensor with its $(i_1, \ldots, i_N)$th element being $m_{i_1,\ldots,i_N}$, and $\mathfrak{Re}(\cdot)$ denotes the real part of its argument.

7.3.5 Summary of the Iterative Algorithm

From the expressions for $Q(\Theta_k)$ evaluated above, it is seen that the calculation of a particular $Q(\Theta_k)$ relies on the statistics of other variables in $\boldsymbol{\Theta}$. As a result, the variational distribution for each variable in $\boldsymbol{\Theta}$ should be iteratively updated. The iterative algorithm is summarized in **Algorithm 10**.

7.3.6 Further Discussions

To gain more insights from **Algorithm 10**, discussions on its convergence property, automatic rank determination, relationship to the OALS algorithm, and computational complexity are presented in the following.

7.3.6.1 Convergence Property

Although the functional minimization of the KL divergence is non-convex over the mean-field family $Q(\boldsymbol{\Theta}) = \prod_k Q(\Theta_k)$, it is convex with respect to a single variational density $Q(\Theta_k)$ when the others $\{Q(\Theta_j)|j \neq k\}$ are fixed. Therefore, **Algorithm 10**, which iteratively updates the optimal solution for each Θ_k, is essentially a coordinate descent algorithm in the functional space of variational distributions with each update solving a convex problem. This guarantees a monotonic decrease of the KL divergence, and **Algorithm 10** is guaranteed to converge to at least a stationary point.

Algorithm 10 Probabilistic Tensor CPD with Orthogonal Factors

Initializations:
Choose $L > R$ and initial values $\{\hat{\Xi}^{(n,0)}\}_{n=1}^{P}$, $\{\mathbf{M}^{(n,0)}, \Sigma^{(n,0)}\}_{n=P+1}^{N}$, $\tilde{b}_l^0$, $\{\tilde{c}_{i_1,\ldots,i_N}^0, \tilde{d}_{i_1,\ldots,i_N}^0\}$ and $\tilde{f}^0$ for all l and $i_1, \ldots, i_N$. Let $\tilde{a}_l = \sum_{n=P+1}^{N} I_n$ and $\tilde{e} = \prod_{n=1}^{N} I_n$.
Iterations: For the tth iteration ($t \geq 1$),
Update the statistics of outliers: $\left\{p_{i_1,\ldots,i_N}, m_{i_1,\ldots,i_N}\right\}_{i_1=1,\ldots,i_N=1}^{I_1,\ldots,I_N}$:

$$p_{i_1,\ldots,i_N}^t = \frac{\tilde{e}}{\tilde{f}^{t-1}} + \frac{\tilde{c}_{i_1,\ldots,i_N}^{t-1}}{\tilde{d}_{i_1,\ldots,i_N}^{t-1}}, \tag{7.27}$$

$$m_{i_1,\ldots,i_N}^t = \frac{\tilde{e}}{\tilde{f}^{t-1} p_{i_1,\ldots,i_N}^t}\Big(\mathcal{Y}_{i_1,\ldots,i_N} - \sum_{l=1}^{L}\Big[\Big(\prod_{n=1}^{P} \hat{\Xi}_{i_n,l}^{(n,t-1)}\Big)\Big(\prod_{n=P+1}^{N} \mathbf{M}_{i_n,l}^{(n,t-1)}\Big)\Big]\Big). \tag{7.28}$$

Update the statistics of factor matrices: $\{\mathbf{M}^{(k)}, \Sigma^{(k)}\}_{k=P+1}^{N}$:

$$\Sigma^{(k,t)} = \Big(\frac{\tilde{e}}{\tilde{f}^{t-1}}\mathfrak{D}\Big[\bigodot_{n=P+1,n\neq k}^{N}\Big(\mathbf{M}^{(n,t-1)H}\mathbf{M}^{(n,t-1)} + I_n\Sigma^{(n,t-1)}\Big)^*\Big] + \mathrm{diag}\Big\{\frac{\tilde{a}_1}{\tilde{b}_1^{t-1}}, \ldots, \frac{\tilde{a}_L}{\tilde{b}_L^{t-1}}\Big\}\Big)^{-1}, \tag{7.29}$$

$$\mathbf{M}^{(k,t)} = \frac{\tilde{e}}{\tilde{f}^{t-1}}\Big([\mathcal{Y} - \mathcal{M}^t]^{(k)}\Big)\Big[\Big(\mathop{\diamond}_{n=P+1,n\neq k}^{N} \mathbf{M}^{(n,t-1)}\Big) \diamond \Big(\mathop{\diamond}_{n=1}^{P} \hat{\Xi}^{(n,t-1)}\Big)\Big]^* \Sigma^{(k,t)}. \tag{7.30}$$

Update the orthonormal factor matrices $\{\hat{\Xi}^{(k)}\}_{k=1}^{P}$:

$$\begin{aligned}\left[\Upsilon^{(k,t)}, \Pi^{(k,t)}\right] &= \mathrm{SVD}\Big[\frac{\tilde{e}}{\tilde{f}^{t-1}}\Big[\mathcal{Y} - \mathcal{M}^t\Big]^{(k)}\Big[\Big(\mathop{\diamond}_{n=P+1}^{N} \mathbf{M}^{(n,t)}\Big) \diamond \Big(\mathop{\diamond}_{n=1,n\neq k}^{P} \hat{\Xi}^{(n,t-1)}\Big)\Big]^*\Big],\\ \hat{\Xi}^{(k,t)} &= \Upsilon^{(k,t)}\Pi^{(k,t)H}.\end{aligned} \tag{7.31}$$

Update $\{\tilde{b}_l\}_{l=1}^{L}$, $\{\tilde{d}_{i_1,\ldots,i_N}\}_{i_1=1,\ldots,i_N=1}^{I_1,\ldots,I_N}$ and $\tilde{f}$

$$\tilde{b}_l^t = \sum_{n=P+1}^{N} \mathbf{M}_{:,l}^{(n,t)H}\mathbf{M}_{:,l}^{(n,t)} + I_n\Sigma_{l,l}^{(n,t)}, \tag{7.32}$$

$$\tilde{c}_{i_1,\ldots,i_N}^t = \tilde{c}_{i_1,\ldots,i_N}^0 + 1, \tag{7.33}$$

$$\tilde{d}_{i_1,\ldots,i_N}^t = \tilde{d}_{i_1,\ldots,i_N}^0 + \Big(m_{i_1,\ldots,i_N}^t\Big)^* m_{i_1,\ldots,i_N}^t + 1/p_{i_1,\ldots,i_N}^t, \tag{7.34}$$

$$\begin{aligned}\tilde{f}^t &= \| \mathcal{Y} - \mathcal{M}^t \|_F^2 + \sum_{i_1=1}^{I_1}\cdots\sum_{i_n=1}^{I_N}(p_{i_1,\ldots,i_N}^t)^{-1} + \mathrm{Tr}\Big(\mathfrak{D}\Big[\bigodot_{n=P+1}^{N}\Big(\mathbf{M}^{(n,t)H}\mathbf{M}^{(n,t)} + I_n\Sigma^{(n,t)}\Big)^*\Big]\Big)\\ &\quad - 2\Re\mathfrak{e}\Big(\mathrm{Tr}\Big(\Big[\mathcal{Y} - \mathcal{M}^t\Big]^{(1)}\Big[\Big(\mathop{\diamond}_{n=P+1}^{N} \mathbf{M}^{(n,t)}\Big) \diamond \Big(\mathop{\diamond}_{n=2}^{P} \hat{\Xi}^{(n,t)}\Big)\Big]^* \hat{\Xi}^{(1,t)H}\Big)\Big).\end{aligned} \tag{7.35}$$

Until Convergence

7.3.6.2 Automatic Rank Determination

The automatic rank determination for the tensor CPD uses an idea from the Bayesian model selection (or Bayesian Occam's razor). More specifically, the parameters

$\{\gamma_l\}_{l=1}^{L}$ control the model complexity, and their optimal variational densities are obtained together with those of other parameters by minimizing the KL divergence. After convergence, if some $\mathbb{E}[\gamma_l]$ are very large, e.g., 10^6, this indicates that their corresponding columns in $\{\mathbf{M}^{(n)}\}_{n=P+1}^{N}$ can be "switched off", as they play no role in explaining the data. Furthermore, according to the definition of the tensor CPD, the corresponding columns in $\{\hat{\Xi}^{(n)}\}_{n=1}^{P}$ should also be pruned accordingly. Finally, the learned tensor rank R is the number of remaining columns in each estimated factor matrix $\hat{\Xi}^{(n)}$.

7.3.6.3 Computational Complexity

For each iteration, the complexity is dominated by updating each factor matrix, costing $O(\sum_{n=1}^{N} I_n L^2 + N \prod_{n=1}^{N} I_n L)$. Thus, the overall complexity is about $O(q(\sum_{n=1}^{N} I_n L^2 + N \prod_{n=1}^{N} I_n L))$ where q is the number of iterations needed for convergence. On the other hand, for the OALS algorithm with exact tensor rank R, its complexity is $O(m(\sum_{n=1}^{N} I_n R^2 + N \prod_{n=1}^{N} I_n R))$ where m is the number of iterations needed for convergence. Therefore, for each iteration, the complexity of **Algorithm 10** is comparable to that of the OALS algorithm.

7.3.6.4 Reduction to Special Cases

The model and inference algorithm are general in the sense that they include various special cases. For example, if there is no orthogonal factor matrix, we can set $P = 0$; if the tensor is in real-valued, we can simply replace hermitian with transpose; if we believe there are no outliers, we can skip the steps related to $\mathcal{E}$.

7.4 Simulation Results and Discussions

In this section, numerical simulations are presented to assess the performance of the developed algorithm (labeled as VB) using synthetic data and two applications, in comparison with various state-of-the-art tensor CPD algorithms. The algorithms being compared include the ALS, the simultaneous diagonalization method for coupled tensor CPD (labeled as SD) [11], the direct algorithm for CPD followed by enhanced ALS (labeled as DIAG-A) [12], the Bayesian tensor CPD (**Algorithm 5**, labeled as BCPD) [4], the robust iteratively reweighed ALS (labeled as IRALS) [13], and the OALS algorithm (labeled as OALS) [3]. Note that some of these algorithms were not originally derived for complex-valued data. In that case, they are extended to handle complex-valued data for comparison.

In all experiments, three outlier models are considered, and they are listed in Table 7.1. For all the simulated algorithms, the initial factor matrix $\hat{\Xi}^{(n,0)}$ is

Table 7.1 Three different outlier models

Scenario	Variable description
Bernoulli–Gaussian	$\mathcal{E}_{i_1,\ldots,i_N} \sim \mathcal{CN}(0, \sigma_e^2)$ with a probability π
Bernoulli-Uniform	$\mathcal{E}_{i_1,\ldots,i_N} \sim \mathcal{U}(-H, H)$ with a probability π
Bernoulli-Student's t	$\mathcal{E}_{i_1,\ldots,i_N} \sim \mathcal{T}(\mu, \lambda, \nu)$ with a probability π

set as the matrix consisting of L leading left singular vectors of $[\mathcal{Y}]^{(n)}$ where $L = \max\{I_1, I_2, \ldots, I_N\}$ for **Algorithm 10** and the BCPD, and $L = R$ for other algorithms. The initial parameters of the algorithm in this chapter are set as $\{\tilde{c}^0_{i_1,\ldots,i_N}, \tilde{d}^0_{i_1,\ldots,i_N}\} = 10^{-6}$ for all $i_1, \ldots, i_N, \tilde{b}^0_l = \sum_{n=P+1}^{N} I_n$ for all l, $\tilde{f}^0 = \prod_{n=1}^{N} I_n$, and $\{\boldsymbol{\Sigma}^{(n,0)}\}_{n=P+1}^{N}$ are all set to be $\mathbf{I}_L$. All the algorithms terminate at the tth iteration when $\| [\![\mathbf{A}^{(1,t)}, \mathbf{A}^{(2,t)}, \ldots, \mathbf{A}^{(N,t)}]\!] - [\![\mathbf{A}^{(1,t-1)}, \mathbf{A}^{(2,t-1)}, \ldots, \mathbf{A}^{(N,t-1)}]\!] \|_F^2 < 10^{-6}$ or the iteration number exceeds 2000.

7.4.1 Validation on Synthetic Data

Synthetic tensors are used in this subsection to assess the performance of **Algorithm 10** on convergence, rank learning ability, and factor matrix recovery under different outlier models. A complex-valued third-order tensor $[\![\mathbf{A}^{(1)}, \mathbf{A}^{(2)}, \mathbf{A}^{(3)}]\!] \in \mathbb{C}^{12\times12\times12}$ with rank $R = 5$ is considered, where the orthogonal factor matrix $\mathbf{A}^{(1)}$ is constructed from the R leading left singular vectors of a matrix drawn from $\mathcal{CMN}(\mathbf{A}|\mathbf{0}_{12\times5,}, \mathbf{I}_{12\times12}, \mathbf{I}_{5\times5})$, and the factor matrices $\{\mathbf{A}^{(n)}\}_{n=2}^{3}$ are drawn from $\mathcal{CMN}(\mathbf{A}|\ \mathbf{0}_{12\times5,}, \mathbf{I}_{12\times12}, \mathbf{I}_{5\times5})$. Parameters for outlier models are set as $\pi = 0.05$, $\sigma_e^2 = 100$, $H = 10 \arg\max_{i_1,\ldots,i_N} |[\![\mathbf{A}^{(1)}, \mathbf{A}^{(2)}, \mathbf{A}^{(3)}]\!]_{i_1,\ldots,i_N}|$, $\mu = 3$, $\lambda = 1/50$, and $\nu = 10$. The signal-to-noise ratio (SNR) is defined as $10\log_{10}(\| [\![\mathbf{A}^{(1)}, \mathbf{A}^{(2)}, \mathbf{A}^{(3)}]\!] \|_F^2 / \| \mathcal{W} \|_F^2)$. Each result in this subsection is obtained by averaging 500 Monte Carlo runs.

Figure 7.2 presents the convergence performance of **Algorithm 10** under different outlier models, where the mean square error (MSE) $\| [\![\hat{\boldsymbol{\Xi}}^{(1)}, \hat{\boldsymbol{\Xi}}^{(2)}, \hat{\boldsymbol{\Xi}}^{(3)}]\!] - [\![\mathbf{A}^{(1)}, \mathbf{A}^{(2)}, \mathbf{A}^{(3)}]\!] \|_F^2$ is chosen as the assessment criterion. From Fig. 7.2, it can be seen that the MSEs decrease significantly in the first few iterations and converge to stable values quickly, demonstrating the rapid convergence property. Furthermore, by comparing the simulation results with outliers to that without outlier, it is clear that **Algorithm 10** is effective in mitigating outliers.

For tensor rank learning, the simulation results of **Algorithm 10** are shown in Fig. 7.3a, while those of the Bayesian tensor CPD algorithm are shown in Fig. 7.3b. Each vertical bar in the figures shows the mean and standard deviation of rank estimates, with the red horizontal dotted lines indicating the true tensor rank. The percentages of correct estimates are also shown on top of the figures. From Fig. 7.3a, it is seen that **Algorithm 10** can recover the true tensor rank with 100% accuracy when

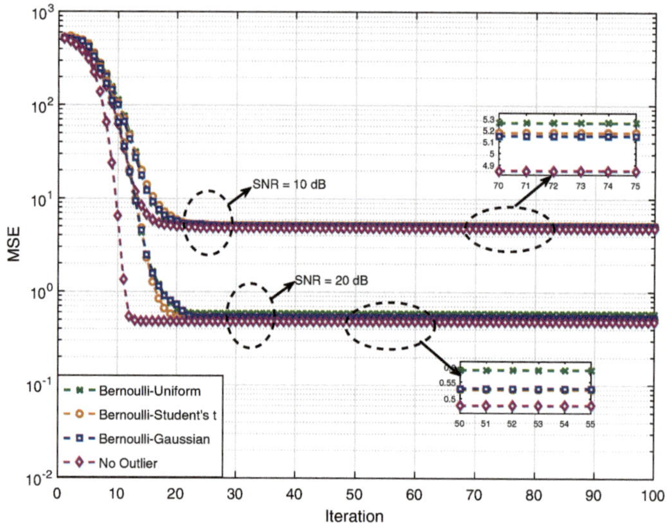

Fig. 7.2 Convergence of **Algorithm 10** under different outlier models

SNR $\geq$ 5 dB, both with or without outliers. This shows the accuracy and robustness of **Algorithm 10** when the noise power is moderate. Even though the performance at low SNRs is not as impressive as that at high SNRs, it can be observed that **Algorithm 10** still gives estimates close to the true tensor rank with the true rank lying mostly within one standard deviation from the mean estimate. On the other hand, in Fig. 7.3b, it is observed that while the Bayesian tensor CPD algorithm performs nearly the same as **Algorithm 10** without outliers, it gives tensor rank estimates very far away from the true value when outliers are present.

Figure 7.4 compares **Algorithm 10** to other state-of-the-art CPD algorithms in terms of recovery accuracy of the orthogonal factor matrix $\mathbf{A}^{(1)}$ under different outlier models. The criterion is set as the best congruence ratio defined as $\min_{\boldsymbol{\Delta}\mathbf{P}} \parallel \mathbf{A}^{(1)} - \hat{\boldsymbol{\Xi}}^{(1)}\mathbf{P}\boldsymbol{\Delta} \parallel_F / \parallel \mathbf{A}^{(1)} \parallel_F$, where the diagonal matrix $\boldsymbol{\Delta}$ and the permutation matrix $\mathbf{P}$ are found via the greedy least-squares column matching algorithm. From Fig. 7.4a, it is seen that both **Algorithm 10** and OALS perform better than other algorithms when outliers are absent. This shows the importance of incorporating the orthogonality information of the factor matrix. On the other hand, while OALS offers the same performance as **Algorithm 10** when there is no outlier, its performance is significantly different in the presence of outliers, as presented in Fig. 7.4b–d. Furthermore, except **Algorithm 10** and IRALS, all other algorithms do not take the outliers into account, thus their performances degrade significantly as shown in Fig. 7.4b–d. Even though the IRALS uses the robust l_p $(0 < p \leq 1)$ norm optimization to alleviate the effects of outliers, it cannot learn the statistical information of the outliers, leading to its worse performance than that of **Algorithm 10** in outliers mitigation.

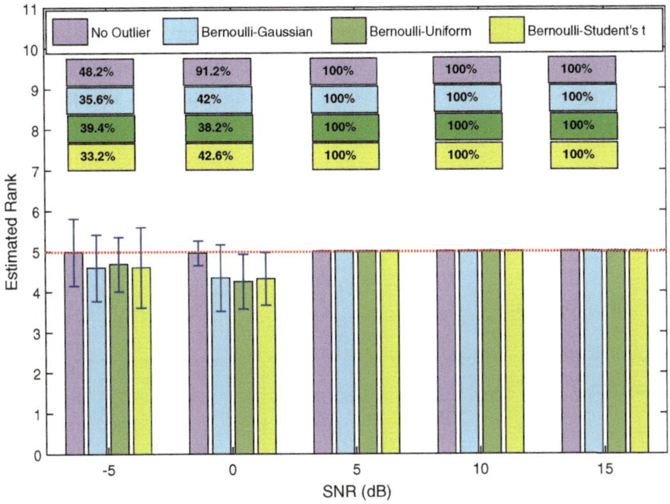

(a)

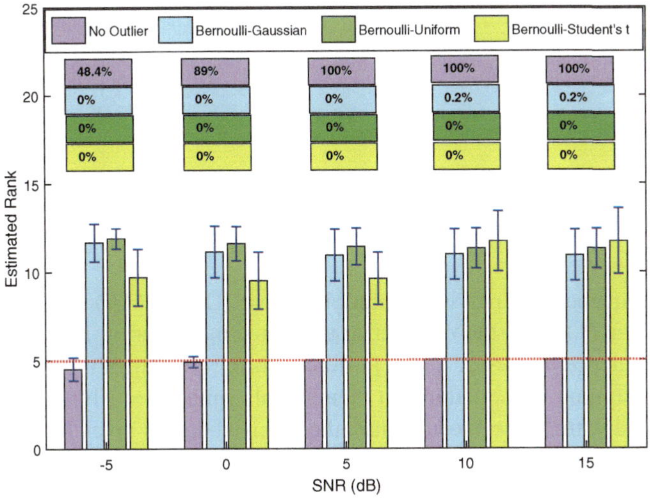

(b)

Fig. 7.3 Rank determination using **a Algorithm 10** and **b** the Bayesian tensor CPD

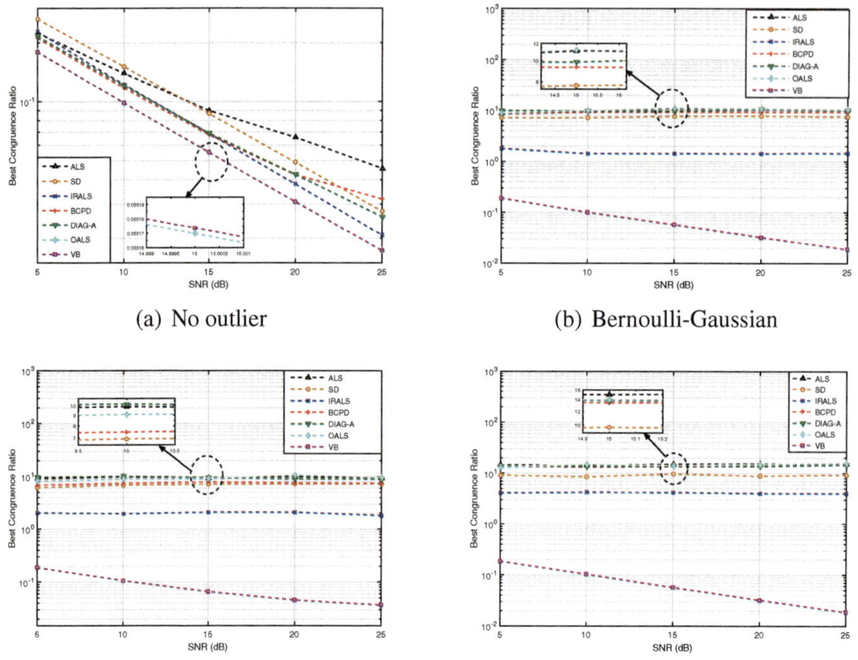

(a) No outlier (b) Bernoulli-Gaussian

(c) Bernoulli-Uniform (d) Bernoulli-Student's t

Fig. 7.4 Performance of factor matrix recovery versus SNR under different outlier models

7.4.2 Blind Data Detection for DS-CDMA Systems

In a direct-sequence code division multiple access (DS-CDMA) system, the transmitted signal $s_r(k)$ from the rth user at the kth symbol period is multiplied by a spreading sequence $[c_{1r}, c_{2r}, \ldots, c_{Zr}]$ where c_{zr} is the zth chip of the applied spreading code. Assuming R users transmit their signals simultaneously to a base station (BS) equipped with M receive antennas, the received data is given by

$$y_{mz}(k) = \sum_{r=1}^{R} h_{mr} c_{zr} s_r(k) + w_{mz}(k), \quad 1 \le m \le M, \quad 1 \le z \le Z, \tag{7.36}$$

where h_{mr} denotes the flat fading channel between the rth user and the mth receive antenna at the base station, and $w_{mz}(k)$ denotes white Gaussian noise. By introducing $\mathbf{H} \in \mathbb{C}^{M\times R}$ with its (m, r)th element being h_{mr}, and $\mathbf{C} \in \mathbb{C}^{Z\times R}$ with its (z, r)th element being c_{zr}, the model (7.36) can be written in matrix form as $\mathbf{Y}(k) = \sum_{r=1}^{R} \mathbf{H}_{:,r} \circ \mathbf{C}_{:,r} s_r(k) + \mathbf{W}(k)$, where $\mathbf{Y}(k), \mathbf{W}(k) \in \mathbb{C}^{M\times Z}$ are matrices with their (m, z)th elements being $y_{mz}(k)$ and $w_{mz}(k)$, respectively. After collecting

T samples along the time dimension and defining $\mathbf{S} \in \mathbb{C}^{T \times R}$ with its (k, r)th element being $s_r(k)$, the system model can be further written in the tensor form as [1]

$$\mathcal{Y} = \sum_{r=1}^{R} \mathbf{H}_{:,r} \circ \mathbf{C}_{:,r} \circ \mathbf{S}_{:,r} + \mathcal{W} = [\![\mathbf{H}, \mathbf{C}, \mathbf{S}]\!] + \mathcal{W} \tag{7.37}$$

where $\mathcal{Y} \in \mathbb{C}^{M \times Z \times T}$ and $\mathcal{W} \in \mathbb{C}^{M \times Z \times T}$ are third-order tensors, which take $y_{mz}(k)$ and $w_{mz}(k)$ as their (m, z, k)th elements, respectively.

It is shown in [1] that under certain mild conditions, the CPD of tensor $\mathcal{Y}$, which solves $\min_{\mathbf{H},\mathbf{C},\mathbf{S}} \parallel \mathcal{Y} - [\![\mathbf{H}, \mathbf{C}, \mathbf{S}]\!] \parallel_F^2$, can blindly recover the transmitted signals $\mathbf{S}$. Furthermore, since the transmitted signals are usually uncorrelated and with zero mean, the orthogonality structure[1] of $\mathbf{S}$ can further be taken into account to give better performance for blind signal recovery [2]. Similar models can also be found in blind data detection for cooperative communication systems [15, 16], and in topology learning for wireless sensor networks (WSNs) [17].

In this simulation, we set $R = 5$ users and the BS is equipped with $M = 8$ antennas. The channel between the rth user and the mth antenna is a flat fading channels $h_{mr} \sim \mathcal{CN}(h_{mr}|0, 1)$. The transmitted data $s_r(k)$ are random binary phase-shift keying (BPSK) symbols. The spreading code is of length $Z = 6$, and with each code element $c_{zr} \sim \mathcal{CN}(c_{zr}|0, 1)$. After observing the received tensor $\mathcal{Y} \in \mathbb{C}^{8 \times 6 \times 100}$, **Algorithm 10** and other state-of-the-art tensor CPD algorithms, combined with ambiguity removal and constellation mapping [1, 2], are executed to blindly detect the transmitted data. Their performance is measured in terms of bit error rate (BER).

The BERs versus SNR under different outlier models are presented in Fig. 7.5, which are averaged over 10000 independent trials. The parameter settings for different outlier models are the same as those in the last subsection. It is seen from Fig. 7.5a that when there is no outlier, **Algorithm 10** and OALS behave the same, and both outperform other CPDs. However, when outliers exist, it is seen from Fig. 7.5b–d that **Algorithm 10** performs significantly better than other algorithms.

7.4.3 Linear Image Coding for a Collection of Images

Given a collection of images representing a class of objects, linear image coding extracts the commonalities of these images, which is important in image compression and recognition [18, 19]. The kth image of size $M \times Z$ naturally corresponds to a matrix $\mathbf{B}(k)$ with its (m, z)th element being the image's intensity at that position. Linear image coding seeks the orthogonal basis matrices $\mathbf{U} \in \mathbb{C}^{M \times R}$ and $\mathbf{V} \in \mathbb{C}^{Z \times R}$ that capture the directions of the largest R variances in the image data, and this problem can be written as [18, 19]

---

[1] Strictly speaking, $\mathbf{S}$ is only approximately orthogonal. But the approximation gets better and better when observation length T increases.

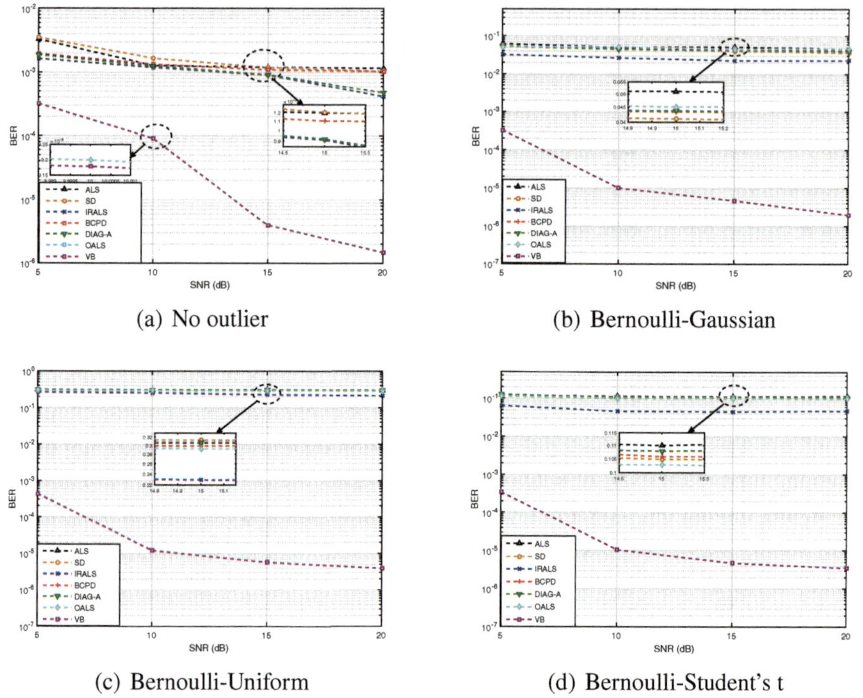

Fig. 7.5 BER versus SNR under different outlier models

$$\min_{\mathbf{U},\mathbf{V},\{d_r(k)\}_{r=1}^R} \sum_{k=1}^{K} \| \mathbf{B}(k) - \mathbf{U}\mathrm{diag}\{d_1(k), \ldots, d_R(k)\}\mathbf{V}^T \|_F^2$$
$$\text{s.t.} \quad \mathbf{U}^H\mathbf{U} = \mathbf{I}_R, \, \mathbf{V}^H\mathbf{V} = \mathbf{I}_R. \tag{7.38}$$

Obviously, if there is only one image (i.e., $K = 1$), problem (7.38) is equivalent to the well-studied SVD problem. Notice that the expression inside the Frobenius norm in (7.38) can be written as $\mathbf{B}(k) - \sum_{r=1}^{R} \mathbf{U}_{:,r} \circ \mathbf{V}_{:,r} d_r(k)$. Further introducing the matrix $\mathbf{D}$ with its (k, r)th element being $d_r(k)$, it is easy to see that problem (7.38) can be rewritten in tensor form as

$$\min_{\mathbf{U},\mathbf{V},\mathbf{D}} \| \mathcal{B} - \underbrace{\sum_{r=1}^{R} \mathbf{U}_{:,r} \circ \mathbf{V}_{:,r} \circ \mathbf{D}_{:,r}}_{=[\![\mathbf{U},\mathbf{V},\mathbf{D}]\!]} \|_F^2$$
$$\text{s.t.} \; \mathbf{U}^H\mathbf{U} = \mathbf{I}_R, \, \mathbf{V}^H\mathbf{V} = \mathbf{I}_R, \tag{7.39}$$

where $\mathcal{B} \in \mathbb{C}^{M \times Z \times K}$ is a third-order tensor with $\mathbf{B}(k)$ as its kth slice. Therefore, linear image coding for a collection of images is equivalent to solving a tensor CPD with two orthonormal factor matrices.

Table 7.2 Classification error and CPD computation time in face recognition

Algorithm	No outlier		Bernoulli–Gaussian		Bernoulli-Uniform		Bernoulli-Student's t	
	Classification error (%)	CPD time (s)	Classification error (%)	CPD time (s)	Classification error (%)	CPD time (s)	Classification error (%)	CPD time (s)
ALS	9	1.9635	51	46.3041	47	63.6765	48	58.1047
SD	12	0.5736	49	0.6287	44	0.6181	49	0.6594
IRALS	9	4.3527	43	15.8361	27	16.6151	36	17.5181
BCPD	2	4.1546	53	20.7338	35	17.9896	48	17.1151
DIAG-A	11	3.7384	50	28.9961	41	30.6317	51	22.5127
OALS	11	1.0174	58	40.2731	34	38.1754	47	24.0162
VB	2	2.4827	10	2.7806	6	2.4912	7	2.8895

We conduct experiments on 165 face images from the Yale Face Database[2] [10], representing different facial expressions (also with or without sunglasses) of 15 people (11 images for each person). In each classification experiment, we randomly pick two people's images. Among these 22 images, 12 (6 from each person) are used for training. In particular, each image is of size 240×320, and the training data can be naturally represented by a third-order tensor $\mathcal{Y} \in \mathbb{R}^{240\times 320\times 12}$. Various algorithms are run to learn the two orthogonal basis matrices.[3] Then, the feature vectors of these 12 training images, which are obtained by projecting them onto the multi-linear subspaces spanned by the two orthogonal basis matrices, are used to train a support vector machine (SVM) classifier. For the 10 testing images, their feature vectors are fed into the SVM classifier to determine which person is in each image. The parameters of various outlier models are $\pi = 0.05$, $\sigma_e = 100$, $H = 100$, $\mu = 1$, $\lambda = 1/1000$, and $\nu = 20$.

Since the tensor rank is not known in the image data, it should be carefully chosen. For the algorithms (ALS, SD, IRALS, DIAG-A, and OALS) that cannot automatically determine the rank, it can be obtained by first running these algorithms with tensor rank ranges from 1 to 12 and then finding the knee point of the reconstruction error decrement [5]. When there is no outlier, it is able to find the appropriate tensor rank. However, when outliers exist, the knee point cannot be found and we set the rank as the upper bound 12. For the BCPD, although it learns the appropriate rank when there are no outliers, it learns the rank as 12 when outliers exist. On the other hand, no matter whether there are outliers or not, **Algorithm 10** automatically learns the appropriate tensor rank without exhaustive search and thus saves considerable computational complexity.

The average classification errors of 10 independent experiments and the corresponding average computation times (benchmarked in Matlab on a personal computer with an i7 CPU) are shown in Table 7.2, and it can be seen that **Algorithm 10** provides the smallest classification error under all considered scenarios.

---

[2] http://vision.ucsd.edu/content/yale-face-database.

[3] Although the image data are real-valued and **Algorithm 10** is derived for complex-valued data, we directly use **Algorithm 10** on image data without modification.

References

1. N.D. Sidiropoulos, G.B. Giannakis, R. Bro, Blind PARAFAC receivers for DS-CDMA systems. IEEE Trans. Signal Process. **48**(3), 810–823 (2000)
2. M. Sorensen, L.D. Lathauwer, L. Deneire, PARAFAC with orthogonality in one mode and applications in DS-CDMA systems, in *Proceedings of the IEEE International Conference on Acoustics, Speech and Signal Processing (ICASSP 2010)*, Dallas, Texas (2010), pp. 4142–4145
3. M. Sorensen, L.D. Lathauwer, P. Comon, S. Icart, L. Deneire, Canonical polyadic decomposition with a columnwise orthonormal factor matrix. SIAM J. Matrix Anal. Appl. **33**(4), 1190–1213 (2012)
4. Q. Zhao, L. Zhang, A. Cichocki, Bayesian CP factorization of incomplete tensors with automatic rank determination. IEEE Trans. Pattern Anal. Mach. Intell. **37**(9), 1751–1753 (2015)
5. K.P. Murphy, *Machine Learning: A Probabilistic Perspective* (MIT Press, Cambridge, 2012)
6. C.G. Khat, K.V. Mardia, The von Mises-Fisher distribution in orientation statistics. J. R. Stat. Soc. **39**, 95–106 (1977)
7. M. West, On scale mixtures of normal distributions. Biometrika **74**(3), 646–648 (1987)
8. A.K. Gupta, D.K. Nagar, *Matrix Variate Distributions* (CRC Press, Boca Raton, 1999)
9. T. Kanamori, A. Takeda, Non-convex optimization on Stiefel manifold and applications to machine learning, in *Neural Information Processing* (2012), pp. 109–116
10. A. Georghiades, D. Kriegman, P. Belhumeur, From few to many: generative models for recognition under variable pose and illumination. IEEE Trans. Pattern Anal. Mach. Intell. **40**, 643–660 (2001)
11. M. Sørensen, D. Ignat, L.D. Lieven, Coupled canonical polyadic decompositions and (coupled) decompositions in multilinear rank-(L_r, n, L_r, n, 1) terms–Part II: Algorithms. SIAM J. Matrix Anal. Appl. **36**(3), 1015–1045 (2015)
12. X. Luciani, L. Albera, Canonical polyadic decomposition based on joint eigenvalue decomposition. Chemom. Intell. Lab. Syst. **132**, 152–167 (2014)
13. X. Fu, K. Huang, W.-K. Ma, N.D. Sidiropoulos, R. Bro, Joint tensor factorization and outlying slab suppression with applications. IEEE Trans. Signal Process. **63**(23), 6315–6328 (2015)
14. L. Cheng, Y.-C. Wu, H.V. Poor, Probabilistic tensor canonical polyadic decomposition with orthogonal factors. IEEE Trans. Signal Process. **65**(3), 663–676 (2017)
15. C.A.R. Fernandes, A.L.F. de Almeida, D.B. da Costa, Unified tensor modeling for blind receivers in multiuser uplink cooperative systems. IEEE Signal Process. Lett. **19**(5), 247–250 (2012)
16. A.Y. Kibangou, A. De Almeida, Distributed PARAFAC based DS-CDMA blind receiver for wireless sensor networks, in *Proceedings of the IEEE International Conference on Signal Processing Advances in Wireless Communications (SPAWC 2010)*, Marrakech, Jun. 20–23 (2010), pp. 1–5
17. A.L.F. de Almeida, A.Y. Kibangou, S. Miron, D.C. Araujo, Joint data and connection topology recovery in collaborative wireless sensor networks, in *Proceedings of the IEEE International Conference on Acoustics, Speech and Signal Processing (ICASSP 2013)*, Vancouver, BC, May 26–31 (2013), pp. 5303–5307
18. B. Pesquet-Popescu, J.-C. Pesquet, A.P. Petropulu, Joint singular value decomposition - a new tool for separable representation of images, in *Proceedings of the IEEE International Conference on Image Processing (ICIP 2001)*, Thessaloniki, Greece (2001), pp. 569–572
19. A. Shashua, A. Levin, Linear image coding for regression and classification using the tensor-rank principle, in *Proceedings of the 2001 IEEE Computer Society Conference on Computer Vision and Pattern Recognition (CVPR 2001)*, Kauai, Hawaii (2001), pp. 42–49

Chapter 8
Handling Missing Value: A Case Study in Direction-of-Arrival Estimation

Abstract In previous chapters, the Bayesian CPDs are derived under fully observed tensors. However, in practice, there are many scenarios where only part of the tensors can be observed. This gives rise to the tensor completion problem. In this chapter, we use subspace identification for direction-of-arrival (DOA) estimation as a case study to elucidate the key idea of the associated Bayesian modeling and inference in data completion. In particular, we firstly introduce how DOA signal subspace recovery is linked to tensor decomposition under missing data. Then, the corresponding probabilistic model is established and the subsequent inference problem is solved by using **Theorem 2.1** in Chap. 2.

8.1 Linking DOA Subspace Estimation to Tensor Completion

Consider an arbitrary sensor array with a total of M sensors as seen in Fig. 8.1, where the mth sensor is located at coordinate (x_m, y_m, z_m). There are R far-field narrow-band radiating sources impinging on this array. With the source at elevation angle θ_r and azimuth angle ϕ_r transmitting a signal $\xi_r(n)$ at the nth snapshot, the discrete-time complex baseband signal received by the mth sensor is [1]

$$y_{x_m,y_m,z_m}(n) = \sum_{r=1}^{R} \xi_r(n)\exp\Big\{j\big[x_m u_r + y_m v_r + z_m p_r\big]\Big\} + w_{x_m,y_m,z_m}(n) \tag{8.1}$$

where $u_r = \frac{2\pi}{\lambda_c}\sin\theta_r\cos\phi_r$, $v_r = \frac{2\pi}{\lambda_c}\sin\theta_r\sin\phi_r$, and $p_r = \frac{2\pi}{\lambda_c}\cos\theta_r$ with λ_c being the wavelength of the carrier signal. It is assumed that the transmitted signal $\xi_r(n)$ is a zero-mean wide-sense stationary random process with correlation $\mathbb{E}\big[\xi_r(k)\xi_{r'}^*(k+\tau)\big] = \delta(r-r')r_r(\tau)$, and the additive noise $w_{x_m,y_m,z_m}(n) \sim \mathcal{CN}\left(w_{x_m,y_m,z_m}(n) \mid 0, \beta^{-1}\right)$ is spatially and temporally independent.

The goal is to find the DOA pairs $\{\theta_r, \phi_r\}_{r=1}^R$ from the received signal $\{y_{x_m,y_m,z_m}(n)\}_{m,n}$, with no knowledge of the source number R, noise power β^{-1}, and the statistics of source signals $\{\xi_r(n)\}_{r,n}$. Since the DOA parameters are

L. Cheng et al., *Bayesian Tensor Decomposition for Signal Processing and Machine Learning*, https://doi.org/10.1007/978-3-031-22438-6_8

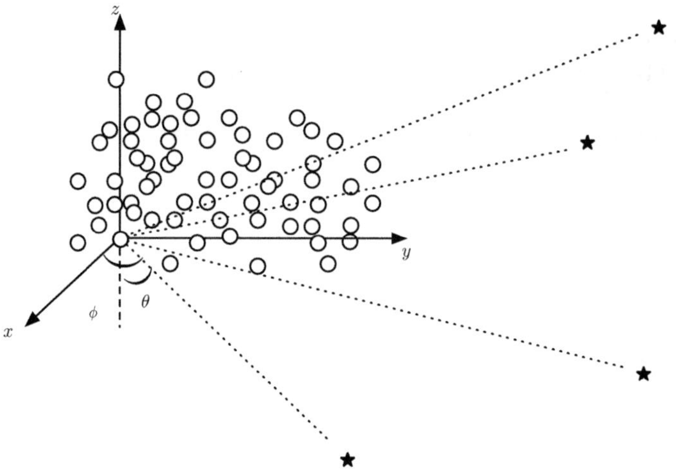

Fig. 8.1 Multiple sources impinging on an arbitrary array (© [2019] IEEE. Reprinted, with permission, from [L. Cheng, C. Xing, and Y.-C. Wu, Irregular Array Manifold Aided Channel Estimation in Massive MIMO Communications, IEEE Journal of Selected Topics in Signal Processing, Sep 2019]. It applies all figures and tables in this chapter)

nonlinearly coupled, it requires an exhaustive search if we employ direct optimization, which however is computationally demanding. As an efficient alternative, subspace methods firstly find the subspace in which the DOA signals lie in, and then the DOA parameters are extracted from the subspace structure, thus bypassing the exhaustive search problem.

Although subspace-based DOA estimation is not new, this chapter treats the signal obtained from the arbitrary array as data from a tensor with missing values. Not only this treatment would lead to a better subspace recovery than its matrix counterpart, but also tensor-based methods allow the subspaces in azimuth domain and elevation domain to be separately estimated, thus further reducing the complexity of subsequent DOA estimations [2]. More importantly, by viewing the subspace estimation problem as a tensor completion problem, it does not require shift-invariant sub-structure of the array. This leads to more general applicability than previous tensor subspace methods [3–6], which all require a shift-invariant sub-structure of the array and are inapplicable when the array shape is arbitrary.

To leverage the power of the tensor framework, we treat the arbitrary array as a cuboid array with missing elements, based on which the subspace estimation problem can be cast as a tensor completion problem. To construct the cuboid array, we project the sensor elements onto the x-axis, y-axis, and z-axis, respectively, as shown in Fig. 8.2. In particular, let S_x denotes the set collecting the projected coordinates on the x-axis, with repeated values eliminated and remaining values arranged from small to large, the value of $S_x(i_1)$ is the i_1th largest number in $\{x_m\}_{m=1}^{M}$. Similarly, sets S_y and S_z collect the projected and ordered coordinates on the y-axis and z-axis, respectively. Denoting the cardinality of sets $|S_x| = I_1$, $|S_y| = I_2$, and $|S_z| = I_3$, the cuboid grid

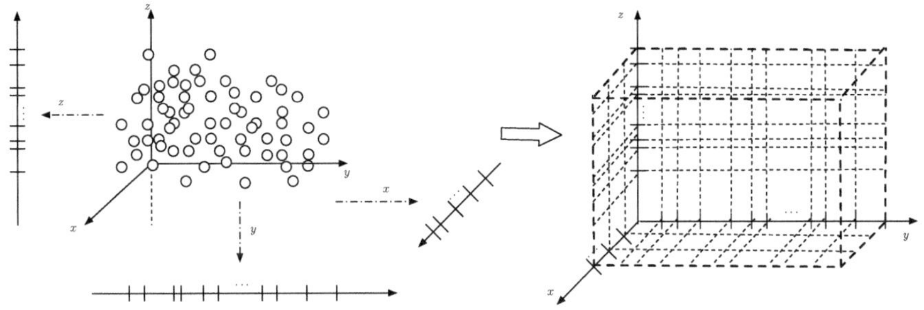

Fig. 8.2 The projected coordinates of sensors are connected to form a cuboid grid

is the collection of coordinates (x, y, z) such that $x \in S_x$, $y \in S_y$, and $z \in S_z$, and it forms a 3D tensor with dimensions I_1, I_2, and I_3. We denote this constructed tensor as $\mathcal{Y}(n)$. For the signal $y_{x_m,y_m,z_m}(n)$ received by the mth sensor, it is naturally assigned to the tensor element $\mathcal{Y}_{i_1,i_2,i_3}(n)$ where the index (i_1, i_2, i_3) satisfies $S_x(i_1) = x_m$, $S_y(i_2) = y_m$, and $S_z(i_3) = z_m$. After assigning the data collected by all M sensors, there is still a portion of tensor elements unknown in the constructed 3D tensor, where the missing ratio is $1 - \frac{M}{I_1 I_2 I_3}$. With each unknown tensor element being assigned a value zero, and using the data model (8.1), the constructed tensor $\mathcal{Y}(n)$ would contain elements

$$\mathcal{Y}_{i_1,i_2,i_3}(n) = O_{i_1,i_2,i_3}(n) \times \left(\sum_{r=1}^{R} \xi_l(n) \exp\left\{ j\left[S_x(i_1)u_r + S_y(i_2)v_r + S_z(i_3)p_r \right] \right\} + w_{S_x(i_1),S_y(i_2),S_z(i_3)}(n) \right) \tag{8.2}$$

where $O_{i_1,i_2,i_3}(n)$ equals one if data $\mathcal{Y}_{i_1,i_2,i_3}(n)$ is available, and zero otherwise.

Using the definition of tensor canonical polyadic decomposition (CPD), it is easy to show that (8.2) can be expressed in a tensor form as

$$\mathcal{Y}(n) = \mathcal{O}(n) \odot \left(\underbrace{\sum_{r=1}^{R} \boldsymbol{a}(u_r) \circ \boldsymbol{a}(v_r) \circ \boldsymbol{a}(p_r) \circ \xi_r(n)}_{\triangleq [\![\boldsymbol{A}[\boldsymbol{u}], \boldsymbol{A}[\boldsymbol{v}], \boldsymbol{A}[\boldsymbol{p}], \boldsymbol{\xi}(n)]\!]} + \mathcal{W}(n) \right), \tag{8.3}$$

where $\odot$ and $\circ$ denote the Hadamard product and the outer product, respectively, while $\mathcal{O}(n)$ is a tensor with (i_1, i_2, i_3)th element given by $O_{i_1,i_2,i_3}(n)$. For the noise tensor $\mathcal{W}(n)$, its (i_1, i_2, i_3)th element is given by $w_{S_x(i_1),S_y(i_2),S_z(i_3)}(n)$ if data $\mathcal{Y}_{i_1,i_2,i_3}(n)$ is available and zero otherwise. In (8.3), $\boldsymbol{\xi}(n) = [\xi_1(n), \xi_2(n), \ldots, \xi_R(n)]$. We also defined $\boldsymbol{A}[\boldsymbol{u}] \in \mathbb{C}^{I_1 \times R}$ with the rth column being $\boldsymbol{a}(u_r) = [\exp(ju_r S_x(1)), \exp(ju_r S_x(2)), \ldots, \exp(ju_r S_x(I_1))]^T$. Matrices $\boldsymbol{A}[\boldsymbol{v}] \in \mathbb{C}^{I_2 \times R}$ and $\boldsymbol{A}\left[\boldsymbol{p}\right] \in \mathbb{C}^{I_3 \times R}$

are defined similarly but with the rth column being $\boldsymbol{a}(v_r) = [\exp(jv_r S_y(1)), \exp(jv_r S_y(2)), \ldots, \exp(jv_r S_y(I_2))]^T$ and $\boldsymbol{a}(p_r) = [\exp(jp_r S_z(1)), \exp(jp_r S_z(2)), \ldots, \exp(jp_r S_z(I_3))]^T$, respectively.

After collecting I_4 snapshots data along the time dimension, we have a fourth-order data tensor $\dot{\mathcal{Y}} \in \mathbb{C}^{I_1 \times I_2 \times I_3 \times I_4}$. Using the tensor CPD definition, it is easy to show the data model can be expressed as

$$\dot{\mathcal{Y}} = \dot{\mathcal{O}} \odot \left(\underbrace{\sum_{r=1}^{R} \boldsymbol{a}(u_r) \circ \boldsymbol{a}(v_r) \circ \boldsymbol{a}(p_r) \circ \boldsymbol{a}(\xi_r)}_{\triangleq [\![\boldsymbol{A}[\boldsymbol{u}], \boldsymbol{A}[\boldsymbol{v}], \boldsymbol{A}[\boldsymbol{p}], \boldsymbol{A}[\boldsymbol{\xi}]]\!]} + \dot{\mathcal{W}} \right), \tag{8.4}$$

where the observation tensor $\dot{\mathcal{O}}$ and noise tensor $\dot{\mathcal{W}}$ are constructed in the same way as data tensor $\dot{\mathcal{Y}}$, and matrix $\boldsymbol{A}\left[\boldsymbol{\xi}\right] \in \mathbb{C}^{I_4 \times R}$ is constructed with the rth column being $\boldsymbol{a}(\xi_r) = [\xi_r(1), \ldots, \xi_r(I_4)]^T$. Note that a more generalized model where $\boldsymbol{A}\left[\boldsymbol{\xi}\right]$ is assumed to be rank-1 can be found in [11], but the subsequent modeling and inference procedure are similar to what is presented in this chapter.

To recover the DOA subspaces from the tensor representation in (8.4), we have the following property.

Property 8.1 *If* $[\![\boldsymbol{A}[\boldsymbol{u}], \boldsymbol{A}[\boldsymbol{v}], \boldsymbol{A}[\boldsymbol{p}], \boldsymbol{A}[\boldsymbol{\xi}]]\!] = [\![\boldsymbol{\Xi}^{(1)}, \boldsymbol{\Xi}^{(2)}, \boldsymbol{\Xi}^{(3)}, \boldsymbol{\Xi}^{(4)}]\!]$ *and* $2 \leq R \leq \min\{I_1, I_2, I_3, I_4\}$, *the following equations hold:* $\boldsymbol{\Xi}^{(1)} = \boldsymbol{A}[\boldsymbol{u}]\boldsymbol{\Delta}\boldsymbol{\Lambda}^{(1)}$, $\boldsymbol{\Xi}^{(2)} = \boldsymbol{A}[\boldsymbol{v}]\boldsymbol{\Delta}\boldsymbol{\Lambda}^{(2)}$, $\boldsymbol{\Xi}^{(3)} = \boldsymbol{A}[\boldsymbol{p}]\boldsymbol{\Delta}\boldsymbol{\Lambda}^{(3)}$, *and* $\boldsymbol{\Xi}^{(4)} = \boldsymbol{A}[\boldsymbol{\xi}]\boldsymbol{\Delta}\boldsymbol{\Lambda}^{(4)}$ *where* $\boldsymbol{\Delta}$ *is a permutation matrix and* $\prod_{k=1}^{4} \boldsymbol{\Lambda}^{(k)} = \boldsymbol{I}_R$.

This property indicates that the columns of the factor matrix $\boldsymbol{\Xi}^{(1)}$ span the same column space as those of matrix $\boldsymbol{A}[\boldsymbol{u}]$, since the permutation matrix $\boldsymbol{\Delta}$ and diagonal matrix $\boldsymbol{\Lambda}^{(1)}$ are invertible. Similarly, columns of $\boldsymbol{\Xi}^{(2)}$ span the same space as that of $\boldsymbol{A}[\boldsymbol{v}]$, and columns of $\boldsymbol{\Xi}^{(3)}$ span the same space as that of $\boldsymbol{A}\left[\boldsymbol{p}\right]$. Therefore, the core problem of subspace estimation is to find the factor matrices $\{\boldsymbol{\Xi}^{(k)}\}_{k=1}^{4}$ under unknown number of sources R, and the problem can be stated as

$$\min_{\{\boldsymbol{\Xi}^{(k)}\}_{k=1}^{4}} \beta \parallel \dot{\mathcal{Y}} - \dot{\mathcal{O}} \odot [\![\boldsymbol{\Xi}^{(1)}, \boldsymbol{\Xi}^{(2)}, \boldsymbol{\Xi}^{(3)}, \boldsymbol{\Xi}^{(4)}]\!] \parallel_F^2 + \sum_{l=1}^{L} \gamma_l \Big(\sum_{k=1}^{4} \boldsymbol{\Xi}_{:,l}^{(k)H} \boldsymbol{\Xi}_{:,l}^{(k)} \Big) \tag{8.5}$$

where L is the maximum possible value of the number of sources R.

This problem is known as complex-valued tensor completion problem, and it is easy to show its non-convexity, since all the factor matrices $\{\boldsymbol{\Xi}^{(k)}\}_{k=1}^{4}$ are coupled via Khatri–Rao products. Furthermore, the tensor rank acquisition is generally NP-hard [7], due to its discrete nature. This issue can be remedied by adding a regularization term $\sum_{l=1}^{L} \gamma_l \left(\sum_{k=1}^{4} \Xi_{:,l}^{(k)H} \Xi_{:,l}^{(k)} \right)$ in order to control the complexity of the model and avoid overfitting of noise. However, determining the optimal regularization parameters $\{\gamma_l\}_{l=1}^{L}$ is difficult, and conventional methods usually rely on computationally demanding search schemes [8]. Fortunately, we can build a probabilistic model, so that the regularization parameters and the factor matrices can be learned from the VI framework.

8.2 Probabilistic Modeling

To use the probabilistic framework, the optimization problem (8.5) needs to be interpreted using probabilistic language. In particular, the squared error term in problem (8.5) can be interpreted as the negative log of a likelihood function given by

$$p\left(\dot{\mathcal{Y}} \mid \{\Xi^{(n)}\}_{n=1}^{4}, \beta\right) \propto \exp\left(-\beta \sum_{i_1,i_2,i_3,i_4} \dot{\mathcal{O}}_{i_1,i_2,i_3,i_4} \left| \dot{\mathcal{Y}}_{i_1,i_2,i_3,i_4} - [\![\Xi^{(1)}, \Xi^{(2)}, \Xi^{(3)}, \Xi^{(4)}]\!]_{i_1,i_2,i_3,i_4} \right|^2 \right). \tag{8.6}$$

On the other hand, the regularization term in problem (8.5) can be interpreted as a zero-mean circularly symmetric complex Gaussian prior distribution over the columns of the factor matrices, i.e.,

$$p(\{\Xi^{(k)}\}_{k=1}^{4} | \{\gamma_l\}_{l=1}^{L}) = \prod_{k=1}^{4} \prod_{l=1}^{L} \mathcal{CN}\left(\Xi_{:,l}^{(k)} | \boldsymbol{m}_c, \gamma_l^{-1} \boldsymbol{I}_{I_k}\right) = \prod_{k=1}^{4} \prod_{\kappa=1}^{I_k} \mathcal{CN}\left(\Xi_{\kappa,:}^{(k)} | \boldsymbol{m}_r, \boldsymbol{\Gamma}\right), \tag{8.7}$$

where $\boldsymbol{m}_c = \boldsymbol{0}_{I_k \times 1}$, $\boldsymbol{m}_r = \boldsymbol{0}_{L \times 1}$, $\boldsymbol{\Gamma} = \text{diag}\{\gamma_1, \gamma_2, \ldots, \gamma_L\}$, and the latter part is an equivalent expression with the unknown changed to the row of $\Xi^{(k)}$. In (8.7), the inverse of the regularization parameter γ_l^{-1} has a physical interpretation of the power of the lth column of various factor matrices. When power γ_l^{-1} goes to zero, it indicates the corresponding columns in various factor matrices play no role and can be pruned out. Since we know nothing about the regularization parameter γ_l and noise power β^{-1} before inference, non-informative gamma prior [9] is imposed on them, i.e., $p(\beta|\boldsymbol{\alpha}_\beta) = \text{gamma}(\beta|10^{-6}, 10^{-6})$ and $p(\{\gamma_l\}_{l=1}^{L}|\boldsymbol{\lambda}_\gamma) = \prod_{l=1}^{L} \text{gamma}(\gamma_l|10^{-6}, 10^{-6})$.

The complete probabilistic model is shown in Fig. 8.3. Compared to the MPCEF model in Chap. 2, it is obvious that $\boldsymbol{\eta}^{(1)} = \{\{\Xi_{\kappa,:}^{(k)}\}_{\kappa,k}, \beta\}$ are the unknown variables

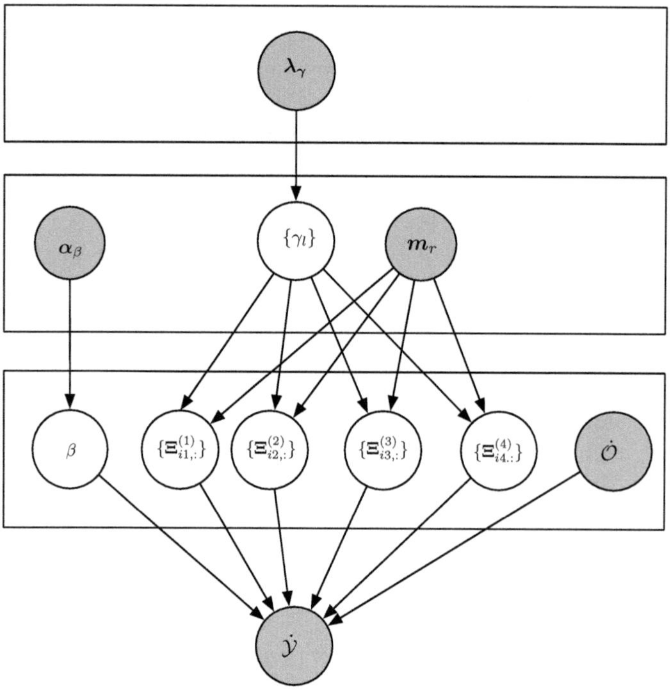

Fig. 8.3 Probabilistic model of tensor subspace estimation

in Layer 1 and $\boldsymbol{\eta}^{(2)} = \{\gamma_l\}_{l=1}^{L}$ are the unknown variables in Layer 2. However, due to complication induced by the missing values, we need to check whether this probabilistic model lies within the MPCEF introduced in Chap. 2.[1]

8.3 MPCEF Model Checking and Optimal Variational Pdfs Derivations

8.3.1 MPCEF Model Checking

For variable $\Xi^{(k)}_{\kappa,:}$ in Layer 1, we need to show that $p\left(\dot{\mathcal{Y}} \mid \Xi^{(k)}_{\kappa,:}, \{\boldsymbol{\eta}^{(1)} \backslash \Xi^{(k)}_{\kappa,:}\}\right)$ takes the form of (2.25). By expanding the square term in (8.6) using $[\![\boldsymbol{A}^{(1)}, \boldsymbol{A}^{(2)}, \ldots, \boldsymbol{A}^{(N)}]\!]_{i_1,i_2,\ldots,i_N} = \boldsymbol{A}^{(k)T}_{i_k,:}\Big(\underset{n=1,n\neq k}{\overset{N}{\diamond}} \boldsymbol{A}^{(n)T}_{i_n,:}\Big)^T$, we arrive at

[1] MPCEF is defined for real-valued model. But we can easily extend the expression to complex-valued.

$$p\left(\dot{\mathcal{Y}}\,|\,\{\Xi^{(k)}\}_{k=1}^{4},\beta\right)=\exp\Bigg\{\Re\mathfrak{e}\Bigg(\sum_{i_1,\ldots,i_4}\dot{O}_{i_1,\ldots,i_4}\ln\left(\frac{\beta}{\pi}\right)-\beta\sum_{i_1,\ldots,i_4}\dot{O}_{i_1,\ldots,i_4}\Bigg(\dot{\mathcal{Y}}_{i_1,\ldots,i_4}\dot{\mathcal{Y}}^{*}_{i_1,\ldots,i_4}$$
$$-2\Bigg(\dot{\mathcal{Y}}^{*}_{i_1,\ldots,i_4}\Xi^{(k)T}_{\kappa,:}\big(\mathop{\diamond}_{n=1,n\neq k}^{4}\Xi^{(n)T}_{i_n,:}\big)^{T}\Bigg)+\Xi^{(k)T}_{\kappa,:}\big(\mathop{\diamond}_{n=1,n\neq k}^{4}\Xi^{(n)T}_{i_n,:}\big)^{T}\big(\mathop{\diamond}_{p=1,p\neq k}^{4}\Xi^{(p)T}_{i_p,:}\big)^{*}\Xi^{(k)*}_{\kappa,:}\Bigg)\Bigg)\Bigg\}.\tag{8.8}$$

By fixing variables other than $\Xi^{(k)}_{\kappa,:}$ in (8.8),

$$p\left(\dot{\mathcal{Y}}\,|\,\Xi^{(k)}_{\kappa,:},\{\boldsymbol{\eta}^{(1)}\backslash\Xi^{(k)}_{\kappa,:}\}\right)=\exp\Bigg\{\Re\mathfrak{e}\Bigg(2\beta\sum_{i_1,\ldots i_k=\kappa,\ldots i_4}\dot{O}_{i_1,\ldots,i_4}\dot{\mathcal{Y}}^{*}_{i_1,\ldots,i_4}\Xi^{(k)T}_{\kappa,:}\big(\mathop{\diamond}_{n=1,n\neq k}^{4}\Xi^{(n)T}_{i_n,:}\big)^{T}$$
$$-\operatorname{vec}\Bigg(\beta\sum_{i_1,\ldots i_k=\kappa,\ldots i_4}\dot{O}_{i_1,\ldots,i_4}\big(\mathop{\diamond}_{n=1,n\neq k}^{4}\Xi^{(n)}_{i_n,:}\big)^{T}\big(\mathop{\diamond}_{p=1,p\neq k}^{4}\Xi^{(p)}_{i_p,:}\big)^{*}\Bigg)^{H}\operatorname{vec}\Big(\Xi^{(k)*}_{\kappa,:}\Xi^{(k)T}_{\kappa,:}\Big)$$
$$-\sum_{i_1,\ldots i_k\neq\kappa,\ldots i_4}\Bigg(2\beta\sum_{i_1,\ldots i_k=\kappa,\ldots i_4}\dot{O}_{i_1,\ldots,i_4}\dot{\mathcal{Y}}^{*}_{i_1,\ldots,i_4}\Xi^{(k)T}_{i_k,:}\big(\mathop{\diamond}_{n=1,n\neq k}^{4}\Xi^{(n)T}_{i_n,:}\big)^{T}$$
$$-\operatorname{vec}\Bigg(\beta\sum_{i_1,\ldots i_k=\kappa,\ldots i_4}\dot{O}_{i_1,\ldots,i_4}\big(\mathop{\diamond}_{n=1,n\neq k}^{4}\Xi^{(n)}_{i_n,:}\big)^{T}\big(\mathop{\diamond}_{p=1,p\neq k}^{4}\Xi^{(p)}_{i_p,:}\big)^{*}\Bigg)^{H}\operatorname{vec}\Big(\Xi^{(k)*}_{i_k,:}\Xi^{(k)T}_{i_k,:}\Big)\Bigg)$$
$$+\sum_{i_1,\ldots,i_4}\dot{O}_{i_1,\ldots,i_4}\ln\left(\frac{\beta}{\pi}\right)-\beta\sum_{i_1,\ldots,i_4}\dot{O}_{i_1,\ldots,i_4}\dot{\mathcal{Y}}_{i_1,\ldots,i_4}\dot{\mathcal{Y}}^{*}_{i_1,\ldots,i_4}\Bigg)\Bigg\}.\tag{8.9}$$

Comparing (8.9) to (2.25), we can see that $\mathbf{t}(\Xi^{(k)}_{\kappa,:})=[\Xi^{(k)*}_{\kappa,:};\ \operatorname{vec}(\Xi^{(k)*}_{\kappa,:}\Xi^{(k)T}_{\kappa,:})]$,

$$\mathbf{n}(\dot{\mathcal{Y}},\{\boldsymbol{\eta}^{(1)}\backslash\Xi^{(k)}_{\kappa,:}\})=\begin{bmatrix}2\beta\sum_{i_1,\ldots i_k=\kappa,\ldots i_4}\dot{O}_{i_1,\ldots,i_4}\dot{\mathcal{Y}}^{*}_{i_1,\ldots,i_4}\big(\mathop{\diamond}_{n=1,n\neq k}^{4}\Xi^{(n)T}_{i_n,:}\big)^{T}\\-\operatorname{vec}\Big(\beta\sum_{i_1,\ldots i_k=\kappa,\ldots i_4}\dot{O}_{i_1,\ldots,i_4}\big(\mathop{\diamond}_{n=1,n\neq k}^{4}\Xi^{(n)}_{i_n,:}\big)^{T}\big(\mathop{\diamond}_{p=1,p\neq k}^{4}\Xi^{(p)}_{i_p,:}\big)^{*}\Big)\end{bmatrix}.\tag{8.10}$$

Furthermore, the prior for $\Xi^{(k)}_{\kappa,:}$ in (8.7) is in the form of (2.26)

$$p\left(\Xi^{(k)}_{\kappa,:}|\mathbf{m}_r,\{\gamma_l\}_{l=1}^{L}\right)=\exp\Bigg\{\Re\mathfrak{e}\Bigg(\begin{bmatrix}\mathbf{0}_{L\times1}\\-\operatorname{vec}(\boldsymbol{\Gamma})\end{bmatrix}^{H}\begin{bmatrix}\Xi^{(k)*}_{\kappa,:}\\\operatorname{vec}(\Xi^{(k)*}_{\kappa,:}\Xi^{(k)T}_{\kappa,:})\end{bmatrix}+\sum_{l=1}^{L}\log\gamma_l-L\log\pi\Bigg)\Bigg\},\tag{8.11}$$

where $\mathbf{n}(\mathbf{m}_r,\{\gamma_l\}_{l=1}^{L})=[\mathbf{0}_{L\times1};\ -\operatorname{vec}(\boldsymbol{\Gamma})]$.

Now, focusing on the variable β in Layer 1, (8.6) can be expressed as

$$p\left(\dot{\mathcal{Y}} \mid \beta, \{\boldsymbol{\eta}^{(1)} \backslash \beta\}\right) = \exp\left\{\Re\mathfrak{e}\left(-\sum_{i_1,\ldots,i_4} \dot{\mathcal{O}}_{i_1,\ldots,i_4} \ln \pi + \sum_{i_1,\ldots,i_4} \dot{\mathcal{O}}_{i_1,\ldots,i_4} \ln \beta \right.\right.$$
$$\left.\left. - \beta \sum_{i_1,\ldots,i_4} \dot{\mathcal{O}}_{i_1,\ldots,i_4} \left| \dot{\mathcal{Y}}_{i_1,\ldots,i_4} - [\![\boldsymbol{\Xi}^{(1)}, \boldsymbol{\Xi}^{(2)}, \boldsymbol{\Xi}^{(3)}, \boldsymbol{\Xi}^{(4)}]\!]_{i_1,\ldots,i_4} \right|^2 \right)\right\}, \quad (8.12)$$

which is in the form of (2.25), where $\mathbf{t}(\beta) = [\beta; \ln \beta]$,

$$\mathbf{n}(\dot{\mathcal{Y}}, \{\boldsymbol{\eta}^{(1)} \backslash \beta\}) = \begin{bmatrix} -\sum_{i_1,\ldots,i_4} \dot{\mathcal{O}}_{i_1,\ldots,i_4} \left| \dot{\mathcal{Y}}_{i_1,\ldots,i_4} - [\![\boldsymbol{\Xi}^{(1)}, \boldsymbol{\Xi}^{(2)}, \boldsymbol{\Xi}^{(3)}, \boldsymbol{\Xi}^{(4)}]\!]_{i_1,\ldots,i_4} \right|^2 \\ \sum_{i_1,\ldots,i_4} \dot{\mathcal{O}}_{i_1,\ldots,i_4} \end{bmatrix}. \quad (8.13)$$

Furthermore, prior $p(\beta|\boldsymbol{\alpha}_\beta) = \text{gamma}(\beta|10^{-6}, 10^{-6})$ is in the form of (2.26), since

$$p(\beta|\boldsymbol{\alpha}_\beta) = \exp\left(\begin{bmatrix} -10^{-6} \\ 10^{-6} - 1 \end{bmatrix}^H \begin{bmatrix} \beta \\ \log \beta \end{bmatrix} + 10^{-6} \ln 10^{-6} - \ln \Gamma(10^{-6}) \right), \quad (8.14)$$

with $\mathbf{n}(\boldsymbol{\alpha}_\beta) = [-10^{-6}; 10^{-6} - 1]$. Therefore, Condition 1 is satisfied.

For variable γ_l in Layer 2 of Fig. 8.3, $p(\{\boldsymbol{\Xi}^{(k)}\}_{k=1}^4 | \gamma_l, \{\boldsymbol{\eta}^{(2)} \backslash \gamma_l\})$ is

$$p(\{\boldsymbol{\Xi}^{(k)}\}_{k=1}^4 | \gamma_l, \{\boldsymbol{\eta}^{(2)} \backslash \gamma_l\}) = \exp\left\{\Re\mathfrak{e}\left(\begin{bmatrix} -\sum_{k=1}^4 \boldsymbol{\Xi}_{:,l}^{(k)H} \boldsymbol{\Xi}_{:,l}^{(k)} \\ \sum_{k=1}^4 I_k \end{bmatrix}^H \begin{bmatrix} \gamma_l \\ \ln \gamma_l \end{bmatrix} \right.\right.$$
$$\left.\left. + \sum_{j=1, j \neq l}^{L} \left(-\sum_{k=1}^4 \boldsymbol{\Xi}_{:,j}^{(k)H} \boldsymbol{\Xi}_{:,j}^{(k)} \right) \gamma_j + \sum_{k=1}^4 I_k \sum_{j=1, j \neq l}^{L} \ln \gamma_j - \sum_{k=1}^4 I_k R \ln \pi \right)\right\}. \quad (8.15)$$

It takes the form of (2.27) if we define $\mathbf{n}(\{\boldsymbol{\Xi}^{(k)}\}_{k=1}^4, \{\boldsymbol{\eta}^{(2)} \backslash \gamma_l\}) = [-\sum_{k=1}^4 \boldsymbol{\Xi}_{:,l}^{(k)H} \boldsymbol{\Xi}_{:,l}^{(k)}; \sum_{k=1}^4 I_k]$ and $\mathbf{t}(\gamma_l) = [\gamma_l; \ln \gamma_l]$. On the other hand, the prior of γ_l takes the form

$$p(\gamma_l|\boldsymbol{\lambda}_\gamma) = \exp\left(\begin{bmatrix} -10^{-6} \\ 10^{-6} - 1 \end{bmatrix}^H \begin{bmatrix} \gamma_l \\ \log \gamma_l \end{bmatrix} + 10^{-6} \ln 10^{-6} - \ln \Gamma(10^{-6}) \right), \quad (8.16)$$

which is consistent with (2.28) if we define $\mathbf{n}(\boldsymbol{\lambda}_\gamma) = [-10^{-6}; 10^{-6} - 1]$. Therefore, Condition 2 is verified. Finally, variables $\{\boldsymbol{\alpha}_\beta, \boldsymbol{\lambda}_\gamma, \mathbf{m}_r, \dot{\mathcal{O}}\}$ are known quantities, and thus Condition 3 holds. To summarize, the Bayesian tensor completion model is in MPCEF.

8.3.2 Optimal Variational Pdfs Derivations

As shown above, the probabilistic model for tensor completion satisfies Conditions 1–3 of MPCEF, and thus the optimal variational pdfs can be directly obtained using **Theorem 2.1** in Chap. 2. More specifically, the optimal variational pdf $Q^*(\Xi^{(k)}_{\kappa,:})$ can be calculated as

$$Q^*(\Xi^{(k)}_{\kappa,:}) \propto \exp\left\{\mathfrak{Re}\left(\mathbb{E}_{\prod_{\theta_j \neq \Xi^{(k)}_{\kappa,:}}}\begin{bmatrix}\mathbf{n}(\dot{\mathcal{Y}}, \boldsymbol{\eta}^{(1)} \backslash \Xi^{(k)}_{\kappa,:})_1 + \mathbf{0}_{L\times 1} \\ \mathbf{n}(\dot{\mathcal{Y}}, \boldsymbol{\eta}^{(1)} \backslash \Xi^{(k)}_{\kappa,:})_2 - \mathrm{vec}(\boldsymbol{\Gamma})\end{bmatrix}^H \begin{bmatrix}\Xi^{(k)*}_{\kappa,:} \\ \mathrm{vec}(\Xi^{(k)*}_{\kappa,:}\Xi^{(k)T}_{\kappa,:})\end{bmatrix}\right)\right\}. \tag{8.17}$$

Its functional form coincides with the complex Gaussian distribution $\mathcal{CN}(\Xi^{(k)}_{\kappa,:} | \mathbf{M}^{(k)}_{\kappa,:}, \boldsymbol{\Sigma}^{(k)}_{\kappa})$, where

$$\boldsymbol{\Sigma}^{(k)}_{\kappa} = \left[\mathbb{E}_{Q(\gamma_l)}[\boldsymbol{\Gamma}] + \mathbb{E}_{Q(\beta)}[\beta] \sum_{i_1,\dots i_k=\kappa,\dots i_4} \dot{O}_{i_1,\dots,i_4} \mathbb{E}_{\prod_n Q(\Xi^{(n)})}\left[\left(\mathop{\diamond}_{n=1,n\neq k}^{4} \Xi^{(n)T}_{i_n,:}\right)^T \left(\mathop{\diamond}_{p=1,p\neq k}^{4} \Xi^{(p)T}_{i_p,:}\right)^*\right]\right]^{-1}, \tag{8.18}$$

$$\mathbf{M}^{(k)}_{\kappa,:} = \boldsymbol{\Sigma}^{(k)}_{\kappa} \sum_{i_1,\dots i_k=\kappa,\dots i_4} \mathbb{E}_{Q(\beta)}[\beta] \dot{O}_{i_1,\dots,i_4} \dot{\mathcal{Y}}_{i_1,\dots,i_4} \left(\mathop{\diamond}_{n=1,n\neq k}^{4} \mathbb{E}_{Q(\Xi^{(n)}_{i_n,:})}[\Xi^{(n)T}_{i_n,:}]\right)^T. \tag{8.19}$$

Similarly, the optimal variational pdf $Q^*(\beta)$ is derived to be

$$Q^*(\beta) \propto \exp\left\{\mathfrak{Re}\left(\mathbb{E}_{\prod_{\theta_j \neq \beta}}\begin{bmatrix}\mathbf{n}(\dot{\mathcal{Y}}, \boldsymbol{\eta}^{(1)} \backslash \beta)_1 - 10^{-6} \\ \mathbf{n}(\dot{\mathcal{Y}}, \boldsymbol{\eta}^{(1)} \backslash \beta)_2 + 10^{-6} - 1\end{bmatrix}^H \begin{bmatrix}\beta \\ \ln \beta\end{bmatrix}\right)\right\}, \tag{8.20}$$

which is a gamma distribution gamma$(\beta | c, d)$ with

$$c = 10^{-6} + \sum_{i_1,\dots,i_4} \dot{O}_{i_1,\dots,i_4}, \tag{8.21}$$

$$d = 10^{-6} + \mathbb{E}_{\prod_n Q(\Xi^{(n)})}\left[\sum_{i_1,\dots,i_4} \dot{O}_{i_1,\dots,i_4} \left|\dot{\mathcal{Y}}_{i_1,\dots,i_4} - [\![\Xi^{(1)}, \Xi^{(2)}, \Xi^{(3)}, \Xi^{(4)}]\!]_{i_1,\dots,i_4}\right|^2\right]. \tag{8.22}$$

For variable γ_l, its optimal variational pdf $Q^*(\gamma_l)$ is

$$Q^*(\gamma_l) \propto \exp\left\{\mathfrak{Re}\left(\mathbb{E}_{\prod_{\theta_j \neq \gamma_l}}\begin{bmatrix}-\sum_{k=1}^{4} \Xi^{(k)H}_{:,l} \Xi^{(k)}_{:,l} - 10^{-6} \\ \sum_{k=1}^{4} I_k + 10^{-6} - 1\end{bmatrix}^H \begin{bmatrix}\gamma_l \\ \ln \gamma_l\end{bmatrix}\right)\right\}. \tag{8.23}$$

It can be seen that $Q^*(\gamma_l)$ follows a gamma distribution gamma$(\beta|a_l, b_l)$, where

$$a_l = 10^{-6} + \sum_{k=1}^{4} I_k \tag{8.24}$$

$$b_l = 10^{-6} + \sum_{k=1}^{4} \mathbb{E}_{\prod_n Q(\Xi^{(n)})}\left[\Xi_{:,l}^{(k)H}\Xi_{:,l}^{(k)}\right]. \tag{8.25}$$

In (8.17)–(8.25), there are several expectations to be computed. Simple expectations $\mathbb{E}_{Q(\Xi_{\kappa,:}^{(k)})}[\Xi_{\kappa,:}^{(k)}]$, $\mathbb{E}_{Q(\gamma_l)}[\gamma_l]$, and $\mathbb{E}_{Q(\beta)}[\beta]$ can be obtained using results presented in previous chapters. However, there are two expectations $\mathbb{E}_{\prod_n Q(\Xi^{(n)})}\left[\left(\underset{n=1,n\neq k}{\overset{4}{\diamond}} \Xi_{i_n,:}^{(n)T}\right)^T \left(\underset{p=1,p\neq k}{\overset{4}{\diamond}} \Xi_{i_p,:}^{(p)T}\right)^*\right]$ and $\mathbb{E}_{\prod_n Q(\Xi^{(n)})}\left[\left|\dot{\mathcal{Y}}_{i_1,\ldots,i_4} - [\![\Xi^{(1)}, \Xi^{(2)}, \Xi^{(3)}, \Xi^{(4)}]\!]_{i_1,\ldots,i_4}\right|^2\right]$ that are more difficult to compute. By **Property 6.2**, they are derived as

$$\mathbb{E}_{\prod_n Q(\Xi^{(n)})}\left[\left(\underset{n=1,n\neq k}{\overset{4}{\diamond}} \Xi_{i_n,:}^{(n)T}\right)^T \left(\underset{p=1,p\neq k}{\overset{4}{\diamond}} \Xi_{i_p,:}^{(p)T}\right)^*\right] = \underset{n=1,n\neq k}{\overset{4}{\odot}} \left[\boldsymbol{M}_{i_n,:}^{(n)}\boldsymbol{M}_{i_n,:}^{(n)H} + \boldsymbol{\Sigma}_{i_n}^{(n)*}\right], \tag{8.26}$$

$$\begin{aligned}
&\mathbb{E}_{\prod_n Q(\Xi^{(n)})}\left[\left|\dot{\mathcal{Y}}_{i_1,\ldots,i_4} - [\![\Xi^{(1)}, \Xi^{(2)}, \Xi^{(3)}, \Xi^{(4)}]\!]_{i_1,\ldots,i_4}\right|^2\right] \\
&= \left|\dot{\mathcal{Y}}_{i_1,\ldots,i_4}\right|^2 - 2\mathfrak{Re}\left(\dot{\mathcal{Y}}_{i_1,\ldots,i_4}^{*}\boldsymbol{M}_{i_1,:}^{(1)T}\left(\overset{4}{\underset{n=2}{\diamond}} \boldsymbol{M}_{i_n,:}^{(n)T}\right)^T\right) \\
&+ \mathrm{Tr}\left(\left[\boldsymbol{M}_{i_1,:}^{(1)}\boldsymbol{M}_{i_1,:}^{(1)H} + \boldsymbol{\Sigma}_{i_1}^{(1)}\right] \overset{4}{\underset{n=2}{\odot}} \left[\boldsymbol{M}_{i_n,:}^{(n)}\boldsymbol{M}_{i_n,:}^{(n)H} + \boldsymbol{\Sigma}_{i_n}^{(n)*}\right]\right).
\end{aligned} \tag{8.27}$$

8.4 Algorithm Summary and Remarks

The variational inference algorithm for the proposed model is summarized in **Algorithm 11**. After convergence, some $\mathbb{E}_{Q^*(\gamma_l)}[\gamma_l] = a_l^t/b_l^t$ will be very large, e.g., 10^6, indicating that the power of corresponding columns goes to zero. Then, these columns can be safely pruned out, and the remaining column number is the estimate of the path number R. Meanwhile, since $Q^*(\Xi_{\kappa,:}^{(k)})$ is Gaussian distribution, factor matrices $\{\Xi^{(k)}\}_{k=1}^4$ are estimated by the $\{\boldsymbol{M}^{(k,t)}\}_{k=1}^4$ at convergence.

According to **Property 8.1**, the subspaces spanned by the columns in $\boldsymbol{A}[\boldsymbol{u}]$, $\boldsymbol{A}[\boldsymbol{v}]$, and $\boldsymbol{A}[\boldsymbol{p}]$ are estimated by the range spaces of $\boldsymbol{M}^{(1,t)}$, $\boldsymbol{M}^{(2,t)}$, and $\boldsymbol{M}^{(3,t)}$, respectively. Thereby, 1D subspace-based DOA estimation methods, such as multiple signal classification (MUSIC) and estimation of signal parameters via rotation invariance technique (ESPRIT), can be applied to the range spaces of $\{\boldsymbol{M}^{(k,t)}\}_{k=1}^3$ to separately estimate the DOAs θ_r and ϕ_r. Details of applying 1D DOA methods in 2D DOA estimation can be found in [2].

Algorithm 11 Probabilistic Tensor CPD for DOA Estimation

Initialization: Choose $L > R$, and initial values $\{\boldsymbol{M}^{(k,0)} \in \mathbb{C}^{I_k \times L}, \{\boldsymbol{\Sigma}_{\kappa}^{(k,0)} \in \mathbb{C}^{L\times L}\}_{\kappa=1}^{I_k}\}_{k=1}^{4}$. Let $a_l^0 = 10^{-6}, b_l^0 = 10^{-6}$ for $l = 1, 2, 3, ..., L$; $c^0 = d^0 = 10^{-6}$.
Iterations: For the tth iteration ($t \geq 1$),
Update the parameter of each $Q^*(\Xi_{\kappa,:}^{(k)})^t$:

$$\boldsymbol{\Sigma}_{\kappa}^{(k,t)} = \left[\frac{c^{t-1}}{d^{t-1}} \sum_{i_1,...,i_k=\kappa,...i_4} \dot{O}_{i_1,...,i_4} \bigodot_{n=1,n\neq k}^{4} \left[\boldsymbol{M}_{i_n,:}^{(n,t-1)} \boldsymbol{M}_{i_n,:}^{(n,t-1)H} + \boldsymbol{\Sigma}_{i_n}^{(n,t-1)*}\right] + \text{diag}\left\{\frac{a_1^{t-1}}{b_1^{t-1}}, \ldots, \frac{a_L^{t-1}}{b_L^{t-1}}\right\}\right]^{-1}, \tag{8.28}$$

$$\boldsymbol{M}_{\kappa,:}^{(k,t)} = \frac{c^{t-1}}{d^{t-1}} \boldsymbol{\Sigma}_{\kappa}^{(k,t)} \sum_{i_1,...i_k=\kappa,...i_4} \left(\dot{O}_{i_1,...,i_4} \dot{\mathcal{Y}}_{i_1,...,i_4} \left(\mathop{\diamond}_{n=1,n\neq k}^{4} \boldsymbol{M}_{i_n,:}^{(n,t-1)T}\right)^T\right), \tag{8.29}$$

Notice that for each $k \in \{1, 2, 3, 4\}$, the update for $\kappa = 1, 2, \ldots, I_k$ can be updated in parallel.
Update the parameter of $Q^*(\{\gamma_l\}_{l=1}^{L})^t$:

$$a_l^t = 10^{-6} + \sum_{k=1}^{4} I_k, \tag{8.30}$$

$$b_l^t = 10^{-6} + \sum_{k=1}^{4}\sum_{\kappa=1}^{I_k} \left|\boldsymbol{M}_{\kappa,l}^{(k,t)}\right|^2 + \left[\boldsymbol{\Sigma}_{\kappa}^{(k,t)}\right]_{l,l}. \tag{8.31}$$

Update the parameter of $Q^*(\beta)^t$:

$$c^t = 10^{-6} + \sum_{i_1,...,i_4} \dot{O}_{i_1,...,i_4}, \tag{8.32}$$

$$d^t = 10^{-6} + \sum_{i_1,...,i_4} \dot{O}_{i_1,...,i_4} \left(\left|\dot{\mathcal{Y}}_{i_1,...,i_4}\right|^2 - 2\Re\mathfrak{e}\left(\dot{\mathcal{Y}}_{i_1,...,i_4}^{*} \boldsymbol{M}_{i_1,:}^{(1,t)T} \left(\mathop{\diamond}_{n=2}^{4} \boldsymbol{M}_{i_n,:}^{(n,t)T}\right)^T\right)\right.$$
$$\left. + \text{Tr}\left(\left[\boldsymbol{M}_{i_1,:}^{(1,t)} \boldsymbol{M}_{i_1,:}^{(1,t)H} + \boldsymbol{\Sigma}_{i_1}^{(1,t)}\right] \bigodot_{n=2}^{4} \left[\boldsymbol{M}_{i_n,:}^{(n,t)} \boldsymbol{M}_{i_n,:}^{(n,t)H} + \boldsymbol{\Sigma}_{i_n}^{(n,t)*}\right]\right)\right). \tag{8.33}$$

Until Convergence

8.5 Simulation Results and Discussions

In this subsection, numerical results are presented to assess the performance of the proposed method (labeled as VI) for subspace estimation over $I_4 = 100$ snapshots. The arbitrary array was generated by randomly deploying M sensor elements so that the projected coordinates form a 3D grid with dimensions $I_1 = 8$, $I_2 = 10$, $I_3 = 12$ and the inter-grid spacing $d_x = \lambda_c/2$, $d_y = \lambda_c/4$, $d_z = \lambda_c/8$. We consider two scenarios for the arbitrary array: $M = 288$ and $M = 576$, corresponding to $\pi = 0.3$ and $\pi = 0.6$ of the grid points being occupied by sensors, respectively. There are 3 sources with elevation DOAs $\{15°, 25°, 120°\}$ and azimuth DOAs $\{-50°, 10°, 70°\}$, respectively. The transmitted signal of each source $\xi_r(n)$ is drawn from a zero-mean circularly symmetric complex Gaussian distribution with unit variance, and

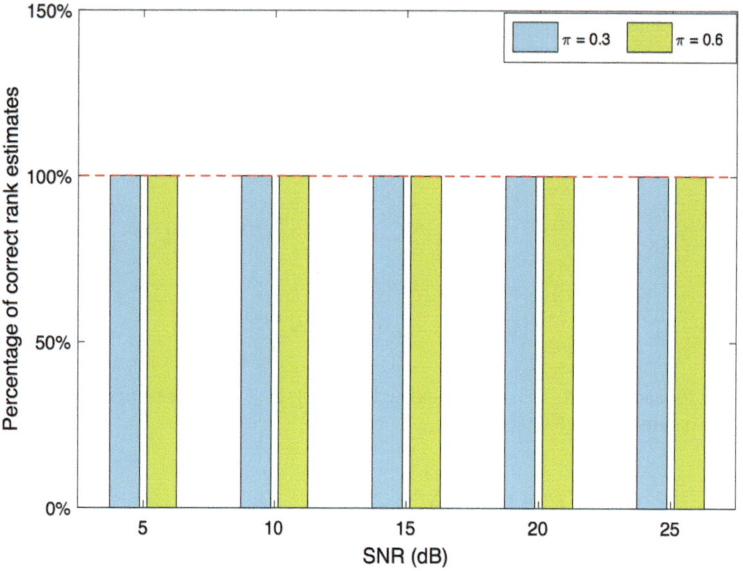

Fig. 8.4 Percentage of correct rank estimates

without any correlation across r and n. The signal-to-noise ratio (SNR) is defined as $10\log_{10}\left(\frac{\|\dot{\mathcal{O}}\odot[\![\boldsymbol{A}[\boldsymbol{u}],\boldsymbol{A}[\boldsymbol{v}],\boldsymbol{A}[\boldsymbol{p}],\boldsymbol{A}[\boldsymbol{\xi}]]\!]\|_F^2}{\|\mathcal{W}\|_F^2}\right)$. For **Algorithm 11**, initial mean $\boldsymbol{M}^{(k,0)}$ for each matrix $\boldsymbol{\Xi}^{(k)}$ is set as the singular value decomposition (SVD) approximation $\boldsymbol{U}_{:,1:R}\left(\boldsymbol{S}_{1:R,1:R}\right)^{\frac{1}{2}}$ where $[\boldsymbol{U},\boldsymbol{S},\boldsymbol{V}] = \mathrm{SVD}[[\dot{\mathcal{Y}}]^{(k)}]$, $L = \max\{I_1, I_2, I_3, I_4\}$. The initial covariance matrix $\boldsymbol{\Sigma}_{\kappa}^{(k,0)} = \boldsymbol{I}_{L\times L}$. Each point in the following figures is an average of 1000 Monte Carlo runs with different realizations of the arbitrary arrays, signals, and noises.

To access the ability of learning the number of sources, the subspace rank learned by the proposed algorithm is shown in Fig. 8.4, with each vertical bar showing the percentages of correct estimates. From Fig. 8.4, it is seen that **Algorithm 11** can recover the true tensor rank with 100% accuracy for a wide range of SNRs. Notice that **Algorithm 11** can only recover the correct tensor rank with 100% accuracy in moderate and high-SNR regions, and particularly in the region SNR > 5 dB for this application.

Finally, the performance of DOA subspace estimation in terms of the averaged largest principal angle $\frac{1}{3}\left[\mathrm{LPA}(\boldsymbol{M}^{(1)}, \boldsymbol{A}[\boldsymbol{u}]) + \mathrm{LPA}(\boldsymbol{M}^{(2)}, \boldsymbol{A}[\boldsymbol{v}]) + \mathrm{LPA}(\boldsymbol{M}^{(3)}, \boldsymbol{A}[\boldsymbol{p}])\right]$ is shown in Fig. 8.5. $\mathrm{LPA}(\boldsymbol{A}, \boldsymbol{B})$ is a measure of the "distance" between two subspaces spanned by the columns of matrices $\boldsymbol{A}$ and $\boldsymbol{B}$, defined as $\cos^{-1}\{\sigma_{\min}\{\mathrm{orth}\{\boldsymbol{A}\}^H\ \mathrm{orth}\{\boldsymbol{B}\}\}\}$, where the operator $\sigma_{\min}\{\boldsymbol{Q}\}$ denotes the smallest singular value of the matrix $\boldsymbol{Q}$ and $\mathrm{orth}\{\boldsymbol{Q}\}$ is an orthonormal basis for the subspace spanned by the columns of $\boldsymbol{Q}$. The range of LPA is from 0° to 90°, where LPA

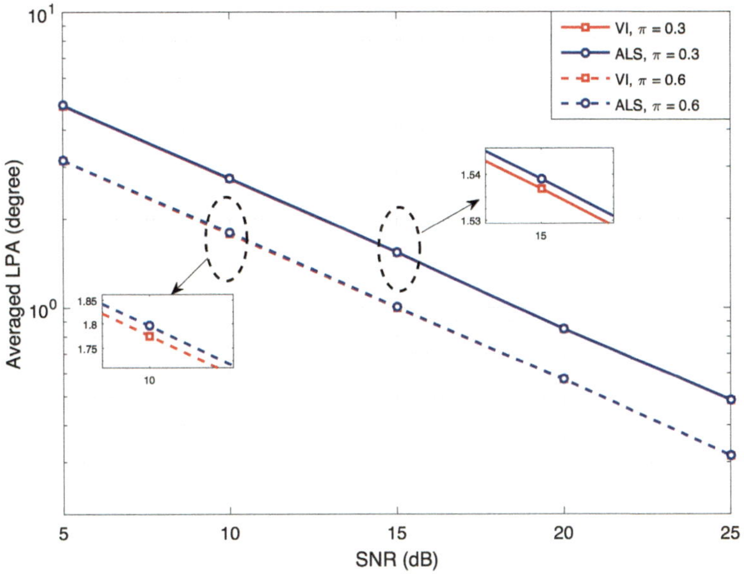

Fig. 8.5 Averaged largest principal angle (LPA) of subspace estimation

0° means that the two subspaces are identical while LPA 90° means that the two subspaces are orthogonal.

From Fig. 8.5, it is seen that **Algorithm 11** gives averaged LPA less than 5° when SNR is larger than 5 dB, no matter $\pi = 0.3$ or $\pi = 0.6$. It shows that **Algorithm 11** gives relatively good subspace estimations. On the other hand, for comparison, a recently proposed alternating least square (ALS)-based tensor completion method [10] (labeled as ALS) is also simulated. It assumes knowing the exact tensor rank $R = 3$, but takes the same initial estimates of factor matrices and termination criterion as that in **Algorithm 11**. From Fig. 8.5, it is apparent that the ALS-based algorithm performs almost indistinguishably from **Algorithm 11**. However, this result is obtained with the ALS-based algorithm knowing the tensor rank, while **Algorithm 11** without knowing the tensor rank. This shows that **Algorithm 11** supersedes the existing algorithm [10] by providing additional rank learning ability.

References

1. Van Trees, *Detection, Estimation, and Modulation Theory, Optimum Array Processing* (Wiley, New York, 2004)
2. L. Cheng, Y.-C. Wu, J. (Charlie) Zhang, L. Liu, Subspace identification for DOA estimation in massive/full-dimension mimo system: bad data mitigation and automatic source enumeration. IEEE Trans. Signal Process. **63**(22), 5897–5909 (2015)

3. M. Haardt, F. Roemer, G.D. Galdo, Higher-order SVD based subspace estimation to improve the parameter estimation accuracy in multi-dimensional harmonic retrieval problems. IEEE Trans. Signal Process. **56**(7), 3198–3213 (2008)
4. F. Roemer, M. Haardt, G.D. Galdo, Analytical performance assessment of multi-dimensional matrix-and tensor-based ESPRIT-type algorithms. IEEE Trans. Signal Process. **62**(10), 2611–2625 (2014)
5. W. Sun, H.C. So, F.K.W. Chan, L. Huang, Tensor approach for eigenvector-based multi-dimensional harmonic retrieval. IEEE Trans. Signal Process. **61**(13), 3378–3388 (2013)
6. X. Guo, S. Miron, D. Brie, S. Zhu, X. Liao, A CANDECOMP/PARAFAC perspective on uniqueness of DOA estimation using a vector sensor array. IEEE Trans. Signal Process. **59**(9), 3475–3481 (2011)
7. Q. Zhao, L. Zhang, A. Cichocki, Bayesian CP factorization of incomplete tensors with automatic rank determination. IEEE Trans. Pattern Anal. Mach. Intell. **37**(9), 1751–1753 (2015)
8. P.C. Hansen, *Rank Deficient and Discrete Ill-Posed Problems-Numerical Aspects of Linear Inversion* (SIAM, Philadelphia, PA, 1998)
9. K.P. Murphy, *Machine Learning: A Probabilistic Perspective* (MIT Press, Cambridge, 2012)
10. L. Karlsson, D. Kressner, A. Uschmajew, Parallel algorithms for tensor completion in the CP format. Parallel Comput. **57**, 222–234 (2016)
11. L. Cheng, C. Xing, Y.-C. Wu, Irregular array manifold aided channel estimation in massive MIMO communications. IEEE J. Selected Topics Signal Process. **13**(5), 974–988 (2019)

Chapter 9
From CPD to Other Tensor Decompositions

Abstract In previous chapters, we have introduced tensor rank learning using GSM priors for tensor CPD and its extensions to scenarios where additional prior information exists or the data structure is altered. In this chapter, we present tensor rank learning for other tensor decomposition formats. It turns out that what has been presented for CPD is instrumental for other Bayesian tensor modelings, as they share many common characteristics.

9.1 Tucker Decomposition (TuckerD)

Tucker decomposition (TuckerD) is introduced in Chap. 1,

$$\mathcal{X} = \mathcal{G} \times_1 \mathbf{U}^{(1)} \times_2 \mathbf{U}^{(2)} \times_3 \cdots \times_N \mathbf{U}^{(N)}, \tag{9.1}$$

where $\mathcal{X} \in \mathbb{R}^{I_1 \times \cdots \times I_N}$ is the target tensor, $\mathcal{G} \in \mathbb{R}^{R_1 \times \cdots \times R_N}$ is a core tensor, and $\{\mathbf{U}^{(n)} \in \mathbb{R}^{I_n \times R_n}\}_{n=1}^N$ are factor matrices. The tuple $(R_1, \ldots, R_N)$ is known as *multi-linear rank*. A graphical illustration of Tucker format is provided in Fig. 1.6 and its deterministic algorithm is summarized in **Algorithm 2**. However, **Algorithm 2** requires the knowledge of the multi-linear rank, which is typically unknown in practice.

To avoid extensive multi-linear rank tuning, a probabilistic approach using Gaussian-gamma prior is developed in [1]. Notice that rank R_n is the column number of factor matrix $\mathbf{U}^{(n)}$. Following the general philosophy introduced in Sect. 3.1, GSM priors can be placed on the columns of $\mathbf{U}^{(n)}$ to learn rank R_n. In particular, the Gaussian-gamma prior is chosen in [1]. Given the multi-linear rank upper bounds $\{L_n\}_{n=1}^N$, the Gaussian-gamma prior for the factor matrices $\{\mathbf{U}^{(n)} \in \mathbb{R}^{I_n \times L_n}\}_{n=1}^N$ is

$$p(\{\mathbf{U}^{(n)}\}_{n=1}^N | \{\boldsymbol{\lambda}^{(n)}\}_{n=1}^N) = \prod_{n=1}^N p(\{\mathbf{U}^{(n)}\}_{n=1}^N | \boldsymbol{\lambda}^{(n)}) = \prod_{n=1}^N \prod_{l_n=1}^{L_n} \mathcal{N}(\mathbf{U}^{(n)}_{:,l_n} | \mathbf{0}_{I_n \times 1}, (\lambda^{(n)}_{l_n})^{-1} \mathbf{I}_{I_n}), \tag{9.2}$$

L. Cheng et al., *Bayesian Tensor Decomposition for Signal Processing and Machine Learning*, https://doi.org/10.1007/978-3-031-22438-6_9

$$p(\{\boldsymbol{\lambda}^{(n)}\}_{n=1}^{N}|a_0^{\lambda}, b_0^{\lambda}) = \prod_{n=1}^{N} p(\boldsymbol{\lambda}^{(n)}|a_0^{\lambda}, b_0^{\lambda}) = \prod_{n=1}^{N}\prod_{l=1}^{L} \text{gamma}(\lambda_{l_n}^{(n)}|a_0^{\lambda}, b_0^{\lambda}), \tag{9.3}$$

where $\boldsymbol{\lambda}^{(n)} \in \mathbb{R}^{L\times 1}$ collects precisions of columns in factor matrix $\mathbf{U}^{(n)}$ and $a_0^{\lambda}, b_0^{\lambda}$ are pre-determined hyper-parameters shared by all random vector $\{\boldsymbol{\lambda}^{(n)}\}_{n=1}^{N}$. Compared to Bayesian CPD modeling, Bayesian Tucker has a distinct vector $\boldsymbol{\lambda}^{(n)}$ for each factor matrix. If certain precision $\lambda_{l_n}^{(n)}$ is learned to be a very large value, it indicates that the corresponding column $\mathbf{U}_{:,l_n}^{(n)}$ is zero and thus tensor rank learning is achieved in each factor matrix (i.e., multi-linear rank).

While an independent sparsity-promoting prior can be placed on the core tensor $\mathcal{G}$, it ignores the relation between the core tensor $\mathcal{G}$ and the factor matrices $\{\mathbf{U}^{(n)} \in \mathbb{R}^{I_n\times R_n}\}_{n=1}^{N}$. To see this, we rewrite TuckerD (9.1) as

$$\mathcal{X} = \sum_{r_1=1}^{R_1}\sum_{r_2=1}^{R_2}\cdots\sum_{r_N=1}^{R_N} \mathcal{G}_{r_1,r_2,\ldots,r_N} \mathbf{U}_{:,r_1}^{(1)} \circ \mathbf{U}_{:,r_2}^{(2)} \circ \cdots \circ \mathbf{U}_{:,r_N}^{(N)}, \tag{9.4}$$

which is a weighted sum of rank-1 tensors, where each rank-1 tensor is $\mathbf{U}_{:,r_1}^{(1)} \circ \mathbf{U}_{:,r_2}^{(2)} \circ \cdots \circ \mathbf{U}_{:,r_N}^{(N)}$ with coefficient $\mathcal{G}_{r_1,r_2,\ldots,r_N}$. Equation (9.4) demonstrates the relation between the core tensor $\mathcal{G}$ and factor matrices $\{\mathbf{U}^{(n)}\}_{n=1}^{N}$. That is, once $\mathbf{U}_{:,l_n}^{(n)}$ is determined to be zero, the corresponding core tensor slice $\mathcal{G}_{\ldots,l_n,\ldots}$ can also be enforced to zero. To model such a relation, a Gaussian-gamma prior is placed on each element of the core tensor as [1]

$$p(\mathcal{G}|\{\boldsymbol{\lambda}^{(n)}\}_{n=1}^{N}, \beta) = \prod_{l_1=1}^{L_1}\cdots\prod_{l_N=1}^{L_N} \mathcal{N}\Big(\mathcal{G}_{l_1,l_2,\ldots,l_N}|0, \Big(\beta\prod_{n=1}^{N}\lambda_{l_n}^{(n)}\Big)^{-1}\Big). \tag{9.5}$$

Notice that the precision of a particular tensor core element includes a product of precisions $\prod_{n=1}^{N}\lambda_{l_n}^{(n)}$. Therefore, if some $\lambda_{l_n}^{(n)}$ are learned to be a very large value, the core tensor element is also zero. In (9.5), β is a scale parameter related to the magnitude of $\mathcal{G}$ and its prior is a gamma distribution $\text{gamma}(\beta|a_0^{\beta}, b_0^{\beta})$.

Since it is assumed that the observation noise is Gaussian, the likelihood function is specified as

$$p(\mathcal{Y}|\mathcal{G}, \{\mathbf{U}^{(n)}\}_{n=1}^{N}, \tau) \propto \exp\Big(-\frac{\tau}{2}\|\mathcal{Y} - \mathcal{G}\times_1\mathbf{U}^{(1)}\times_2\mathbf{U}^{(2)}\times_3\cdots\times_N\mathbf{U}^{(N)}\|_F^2\Big), \tag{9.6}$$

where τ is the noise precisions and it is assigned with a Gamma prior distribution $\text{gamma}(\tau|a_0^{\tau}, b_0^{\tau})$. To summarize, the joint distribution of tensor data $\mathcal{Y}$ and model parameters $\boldsymbol{\Theta} = \{\{\mathbf{U}^{(n)}\}_{n=1}^{N}, \mathcal{G}, \{\boldsymbol{\lambda}^{(n)}\}_{n=1}^{N}, \beta, \tau\}$ is

$$p(\mathcal{Y}, \Theta) = p(\mathcal{Y}|\mathcal{G}, \{\mathbf{U}^{(n)}\}_{n=1}^N, \tau) p(\{\mathbf{U}^{(n)}\}_{n=1}^N | \{\boldsymbol{\lambda}^{(n)}\}_{n=1}^N) p(\mathcal{G}|\{\boldsymbol{\lambda}^{(n)}\}_{n=1}^N, \beta)$$
$$\times\, p(\{\boldsymbol{\lambda}^{(n)}\}_{n=1}^N | a_0^\lambda, b_0^\lambda) p(\beta|a_0^\beta, b_0^\beta) p(\tau|a_0^\tau, b_0^\tau). \tag{9.7}$$

The mean-field variational inference (MF-VI) algorithm is derived in [1] to learn model parameters Θ from tensor data $\mathcal{Y}$. The mean field employed is $Q(\Theta) = Q(\mathcal{G})Q(\beta)\prod_{n=1}^N Q(\mathbf{U}^{(n)})\prod_{n=1}^N Q(\boldsymbol{\lambda}^{(n)})Q(\tau)$, and the resultant Bayesian algorithm is summarized in **Algorithm 12**. Interested readers can refer to [1] for more details.

9.2 Tensor Train Decomposition (TTD)

Tensor Train Decomposition (TTD) decomposes tensor data $\mathcal{X} \in \mathbb{R}^{I_1 \times \cdots \times I_N}$ into a set of core tensors $\{\mathcal{G}^{(n)} \in \mathbb{R}^{R_n \times I_n \times R_{n+1}}\}_{n=1}^N$, where each element is in the form

$$\mathcal{X}_{i_1,\ldots,i_N} = \mathcal{G}^{(1)}_{:,i_1,:} \times \cdots \times \mathcal{G}^{(N)}_{:,i_N,:}. \tag{9.8}$$

The tuple $(R_1, \ldots, R_{N+1})$ is termed as *TT-rank*. While a deterministic algorithm for TTD is summarized in **Algorithm 3**, it requires the user to specify the prescribed accuracy, which controls the TT-rank.

To bypass tuning of the prescribed accuracy to obtain satisfactory performance, [2] formally introduces a novel Gaussian-product-gamma prior to model sparsity in TTD. In particular, assuming the TT-rank upper bound is $(1, L_2, \ldots, L_N, 1)$ (in (9.8), each $\mathcal{X}_{i_1,\ldots,i_N}$ is a scalar, and thus R_1 and R_{N+1} are both required to be 1), the Gaussian-product-gamma prior for core tensor $\mathcal{G}^{(n)}$ is

$$p(\mathcal{G}^{(n)}|\boldsymbol{\lambda}^{(n)}, \boldsymbol{\lambda}^{(n+1)}) = \prod_{k=1}^{L_n}\prod_{l=1}^{L_{n+1}} \mathcal{N}(\mathcal{G}^{(n)}_{k,:,l}|\mathbf{0}_{I_n \times 1}, (\lambda_k^{(n)}\lambda_l^{(n+1)})^{-1}\mathbf{I}_{I_n}), \forall n \in \{1, \ldots, N\}, \tag{9.9}$$

$$p(\boldsymbol{\lambda}^{(n)}|\boldsymbol{\alpha}^{(n)}, \boldsymbol{\beta}^{(n)}) = \prod_{k=1}^{L_n} \text{gamma}(\lambda_k^{(n)}|\alpha_k^{(n)}, \beta_k^{(n)}), \forall n \in \{2, \ldots, N\}, \tag{9.10}$$

where $\boldsymbol{\lambda}^{(n)} = [\lambda_1^{(n)}, \ldots, \lambda_{L_n}^{(n)}], \forall n \in \{2, \ldots, N\}$, and $\lambda^{(1)}, \lambda^{(N+1)}$ are set as 1. $\boldsymbol{\alpha}^{(n)} \in \mathbb{R}^{L_n}, \boldsymbol{\beta}^{(n)} \in \mathbb{R}^{L_n}$ are pre-determined hyper-parameters of the gamma distributions. From (9.9), if either $\lambda_k^{(n)}$ or $\lambda_l^{(n+1)}$ is learned to be a very large value, $\mathcal{G}^{(n)}_{k,:,l}$ is zero. It is theoretically proved in [2] that such prior could lead to sparsity in TT slices. Interested readers can refer to [2] for more details of the Gaussian-product-gamma prior. Notice that (9.9) bears some resemblance with the core tensor modeling (9.5) in TuckerD, where the precesions of different factor matrix columns are also coupled together.

Algorithm 12 Probabilistic TuckerD [1]

Initializations: Choose $L_n > R_n, \forall n$ and initial values $\mathrm{vec}(\widetilde{\mathcal{G}})$, Σ_G, $\{\widetilde{\mathbf{U}}^{(n)}, \Psi^{(n)}\}_{n=1}^{N}$, a_M^{β}, b_M^{β}, $\{\tilde{a}_{l_n}^{(n)}, \tilde{b}_{l_n}^{(n)}\}_{l_n=1,n=1}^{L_n,N}$, a_M^{τ}, b_M^{τ}.

Iterations:

Update $\mathcal{G}$

The optimal variational pdf of $\mathcal{G}$ is a Gaussian distribution $Q(\mathcal{G}) = \mathcal{N}\left(\mathrm{vec}(\mathcal{G}) \mid \mathrm{vec}(\widetilde{\mathcal{G}}), \Sigma_G\right)$ with its covariance and mean as

$$\mathrm{vec}(\widetilde{\mathcal{G}}) = \mathbb{E}[\tau]\Sigma_G \left(\bigodot_n \mathbb{E}\left[\mathbf{U}^{(n)T}\right]\right)\mathrm{vec}(\mathcal{Y}), \tag{9.11}$$

$$\Sigma_G = \left\{\mathbb{E}[\beta] \bigodot_n \mathbb{E}\left[\boldsymbol{\Lambda}^{(n)}\right] + \mathbb{E}[\tau] \bigodot_n \mathbb{E}\left[\mathbf{U}^{(n)T}\mathbf{U}^{(n)}\right]\right\}^{-1}. \tag{9.12}$$

Update $U^{(n)}$ from $n = 1$ to N

The optimal variational pdf of $U^{(n)}$ is $Q\left(\mathbf{U}^{(n)}\right) = \prod_{i_n=1}^{I_n} \mathcal{N}\left(\mathbf{U}_{i_n,:}^{(n)} \mid \widetilde{\mathbf{U}}_{i_n,:}^{(n)}, \Psi^{(n)}\right)$, where

$$\widetilde{\mathbf{U}}^{(n)} = \mathbb{E}[\tau]\mathbf{Y}_{(n)}\left(\bigodot_{k\neq n} \mathbb{E}\left[\mathbf{U}^{(k)}\right]\right)\mathbb{E}\left[\mathbf{G}_{(n)}^{T}\right]\Psi^{(n)}, \tag{9.13}$$

$$\Psi^{(n)} = \left\{\mathbb{E}\left[\boldsymbol{\Lambda}^{(n)}\right] + \mathbb{E}[\tau]\mathbb{E}\left[\mathbf{G}_{(n)}\left(\bigodot_{k\neq n} \mathbf{U}^{(k)T}\mathbf{U}^{(k)}\right)\mathbf{G}_{(n)}^{T}\right]\right\}^{-1}. \tag{9.14}$$

Update β

The optimal variational pdf of β is $Q(\beta) = \mathrm{gamma}\left(a_M^{\beta}, b_M^{\beta}\right)$ with parameters

$$a_M^{\beta} = a_0^{\beta} + \frac{1}{2}\prod_n L_n, \tag{9.15}$$

$$b_M^{\beta} = b_0^{\beta} + \frac{1}{2}\mathbb{E}\left[\mathrm{vec}\left(\mathcal{G}^2\right)^T\right]\bigodot_n \mathbb{E}\left[\boldsymbol{\lambda}^{(n)}\right]. \tag{9.16}$$

Update $\boldsymbol{\lambda}^{(n)}$ from $n = 1$ to N

The optimal variational pdf of $\boldsymbol{\lambda}^{(n)}$ is $Q\left(\boldsymbol{\lambda}^{(n)}\right) = \prod_{l_n=1}^{L_n} \mathrm{gamma}\left(\lambda_{l_n}^{(n)} \mid \tilde{a}_{l_n}^{(n)}, \tilde{b}_{l_n}^{(n)}\right)$ with parameters

$$\tilde{a}_{l_n}^{(n)} = a_0^{\lambda} + \frac{1}{2}\left(I_n + \prod_{k\neq n} L_k\right), \tag{9.17}$$

$$\tilde{b}_{l_n}^{(n)} = b_0^{\lambda} + \frac{1}{2}\mathbb{E}\left[\mathbf{u}_{\cdot l_n}^{(n)T}\mathbf{u}_{\cdot l_n}^{(n)}\right] + \frac{1}{2}\mathbb{E}[\beta]\mathbb{E}\left[\mathrm{vec}\left(\mathcal{G}_{\cdots l_n \cdots}^2\right)^T\right]\bigodot_{k\neq n} \mathbb{E}\left[\boldsymbol{\lambda}^{(k)}\right]. \tag{9.18}$$

Update τ

The optimal variational pdf of τ is $Q(\tau) = \mathrm{gamma}\left(a_M^{\tau}, b_M^{\tau}\right)$ with parameters

$$a_M^{\tau} = a_0^{\tau} + \frac{1}{2}\prod_n I_n, \tag{9.19}$$

$$b_M^{\tau} = b_0^{\tau} + \frac{1}{2}\mathbb{E}\left[\left\|\mathrm{vec}(\mathcal{Y}) - \left(\bigodot_n \mathbf{U}^{(n)}\right)\mathrm{vec}(\mathbf{G})\right\|_F^2\right]. \tag{9.20}$$

Until Convergence

The likelihood function is specified as

$$p(\mathcal{Y}|\{\mathcal{G}^{(n)}\}_{n=1}^{N}, \tau) \propto \exp\left(-\frac{\tau}{2}\sum_{i_1=1}^{I_1}\cdots\sum_{i_n=1}^{I_N}(\mathcal{Y}_{i_1,\ldots,i_N} - \mathcal{G}^{(1)}_{:,i_1,:} \times \cdots \times \mathcal{G}^{(N)}_{:,i_N,:})^2\right), \tag{9.21}$$

where τ is the noise precesion and it is assigned with a Gamma prior distribution $\text{gamma}(\tau|a_0^{\tau}, b_0^{\tau})$. The complete probabilistic model is summarized in the joint distribution of tensor data $\mathcal{Y}$ and model parameters $\boldsymbol{\Theta} = \{\{\mathcal{G}^{(n)}\}_{n=1}^{N}, \{\boldsymbol{\lambda}^{(n)}\}_{n=2}^{N}, \tau\}$,

$$p(\mathcal{Y}, \boldsymbol{\Theta}) = p(\mathcal{Y}|\{\mathcal{G}^{(n)}\}_{n=1}^{N}, \tau)\prod_{n=1}^{N} p(\mathcal{G}^{(n)}|\boldsymbol{\lambda}^{(n)}, \boldsymbol{\lambda}^{(n+1)})\prod_{n=2}^{N} p(\boldsymbol{\lambda}^{(n)}|\boldsymbol{\alpha}^{(n)}, \boldsymbol{\beta}^{(n)})p(\tau). \tag{9.22}$$

A MF-VI algorithm is derived for the probabilistic model (9.22) under the mean field $Q(\boldsymbol{\Theta}) = Q(\tau)\prod_{n=1}^{N+1} Q(\boldsymbol{\lambda}^{(n)})\prod_{n=1}^{N}\prod_{k=1}^{L_n}\prod_{\ell=1}^{L_{n+1}} Q(\mathcal{G}^{(n)}_{k,:,\ell})$ in [2] and is summarized in **Algorithm 13**. A similar model and algorithm for tensor ring, which is a variant of TT, is also reported in [3].

9.3 PARAFAC2

Parallel factor analysis 2 (PARAFAC2), which was firstly introduced in [4, 5], has recently gained increasing interest due to its effectiveness in analyzing irregular tensor data [6–16]. In particular, given an irregular tensor data $\mathcal{Y} = \{\mathbf{Y}_k \in \mathbb{R}^{I\times J_k}\}_{k=1}^{K}$, in which each tensor slice $\mathbf{Y}_k$ has a different column number J_k, PARAFAC2 seeks for rank-R factor matrices $\{\mathbf{U}^{(1)} \in \mathbb{R}^{I\times R}, \mathbf{U}^{(3)} \in \mathbb{R}^{K\times R}, \{\mathbf{F}_k \in \mathbb{R}^{J_k\times R}\}_{k=1}^{K}\}$ via solving the following problem [5]:

$$\begin{aligned}\min_{\mathbf{U}^{(1)},\mathbf{U}^{(3)},\{\mathbf{F}_k\}_{k=1}^{K}} \quad & \sum_{k=1}^{K}\|\mathbf{Y}_k - \mathbf{U}^{(1)}\text{diag}(\mathbf{U}^{(3)}_{k,:})\mathbf{F}_k^T\|_F^2,\\ \text{s.t.} \quad & \mathbf{F}_i^T\mathbf{F}_i = \mathbf{F}_j^T\mathbf{F}_j, \quad \forall i, j \in \{1,\ldots,K\}.\end{aligned} \tag{9.23}$$

Notice that in problem (9.23), each tensor slice $\mathbf{Y}_k \in \mathbb{R}^{I\times J_k}$ is decomposed into a common set of factor matrices $\{\mathbf{U}^{(1)} \in \mathbb{R}^{I\times R}, \mathbf{U}^{(3)} \in \mathbb{R}^{K\times R}\}$ while having a dedicated factor matrix $\mathbf{F}_k \in \mathbb{R}^{J_k\times R}$ for the kth slice.

Algorithm 13 Probabilistic Tensor Train [2]

Initializations: Choose $L_n > R_n, \forall n$ and initial values $\{\hat{\alpha}_k^{(n)}, \hat{\beta}_k^{(n)}\}_{k=1,n=2}^{L_n,N}, \hat{\alpha}_\tau, \hat{\beta}_\tau$.
Iterations:
Update $\mathcal{G}^{(n)}$ from $n = 1$ to N

The variational update of the TT core fibers follows a Gaussian distribution $Q(\mathcal{G}_{k,:,\ell}^{(n)}) = \prod_{i_n=1}^{I_n} \mathcal{N}(\mu_{\mathcal{G}_{k,i_n,\ell}^{(n)}}, v_{\mathcal{G}_{k,i_n,\ell}^{(n)}})$, with its variance and mean as

$$v_{\mathcal{G}_{k,i_n,\ell}^{(n)}} = \Big(\mathbb{E}[\tau] \sum_{i_1,i_2,\ldots,i_N \setminus i_n} \mathbb{E}[\boldsymbol{b}_{(k-1)L_n+k}^{(<n)}]\mathbb{E}[\boldsymbol{b}_{(\ell-1)L_{n+1}+\ell}^{(>n)}] + \mathbb{E}[\boldsymbol{\lambda}_k^{(n)}]\mathbb{E}[\boldsymbol{\lambda}_\ell^{(n+1)}]\Big)^{-1}, \tag{9.24}$$

$$\begin{aligned}\mu_{\mathcal{G}_{k,i_n,\ell}^{(n)}} =& v_{\mathcal{G}_{k,i_n,\ell}^{(n)}} \mathbb{E}[\tau] \sum_{i_1,i_2,\ldots,i_N \setminus i_n} \Big(\mathcal{Y}_{i_1 i_2 \ldots i_N} \mathbb{E}[\boldsymbol{t}_k^{(<n)}]\mathbb{E}[\boldsymbol{t}_\ell^{(>n)}] \\ &- \sum_{\substack{k'=1 \\ k' \neq k}}^{L_n} \sum_{\substack{\ell'=1 \\ \ell' \neq \ell}}^{L_{n+1}} \mathbb{E}[\boldsymbol{b}_{(k-1)L_n+k'}^{(<n)}]\mathbb{E}[\mathcal{G}_{k',i_n,\ell'}^{(n)}]\mathbb{E}[\boldsymbol{b}_{(\ell-1)L_{n+1}+\ell'}^{(>n)}]\Big),\end{aligned} \tag{9.25}$$

in which the following notations are adopted to make the expression more concise:

$$\mathbb{E}[\boldsymbol{t}^{(<n)}] = \prod_{d=1}^{n-1} \mathbb{E}[\mathcal{G}_{:,i_d,:}^{(d)}], \tag{9.26}$$

$$\mathbb{E}[\boldsymbol{b}^{(<n)}] = \prod_{d=1}^{n-1} \mathbb{E}[\mathcal{G}_{:,i_d,:}^{(d)} \otimes \mathcal{G}_{:,i_d,:}^{(d)}], \tag{9.27}$$

and $\mathbb{E}[\boldsymbol{t}^{(>n)}]$ and $\mathbb{E}[\boldsymbol{b}^{(>n)}]$ are defined similarly.
Update $\boldsymbol{\lambda}^{(n)}$ from $n = 2$ to N

The variational distribution of $\boldsymbol{\lambda}_k^{(n)}$ is $Q(\boldsymbol{\lambda}_k^{(n)}) = \text{gamma}(\hat{\alpha}_k^{(n)}, \hat{\beta}_k^{(n)})$, with

$$\hat{\alpha}_k^{(n)} = \frac{I_n L_{n+1}}{2} + \frac{I_{n-1} L_{n-1}}{2} + \alpha_k^{(n)}, \tag{9.28}$$

$$\hat{\beta}_k^{(n)} = \frac{1}{2} \sum_{i_n=1}^{I_n} \sum_{\ell=1}^{L_{n+1}} (\mathbb{E}[\mathcal{G}_{k,i_n,\ell}^{(n)}{}^2]\mathbb{E}[\boldsymbol{\lambda}_\ell^{(n+1)}]) + \frac{1}{2} \sum_{i_{n-1}=1}^{I_{n-1}} \sum_{\ell'=1}^{L_{n-1}} (\mathbb{E}[\mathcal{G}_{\ell',i_{n-1},k}^{(n-1)}{}^2]\mathbb{E}[\boldsymbol{\lambda}_{\ell'}^{(n-1)}]) + \beta_k^{(n)}. \tag{9.29}$$

Update τ

The variational distribution of τ is $Q(\tau) = \text{gamma}(\hat{\alpha}_\tau, \hat{\beta}_\tau)$, where

$$\hat{\alpha}_\tau = \frac{\prod_{n=1}^N I_n}{2} + \alpha_\tau, \tag{9.30}$$

$$\hat{\beta}_\tau = \frac{1}{2}\Big(\|\mathcal{Y}\|_F^2 - 2\sum_{i_1=1}^{I_1} \cdots \sum_{i_N=1}^{I_N} \mathcal{Y}_{i_1 \ldots i_N} \prod_{n=1}^{N} \mathbb{E}[\mathcal{G}_{:,i_n,:}^{(n)}] + \sum_{i_1=1}^{I_1} \cdots \sum_{i_N=1}^{I_N} \prod_{n=1}^{N} \mathbb{E}[\mathcal{G}_{:,i_n,:}^{(n)} \otimes \mathcal{G}_{:,i_n,:}^{(n)}]\Big) + \beta_\tau. \tag{9.31}$$

Until Convergence

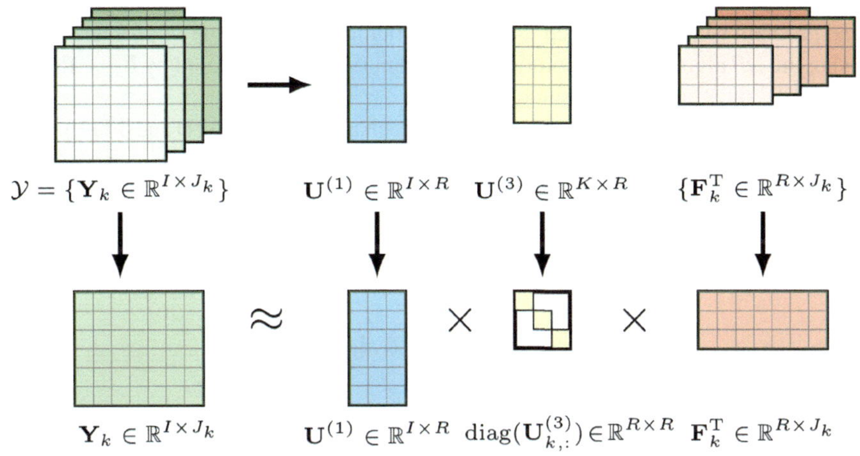

Fig. 9.1 An illustration of PARAFAC2

Even for regular tensor data $\boldsymbol{\mathcal{Y}} = \{\mathbf{Y}_k \in \mathbb{R}^{I \times J}\}_{k=1}^{K}$, PARAFAC2 and tensor CPD differ in the sense that tensor CPD restricts factor matrices $\{\mathbf{F}_k\}_{k=1}^{K}$ to be equal across slices, i.e., $\mathbf{F}_k = \mathbf{F}$, $\forall k$. More specifically, tensor CPD solves the following problem:

$$\min_{\mathbf{U}^{(1)},\mathbf{U}^{(3)},\mathbf{F}} \sum_{k=1}^{K} \|\mathbf{Y}_k - \mathbf{U}^{(1)}\mathrm{diag}(\mathbf{U}^{(3)}_{k,:})\mathbf{F}^T\|_F^2. \tag{9.32}$$

By comparing problem (9.23) with problem (9.32), it can be seen that PARAFAC2 generalizes tensor CPD to deal with irregular tensor data with unaligned size along one dimension, as illustrated in Fig. 9.1.

To simplify the constraints of $\{\mathbf{F}_k\}_{k=1}^{K}$ in problem (9.23), a set of orthogonal matrices $\boldsymbol{\mathcal{P}} = \{\mathbf{P}_k \in \mathbb{R}^{J_k \times R}\}$ and a rank-R factor matrix $\mathbf{U}^{(2)} \in \mathbb{R}^{R \times R}$ are introduced in [5], which transforms problem (9.23) to

$$\begin{aligned} &\min_{\{\mathbf{U}^{(n)}\}_{n=1}^{3},\boldsymbol{\mathcal{P}}} \sum_{k=1}^{K} \|\mathbf{Y}_k - \mathbf{U}^{(1)}\mathrm{diag}(\mathbf{U}^{(3)}_{k,:})\left[\mathbf{U}^{(2)}\right]^T \mathbf{P}_k^T\|_F^2, \\ &\text{s.t.} \quad \mathbf{P}_k^T\mathbf{P}_k = \mathbf{I}_R, \ \forall k \in \{1,\ldots,K\}, \\ &\mathbf{P}_k \in \mathbb{R}^{J_k \times R}, \ \forall k \in \{1,\ldots,K\}, \ \mathbf{U}^{(2)} \in \mathbb{R}^{R \times R}. \end{aligned} \tag{9.33}$$

In [5], the direct fitting (DF) algorithm was developed to solve problem (9.33). Particularly, given the orthogonal matrices in $\boldsymbol{\mathcal{P}}$, problem (9.33) reduces to a tensor CPD problem, which can be solved using **Algorithm 1**. On the other hand, by fixing the factor matrices $\{\mathbf{U}^{(n)}\}_{n=1}^{3}$, each orthogonal matrix in $\boldsymbol{\mathcal{P}}$ can be optimized via singular value decomposition (SVD). Due to the equivalence between problem (9.23) and problem (9.33) established in [5], after obtaining the solution of problem (9.33) using the DF algorithm, the solution to problem (9.23) is recovered by $\mathbf{F}_k = \mathbf{P}_k \mathbf{U}^{(2)}$, with $\{\mathbf{U}^{(1)}, \mathbf{U}^{(3)}\}$ remain the same.

Similar to Bayesian tensor CPD in previous chapters, or TuckerD and TTD, probabilistic PARAFAC2 [17] also employs the sparsity-promoting prior to achieve tensor rank learning. It starts with $L (L \geq R)$ columns in all factor matrices and places Gaussian-gamma prior on such over-parametrized columns to encode sparsity information. To be more specific, the prior design for factor matrices is [17]

$$p(\mathbf{U}^{(1)}) = \prod_{l=1}^{L} \mathcal{N}(\mathbf{U}^{(1)}_{:,l} | \mathbf{0}_{I \times 1}, \mathbf{I}_I), \tag{9.34}$$

$$p(\mathbf{U}^{(2)}) = \prod_{l=1}^{L} \mathcal{N}(\mathbf{U}^{(2)}_{:,l} | \mathbf{0}_{J \times 1}, \mathbf{I}_J), \tag{9.35}$$

$$p(\mathbf{U}^{(3)} | \{\gamma_l\}_{l=1}^{L}) = \prod_{l=1}^{L} \mathcal{N}(\mathbf{U}^{(3)}_{:,l} | \mathbf{0}_{I \times 1}, \gamma_l^{-1} \mathbf{I}_K), \tag{9.36}$$

$$p(\{\gamma_l\}_{l=1}^{L}) = \prod_{l=1}^{L} \text{gamma}(\gamma_l | c_l^0, d_l^0). \tag{9.37}$$

On the other hand, the likelihood function in probabilistic PARAFAC2 [17] is obtained from the objective function of (9.33), by assuming each observation in $\mathcal{Y}$ subject to independent Gaussian noise perturbation. This results in

$$p(\mathcal{Y} | \{\mathbf{U}^{(n)}\}_{n=1}^{3}, \tau; \mathcal{P}) \propto \exp\left\{ -\frac{\tau}{2} \sum_{k=1}^{K} \| \mathbf{Y}_k - \mathbf{U}^{(1)} \text{diag}(\mathbf{U}^{(3)}_{k,:}) \left[\mathbf{U}^{(2)}\right]^T \mathbf{P}_k^T \|_F^2 \right\}. \tag{9.38}$$

For the noise precision τ, a Gamma prior gamma$(\tau | a_0, b_0)$ is assigned. Based on the prior and likelihood functions in (9.34)–(9.38), the joint probability of data $\mathcal{Y}$ and parameters $\boldsymbol{\Theta} = \{\{\mathbf{U}^{(n)}\}_{n=1}^{3}, \{\gamma_l\}_{l=1}^{L}, \tau\}$ is summarized as

$$\begin{aligned} p(\mathcal{Y}, \{\mathbf{U}^{(n)}\}_{n=1}^{3}, \{\gamma_l\}_{l=1}^{L}, \tau; \mathcal{P}) &= p(\mathcal{Y} | \{\mathbf{U}^{(n)}\}_{n=1}^{3}, \tau; \mathcal{P}) p(\mathbf{U}^{(1)}) p(\mathbf{U}^{(2)}) \\ &\times p(\mathbf{U}^{(3)} | \{\gamma_l\}_{l=1}^{L}) p(\{\gamma_l\}_{l=1}^{L}) p(\tau). \end{aligned} \tag{9.39}$$

An inference algorithm was developed in [17] by using variational EM framework under the mean-field assumption $Q(\boldsymbol{\Theta}) = \prod_{n=1}^{N} Q(\mathbf{U}^{(n)}) \prod_{l=1}^{L} Q(\gamma_l) Q(\tau)$. The resultant algorithm is summarized in **Algorithm 14**.

Algorithm 14 Probabilistic PARAFAC2 [17]

Initializations: Choose $L > R$ and initial values $\{\mathbf{M}^{(n)}, \mathbf{\Sigma}^{(n)}\}_{n=1}^{N}$, $\{c_l, d_l\}_{l=1}^{L}$, a, b.
Iterations:
Update $\mathcal{P}$

$$\mathbf{P}_k = \mathbf{\Psi}_k \mathbf{\Xi}_k^T, \quad \forall k \in \{1, \ldots, K\}, \tag{9.40}$$

where $\mathbf{\Xi}_k$ and $\mathbf{\Psi}_k$ are orthogonal matrices obtained from the following singular value decomposition (SVD):

$$\mathbb{E}[\mathbf{U}^{(2)}]\text{diag}(\mathbb{E}[\mathbf{U}_{k,:}^{(3)}]) \left[\mathbb{E}[\mathbf{U}^{(1)}]\right]^T \mathbf{Y}_k = \mathbf{\Xi}_k \mathbf{\Upsilon}_k \mathbf{\Psi}_k^T. \tag{9.41}$$

Update $\mathcal{W}$

$$\mathbf{W}_k = \mathbf{Y}_k \mathbf{P}_k, \quad \forall k \in \{1, \ldots, K\}, \tag{9.42}$$

Update $\mathbf{U}^{(n)}$ from $n = 1$ to 3

The optimal variational pdf $Q^*(\mathbf{U}_{:,l}^{(n)})$ is a Gaussian distribution $\mathcal{N}(\mathbf{U}_{:,l}^{(n)}|\mathbf{M}_{:,l}^{(n)}, \mathbf{\Sigma}^{(n)})$ with its covariance and mean as

$$\mathbf{\Sigma}^{(n)} = \left[\mathbb{E}[\tau]\mathbb{E}\left[\left(\underset{k=1,k\neq n}{\overset{3}{\diamond}} \mathbf{U}^{(k)}\right)^T \left(\underset{k=1,k\neq n}{\overset{3}{\diamond}} \mathbf{U}^{(k)}\right)\right] + \mathbf{I}_L\right]^{-1}, n = 1, 2, \tag{9.43}$$

$$\mathbf{\Sigma}^{(n)} = \left[\mathbb{E}[\tau]\mathbb{E}\left[\left(\underset{k=1,k\neq n}{\overset{3}{\diamond}} \mathbf{U}^{(k)}\right)^T \left(\underset{k=1,k\neq n}{\overset{3}{\diamond}} \mathbf{U}^{(k)}\right)\right] + \text{diag}\{\mathbb{E}[\gamma_1], \ldots, \mathbb{E}[\gamma_L]\}\right]^{-1}, n = 3, \tag{9.44}$$

$$\mathbf{M}^{(n)} = \mathbb{E}[\tau]\, \mathbf{W}^{(n)T}\mathbb{E}\left[\underset{k=1,k\neq n}{\overset{3}{\diamond}} \mathbf{U}^{(k)}\right] \mathbf{\Sigma}^{(n)}, \tag{9.45}$$

where $\mathbf{W}^{(n)}$ is the Mode-n unfolding of the tensor $\mathcal{W}$.
Update γ_l from $l = 1$ to L

The optimal variational pdf $Q^*(\alpha_l)$ coincides with Gamma distribution gamma(c_l, d_l), where

$$c_l = c_l^0 + \frac{K}{2}, \tag{9.46}$$

$$d_l = d_l^0 + \frac{1}{2}\mathbb{E}\left[\left[\mathbf{U}^{(3)}\right]^T \left[\mathbf{U}^{(3)}\right]\right]_{l,l}. \tag{9.47}$$

Update τ

The optimal variational pdf $Q^*(\tau)$ is derived to be a Gamma distribution gamma(a, b), with

$$a = a_0 + \frac{I \times J \times K}{2}, \tag{9.48}$$

$$b = b_0 + \frac{1}{2}\mathbb{E}\left[\sum_{k=1}^{K} \|\mathbf{W}_k - \mathbf{U}^{(1)}\text{diag}(\mathbf{U}_{k,:}^{(3)})\left[\mathbf{U}^{(2)}\right]^T\|_F^2\right]. \tag{9.49}$$

Until Convergence

9.4 Tensor-SVD (T-SVD)

Tensor-SVD (T-SVD) is a relatively new tensor decomposition and first appears in [18]. It has a wide range of applications, especially for color images, videos, and multi-channel audio sequences modeling [19]. To formally define T-SVD, we firstly introduce several definitions that are essential for T-SVD.

Definition 9.1 (*T-product*) Given $\mathcal{X} \in \mathbb{R}^{I_1 \times R \times I_3}$ and $\mathcal{Y} \in \mathbb{R}^{R \times I_2 \times I_3}$, the T-product $\mathcal{X} * \mathcal{Y}$ is the $I_1 \times I_2 \times I_3$ tensor

$$\mathcal{X} * \mathcal{Y} = \text{ fold } (\text{circ}(\mathcal{X}) \text{ unfold } (\mathcal{Y})), \tag{9.50}$$

where $\text{unfold}(\mathcal{X}) = \left[\mathcal{X}_{:,:,1}; \ldots; \mathcal{X}_{:,:,I_3}\right]$, fold is the inverse operator of unfold, and

$$\text{circ}(\mathcal{X}) = \begin{bmatrix} \mathcal{X}_{:,:,1} & \mathcal{X}_{:,:,I_3} & \cdots & \mathcal{X}_{:,:,2} \\ \mathcal{X}_{:,:,2} & \mathcal{X}_{:,:,1} & \cdots & \mathcal{X}_{:,:,3} \\ \vdots & \vdots & \ddots & \vdots \\ \mathcal{X}_{:,:,I_3} & \mathcal{X}_{:,:,I_3-1} & \cdots & \mathcal{X}_{:,:,1} \end{bmatrix}. \tag{9.51}$$

Definition 9.2 (*Identity Tensor*) The identity tensor $\mathcal{I} \in \mathbb{R}^{I \times I \times I_3}$ is defined as the tensor whose first frontal slice is the $I \times I$ identity matrix, and other slices are all zeros, i.e., $\mathcal{I}_{:,:,1} = \mathbf{I}_I$, $\mathcal{I}_{:,:,i} = \mathbf{0}_{I \times I}$, $\forall i = 2, \ldots, I_3$.

Definition 9.3 (*F-diagonal Tensor*) A tensor is called f-diagonal if its frontal slices are all diagonal matrices.

Definition 9.4 (*Conjugate Transpose*) The conjugate transpose of a tensor is defined as the tensor $\mathcal{X}^\dagger \in \mathbb{R}^{I_2 \times I_1 \times I_3}$ constructed by conjugate transposing each frontal slice of $\mathcal{X} \in \mathbb{R}^{I_1 \times I_2 \times I_3}$ and then reversing the order of the transposed frontal slices 2 through I_3, i.e., $\mathcal{X}^\dagger_{:,:,1} = \mathcal{X}^H_{:,:,1}$, $\mathcal{X}^\dagger_{:,:,i} = \mathcal{X}^H_{:,:,I_3-i+2}$, $\forall i = 2, \ldots, I_3$.

Definition 9.5 (*Orthogonality*) A tensor $\mathcal{Q} \in \mathbb{R}^{I \times I \times I_3}$ is called orthogonal, provided that $\mathcal{Q}^\dagger * \mathcal{Q} = \mathcal{Q} * \mathcal{Q}^\dagger = \mathcal{I}$ with $\mathcal{I}$ being an $I \times I \times I_3$ identity tensor.

Now, we are ready for the definition of T-SVD.

Definition 9.6 (*T-SVD*) Let $\mathcal{X}$ be an $I_1 \times I_2 \times I_3$ real-valued tensor. Then $\mathcal{X}$ can be factored as

$$\mathcal{X} = \mathcal{U} * \mathcal{D} * \mathcal{V}^{\dagger}, \tag{9.52}$$

where $\mathcal{U} \in \mathbb{R}^{I_1 \times I_1 \times I_3}$, $\mathcal{V} \in \mathbb{R}^{I_2 \times I_2 \times I_3}$ are orthogonal tensors, and $\mathcal{D} \in \mathbb{R}^{I_1 \times I_2 \times I_3}$ is an f-diagonal tensor. The factorization (9.52) is called the T-SVD (i.e., tensor SVD).

For a matrix $\mathbf{X}$, if we perform SVD, $\mathbf{X} = \mathbf{U}\mathbf{S}\mathbf{V}^T$, the rank of $\mathbf{X}$ is defined as the number of non-zero elements in the diagonal matrix $\mathbf{S}$. Generalizing this to T-SVD, the tubal rank of $\mathcal{X}$ is formally defined as follows.

Definition 9.7 (*Tensor tubal rank*) The tubal rank of $\mathcal{X}$ is the number of non-zero tubes of $\mathcal{D}$ from the T-SVD of $\mathcal{X} = \mathcal{U} * \mathcal{D} * \mathcal{V}^{\dagger}$, i.e.,

$$\text{Rank}_t(\mathcal{X}) = \#\{i, \mathcal{D}(i, i, :) \neq \mathbf{0}\}. \tag{9.53}$$

According to [20], any tensor with a tubal rank up to R can be factorized as

$$\mathcal{X} = \mathcal{U} * \mathcal{V}^{\dagger}, \tag{9.54}$$

for some $\mathcal{U}$ and $\mathcal{V}$ satisfying $\text{Rank}_t(\mathcal{U}) = \text{Rank}_t(\mathcal{V}) = R$. In (9.54), $\mathcal{U} \in \mathbb{R}^{I_1 \times R \times I_3}$, $\mathcal{V} \in \mathbb{R}^{I_2 \times R \times I_3}$, and $R \leq \min(I_1, I_2)$ controls the tubal rank. To infer tubal rank, [21] employs the Gaussian-gamma prior to impose sparsity in $\mathcal{U}$ and $\mathcal{V}$. In particular, assume the low-tubal rank upper bound is $L > R$,

$$p(\mathcal{U} \mid \boldsymbol{\lambda}) = \prod_{i=1}^{I_1} \prod_{l=1}^{L} \prod_{k=1}^{I_3} \mathcal{N}\left(\mathcal{U}_{i,l,k} \mid 0, \lambda_l^{-1}\right), \tag{9.55}$$

$$p(\mathcal{V} \mid \boldsymbol{\lambda}) = \prod_{j=1}^{I_2} \prod_{l=1}^{L} \prod_{k=1}^{I_3} \mathcal{N}\left(\mathcal{V}_{j,l,k} \mid 0, \lambda_l^{-1}\right), \tag{9.56}$$

$$p(\boldsymbol{\lambda}) = \prod_{l=1}^{L} \text{gamma}\left(\lambda_l \mid a_0^{\lambda}, b_0^{\lambda}\right), \tag{9.57}$$

where a_0^λ, b_0^λ in the gamma prior are pre-determined hyper-parameters. In (9.55)–(9.56), each element in the tensor slice $\mathcal{U}_{:,l,:}$ and $\mathcal{V}_{:,l,:}$ follows a zero-mean Gaussian distribution with precesion λ_l, of which the philosophy is similar to the probabilistic tensor CPD presented in Chap. 3.

Prior work [21] also considers the outlier modeling. More specifically, given the tensor data $\mathcal{Y} \in \mathbb{R}^{I_1 \times I_2 \times I_3}$, the likelihood function is specified as

$$p(\mathcal{Y} \mid \mathcal{U}, \mathcal{V}, \mathcal{S}, \tau) = \prod_{i=1}^{I_1}\prod_{j=1}^{I_2}\prod_{k=1}^{I_3} \mathcal{N}\left(\mathcal{Y}_{i,j,k} \mid (\mathcal{U} * \mathcal{V}^\dagger)_{i,j,k} + \mathcal{S}_{i,j,k}, \tau^{-1}\right), \quad (9.58)$$

where the sparse component $\mathcal{S}$ is introduced to model the outlier. Exactly the same as in Chap. 6, independent Gaussian-gamma priors are placed over each element in the additional sparse component $\mathcal{S}$,

$$p(\mathcal{S} \mid \boldsymbol{\beta}) = \prod_{i=1}^{I_1}\prod_{j=1}^{I_2}\prod_{k=1}^{I_3} \mathcal{N}\left(\mathcal{S}_{i,j,k} \mid 0, \beta_{i,j,k}^{-1}\right), \quad (9.59)$$

$$p(\boldsymbol{\beta}) = \prod_{i=1}^{I_1}\prod_{j=1}^{I_2}\prod_{k=1}^{I_3} \text{gamma}\left(\beta_{i,j,k} \mid a_0^\beta, b_0^\beta\right), \quad (9.60)$$

where a_0^β, b_0^β are pre-determined hyper-parameters. To complete the Bayesian modeling, the prior of the noise precision τ is $p(\tau) = \text{gamma}\left(\tau \mid a_0^\tau, b_0^\tau\right)$. The probabilistic modeling of T-SVD is summarized by the joint distribution of $\mathcal{Y}$ and $\boldsymbol{\Theta} = \{\mathcal{U}, \mathcal{V}, \mathcal{S}, \boldsymbol{\lambda}, \boldsymbol{\beta}, \tau\}$,

$$p(\mathcal{Y}, \boldsymbol{\Theta}) = p(\mathcal{Y} \mid \mathcal{U}, \mathcal{V}, \mathcal{S}, \tau) p(\mathcal{U} \mid \boldsymbol{\lambda}) p(\mathcal{V} \mid \boldsymbol{\lambda}) p(\boldsymbol{\lambda}) p(\mathcal{S} \mid \boldsymbol{\beta}) p(\boldsymbol{\beta}) p(\tau). \quad (9.61)$$

A variational inference algorithm under the mean field $Q(\boldsymbol{\Theta}) = Q(\mathcal{U})Q(\mathcal{V})Q(\mathcal{S}) Q(\boldsymbol{\lambda})Q(\boldsymbol{\beta})Q(\tau)$ is developed in [21] and is summarized in **Algorithm 15**. In **Algorithm 15**, $\overrightarrow{\mathbf{x}}_{i\cdot}$ denotes the vector formed by unfolding $\mathcal{X}^\dagger_{i,:,:}$, $\overrightarrow{\mathbf{x}}_{\cdot j}$ denotes the vector formed by unfolding $\mathcal{X}_{:,j,:}$, and $\mathcal{L}$ is the $L \times L \times I_3$ tensor whose first frontal slice is the diagonal matrix $\mathcal{L}_{:,:,1} = \text{diag}\{\lambda_1, \ldots, \lambda_L\}$ and all other slices are zeros.

Algorithm 15 Probabilistic T-SVD [21]

Initializations: Choose $L > R$ and initial values $\{\langle\vec{\mathbf{u}}_{i\cdot}\rangle\}_{i=1}^{I_1}$, $\mathbf{\Sigma}^u$, $\{\langle\vec{\mathbf{v}}_{j\cdot}\rangle\}_{j=1}^{I_2}$, $\mathbf{\Sigma}^v$, $\{a_r^\lambda, b_r^\lambda\}_{l=1}^{L}$, $\{\langle S_{i,j,k}\rangle, \sigma_{i,j,k}^2, a_{i,j,k}^\beta, b_{i,j,k}^\beta\}_{i=1,j=1,k=1}^{I_1,I_2,I_3}$, a^τ, b^τ.

Iterations:

Update $\mathcal{U}$ and $\mathcal{V}$

The optimal variational pdf $Q(\mathcal{U})$ is $Q(\mathcal{U}) = \prod_{i=1}^{I_1} \mathcal{N}\left(\vec{\mathbf{u}}_{i\cdot} \mid \langle\vec{\mathbf{u}}_{i\cdot}\rangle, \mathbf{\Sigma}^u\right)$, whose parameters are given by

$$\langle\vec{\mathbf{u}}_{i\cdot}\rangle = \langle\tau\rangle \mathbf{\Sigma}^u \operatorname{circ}(\langle\mathcal{V}\rangle)^\top \left(\vec{\mathbf{y}}_{i\cdot} - \langle\vec{\mathbf{s}}_{i\cdot}\rangle\right), \tag{9.62}$$

$$\mathbf{\Sigma}^u = \left(\langle\tau\rangle \left\langle\operatorname{circ}(\mathcal{V})^\top \operatorname{circ}(\mathcal{V})\right\rangle + \operatorname{circ}(\langle\mathcal{L}\rangle)\right)^{-1}. \tag{9.63}$$

Similarly, the optimal variational pdf of $\mathcal{V}$ is given by $Q(\mathcal{V}) = \prod_{j=1}^{I_2} \mathcal{N}\left(\vec{\mathbf{v}}_{j\cdot} \mid \langle\vec{\mathbf{v}}_{j\cdot}\rangle, \mathbf{\Sigma}^v\right)$ with the mean and covariance

$$\langle\vec{\mathbf{v}}_{j\cdot}\rangle = \langle\tau\rangle \mathbf{\Sigma}^v \operatorname{circ}(\langle\mathcal{U}\rangle)^\top \left(\vec{\mathbf{y}}_{\cdot j} - \langle\vec{\mathbf{s}}_{\cdot j}\rangle\right), \tag{9.64}$$

$$\mathbf{\Sigma}^v = \left(\langle\tau\rangle \left\langle\operatorname{circ}(\mathcal{U})^\top \operatorname{circ}(\mathcal{U})\right\rangle + \operatorname{circ}(\langle\mathcal{L}\rangle)\right)^{-1}. \tag{9.65}$$

Update λ

The optimal variational pdf is $Q(\lambda) = \prod_{l=1}^{L} \operatorname{gamma}\left(\lambda_l \mid a_l^\lambda, b_l^\lambda\right)$ with parameters

$$a_l^\lambda = a_0^\lambda + \frac{(I_1 + I_2)\, I_3}{2}, \tag{9.66}$$

$$b_l^\lambda = b_0^\lambda + \frac{1}{2}\left\langle\|\vec{\mathbf{u}}_{\cdot l}\|^2 + \|\vec{\mathbf{v}}_{\cdot l}\|^2\right\rangle. \tag{9.67}$$

Update $\mathcal{S}$

The optimal variational pdf of $\mathcal{S}$ can be obtained as $Q(\mathcal{S}) = \prod_{i,j,k} \mathcal{N}\left(S_{ijk} \mid \langle S_{i,j,k}\rangle, \sigma_{i,j,k}^2\right)$ with the parameters

$$\langle S_{i,j,k}\rangle = \langle\tau\rangle \left(\langle\beta_{i,j,k}\rangle + \langle\tau\rangle\right) \left(\mathcal{Y}_{i,j,k} - (\mathcal{U} * \mathcal{V}^\dagger)_{i,j,k}\right), \tag{9.68}$$

$$\sigma_{i,j,k}^2 = \left(\langle\beta_{i,j,k}\rangle + \langle\tau\rangle\right)^{-1} \tag{9.69}$$

Update β

The optimal variational pdf of β is given by $Q\left(\beta_{i,j,k}\right) = \operatorname{gamma}\left(\beta_{i,j,k} \mid a_{i,j,k}^\beta, b_{i,j,k}^\beta\right)$, whose parameters are

$$a_{i,j,k}^\beta = a_0^\beta + \frac{1}{2}, \tag{9.70}$$

$$b_{i,j,k}^\beta = b_0^\beta + \frac{1}{2}\left\langle\beta_{i,j,k}^2\right\rangle. \tag{9.71}$$

Update τ

The noise precision has the following posterior distribution: $Q(\tau) = \text{gamma}\,(\tau \mid a^\tau, b^\tau)$, whose parameters are

$$a^\tau = a_0^\tau + \frac{\prod_{n=1}^N I_n}{2}, \tag{9.72}$$

$$b^\tau = b_0^\tau + \frac{1}{2}\sum_{i,j,k}\left\langle \left\| \mathcal{Y}_{i,j,k} - (\mathcal{U} * \mathcal{V}^\dagger)_{i,j,k} - \mathcal{S}_{i,j,k} \right\|^2 \right\rangle. \tag{9.73}$$

Until Convergence

References

1. Q. Zhao, L. Zhang, A. Cichocki, Bayesian sparse tucker models for dimension reduction and tensor completion (2015). arXiv:1505.02343
2. L. Xu, L. Cheng, N. Wong, Y.-C. Wu, Overfitting avoidance in tensor train factorization and completion: prior analysis and inference, in *2021 IEEE International Conference on Data Mining (ICDM)* (IEEE, 2021), pp. 1439–1444
3. Z. Long, C. Zhu, J. Liu, Y. Liu, Bayesian low rank tensor ring for image recovery. IEEE Trans. Image Process. **30**, 3568–3580 (2021)
4. R.A. Harshman, Parafac2: Mathematical and technical notes, in *UCLA Working Papers in Phonetics*, vol. 22, no. 10, pp. 30–44 (1972)
5. H.A. Kiers, J.M. Ten Berge, R. Bro, Parafac2–part i. a direct fitting algorithm for the parafac2 model. J. Chemom.: J. Chemom. Soc. **13**(3–4), 275–294 (1999)
6. Y. Panagakis, C. Kotropoulos, Automatic music tagging via parafac2, in *2011 IEEE International Conference on Acoustics, Speech and Signal Processing (ICASSP)* (IEEE, 2011), pp. 481–484
7. E. Pantraki, C. Kotropoulos, Automatic image tagging and recommendation via parafac2, in *2015 IEEE 25th International Workshop on Machine Learning for Signal Processing (MLSP)* (IEEE, 2015), pp. 1–6
8. E. Pantraki, C. Kotropoulos, A. Lanitis, Age interval and gender prediction using parafac2 applied to speech utterances, in *2016 4th International Conference on Biometrics and Forensics (IWBF)* (IEEE, 2016), pp. 1–6
9. P.A. Chew, B.W. Bader, T.G. Kolda, A. Abdelali, Cross-language information retrieval using parafac2, in *Proceedings of the 13th ACM SIGKDD international conference on Knowledge discovery and data mining* (2007), pp. 143–152
10. Y. Shin, S.S. Woo, What is in your password? analyzing memorable and secure passwords using a tensor decomposition, in *The World Wide Web Conference* (2019), pp. 3230–3236
11. I. Perros, E.E. Papalexakis, F. Wang, R. Vuduc, E. Searles, M. Thompson, J. Sun, Spartan: scalable parafac2 for large & sparse data, in *Proceedings of the 23rd ACM SIGKDD International Conference on Knowledge Discovery and Data Mining* (2017), pp. 375–384
12. A. Afshar, I. Perros, E.E. Papalexakis, E. Searles, J. Ho, J. Sun, Copa: constrained parafac2 for sparse & large datasets, in *Proceedings of the 27th ACM International Conference on Information and Knowledge Management* (2018), pp. 793–802
13. K. Yin, A. Afshar, J.C. Ho, W.K. Cheung, C. Zhang, J. Sun, Logpar: logistic parafac2 factorization for temporal binary data with missing values, in *Proceedings of the 26th ACM SIGKDD International Conference on Knowledge Discovery & Data Mining* (2020), pp. 1625–1635
14. A. Afshar, I. Perros, H. Park, C. Defilippi, X. Yan, W. Stewart, J. Ho, J. Sun, Taste: temporal and static tensor factorization for phenotyping electronic health records, in *Proceedings of the ACM Conference on Health, Inference, and Learning* (2020), pp. 193–203

15. Y. Ren, J. Lou, L. Xiong, J.C. Ho, Robust irregular tensor factorization and completion for temporal health data analysis, in *Proceedings of the 29th ACM International Conference on Information & Knowledge Management* (2020), pp. 1295–1304
16. I. Perros, X. Yan, J.B. Jones, J. Sun, W.F. Stewart, Using the parafac2 tensor factorization on ehr audit data to understand pcp desktop work. J. Biomed. Inform. **101**, 103312 (2020)
17. P.J. Jørgensen, S.F. Nielsen, J.L. Hinrich, M.N. Schmidt, K.H. Madsen, M. Mørup, Analysis of chromatographic data using the probabilistic parafac2, in *33rd Conference on Neural Information Processing Systems* (2019)
18. K. Braman, Third-order tensors as linear operators on a space of matrices. Linear Algebra Appl. **433**(7), 1241–1253 (2010)
19. M.E. Kilmer, C.D. Martin, Factorization strategies for third-order tensors. Linear Algebra Appl. **435**(3), 641–658 (2011)
20. O. Semerci, N. Hao, M.E. Kilmer, E.L. Miller, Tensor-based formulation and nuclear norm regularization for multienergy computed tomography. IEEE Trans. Image Process. **23**(4), 1678–1693 (2014)
21. Y. Zhou, Y.-M. Cheung, Bayesian low-tubal-rank robust tensor factorization with multi-rank determination. IEEE Trans. Pattern Anal. Mach. Intell. **43**(1), 62–76 (2019)

GPSR Compliance

The European Union's (EU) General Product Safety Regulation (GPSR) is a set of rules that requires consumer products to be safe and our obligations to ensure this.

If you have any concerns about our products, you can contact us on ProductSafety@springernature.com

In case Publisher is established outside the EU, the EU authorized representative is:

Springer Nature Customer Service Center GmbH
Europaplatz 3
69115 Heidelberg, Germany

Batch number: 10726135

Printed by Printforce, the Netherlands